# YOU & YOUR AQUARIUM

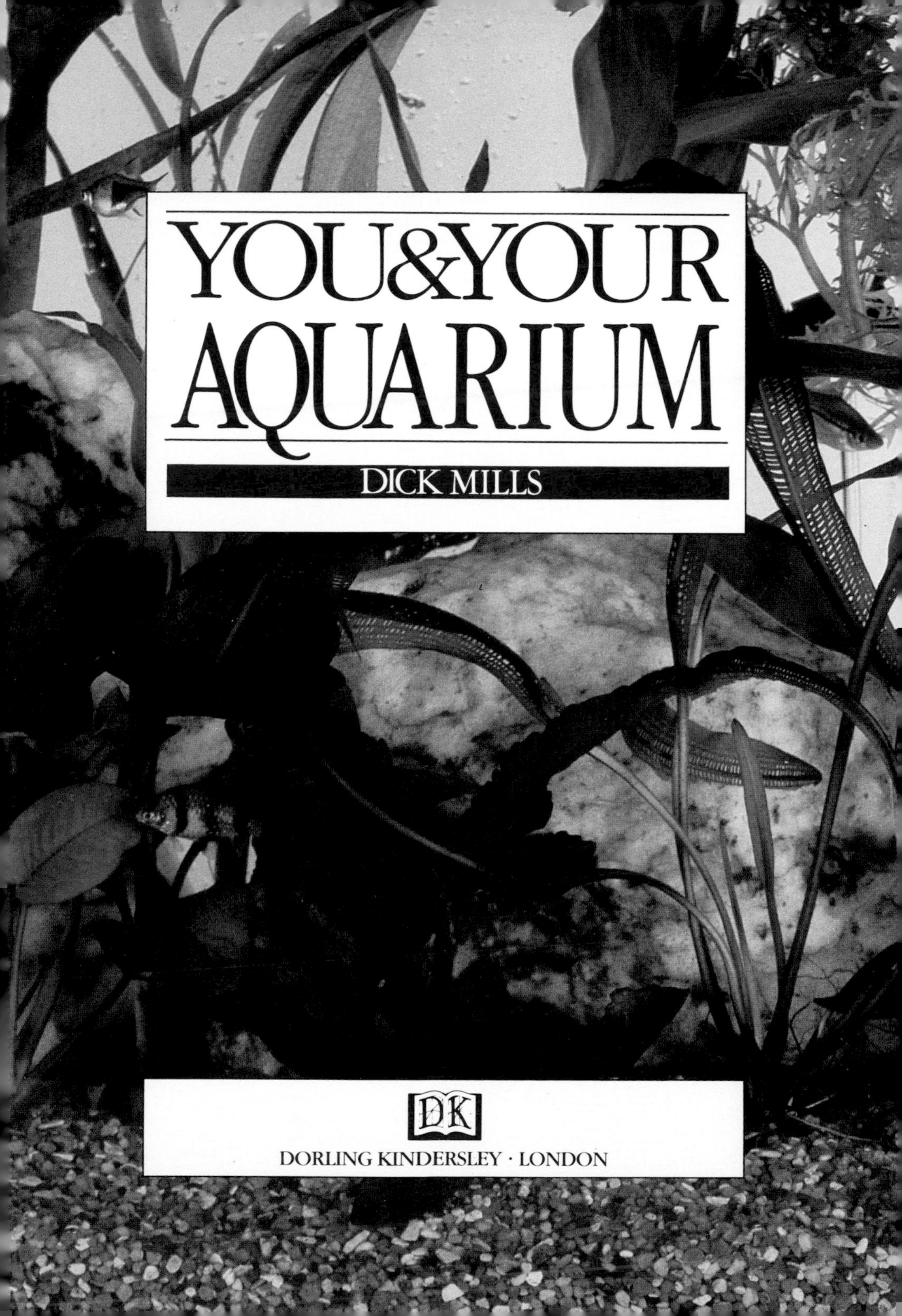

# YOU & YOUR AQUARIUM

DICK MILLS

DORLING KINDERSLEY · LONDON

**Project editor** Judith More
**Art editor** Julia Goodman

**Assistant designers** Jo Martin
Tina Vaughan
**Assistant editor** Janice Lacock

**Senior art editor** Peter Luff
**Managing editor** Alan Buckingham

First published in Great Britain in 1986 by
Dorling Kindersley Publishers Limited,
9 Henrietta Street, London WC2E 8PS.
First published in paperback in 1987 by
Dorling Kindersley Limited
fourth impression 1991
fifth impression 1991
sixth impression 1993
seventh impression 1994


**British Library Cataloguing in Publication Data**

Mills, Dick
You and your aquarium,
1. Aquariums
I. Title
639.3'4 SF457
ISBN 0-86318-086-8 (hardback)
ISBN 0-83618-362-X (paperback)

Printed and bound in Hong Kong by
Wing King Tong

# Contents

6 INTRODUCTION

1
FISH ANATOMY
14 The basic fish
16 The outer covering
17 The fins
19 Bodily functions

2
CHOOSING FISHES
24 Choosing the system
26 Buying fishes
30 Grouping fishes for compatibility

3
SPECIES GUIDE
35 *Tropical freshwater species*
36 Cyprinids
45 Characins
53 Cichlids
60 Anabantoids
66 Livebearers
70 Killifishes
73 Catfishes
78 Loaches
80 Other tropical egglaying species
83 *Coldwater freshwater species*
84 Goldfishes
89 Koi
91 Other coldwater species
95 *Tropical marine species*
96 Anemonefishes and Damselfishes

98 Angelfishes and Butterflyfishes
103 Other tropical marine species
109 *Coldwater marine species*

## 4
## THE INDOOR AQUARIUM

114 *Tanks*
115 Choosing the right tank
117 Tank construction
118 Buying and siting a tank
121 *Aeration and filtration*
122 How do aeration and filtration work?
123 The airpump
124 Air-operated filters
126 Power filters
128 The process of biological filtration
130 *Heating*
131 Aquarium heaters
133 Aquarium thermostats
135 Conserving heat
136 *Water*
137 What is water?
138 Water and habitat
144 Sources of water
145 Water quality
148 *Lighting*
149 Installing lighting

## 5
## SETTING UP AN AQUARIUM

154 *Aquascaping*
155 The base covering
157 Adding decorative materials
158 *Plants*
160 Cultivating and using plants
162 Types of plants
172 *Setting up case studies*
173 Tank designs
174 A tropical freshwater set-up
180 A coldwater freshwater set-up
186 A tropical marine set-up
192 A coldwater marine set-up

## 6
## FEEDING

200 The basic diet
202 Choosing pre-packed foods
204 Collecting live foods
206 Other foods
207 Feeding methods
209 Feeding young fishes

## 7
## HEALTH CARE

212 What is a healthy fish?
213 *Preventing disease*
214 Maintaining a healthy aquarium
220 *Diagnosis charts*
226 *Ailments*
227 How disease affects fishes
229 Internal disorders
230 External disorders
234 Isolation and treatment of sick fishes

## 8
## BREEDING

238 The breeding process
245 Preparing for breeding
246 The spawning
248 Setting up breeding tanks
251 After the spawning
253 Raising fry

## 9
## SHOWING FISHES

256 What is a fish show?
258 What makes a good fish?
262 Guidance on exhibiting fishes

## 10
## PHOTOGRAPHING FISHES

266 The camera and lenses
269 Focusing
270 Lighting
272 Setting up a good picture

274 Glossary
279 Appendices
282 Index
288 Acknowledgements

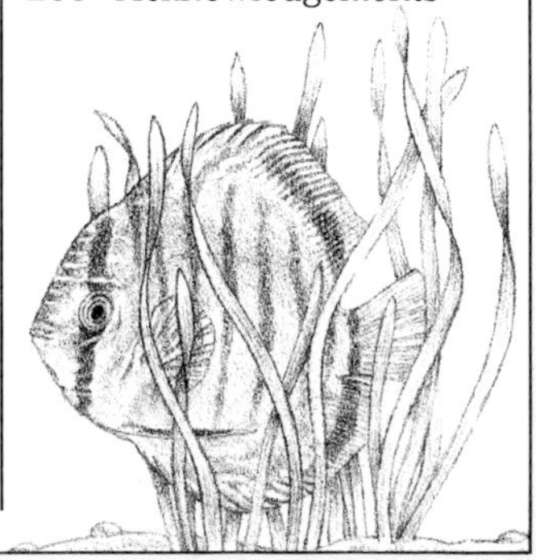

# INTRODUCTION

Interest in fishes can be divided into two separate "streams" – for the table and for ornamentation. Fishkeeping first began as a way of storing live food, and stocks were kept in ponds or moats. The ancient Egyptians may have been the first aquarists – they kept large glass tanks of coldwater fishes for decoration. However, ornamental fish-keeping may have begun in the Far East. Certainly, selective breeding of colour strains started in China, and the cultivation of Goldfishes was well-established in the Sung dynasty (970–1278 AD). By the sixteenth century, the keeping of fishes in glass bowls had spread to Europe. Advances in scientific knowledge in the nineteenth century led to the opening of public aquariums, and by the end of that century the first aquarium societies were formed in the U.S., and the first tropical aquariums were seen. These were "Heath Robinson" affairs by

***Print of Japanese fishkeeping c. 1820 (below)***
*The selective breeding of coldwater freshwater species such as Goldfishes and Koi began in the Far East. Such fishes were usually kept in outdoor ponds, but were sometimes brought indoors.*

***Victorian engraving of a private fishtank (right)***
*Setting up domestic indoor aquariums first became fashionable in the nineteenth century. Early tanks were made from cast-iron and glass, and were often very ornate. Aquariums like this one, which was designed as the centre-piece of a large conservatory, would have housed coldwater freshwater fishes only.*

***Engraving of a public aquarium, 1873 (below)***
*As interest in fishkeeping spread, aquariums such as this one in Brighton, England, were opened for the public.*

today's standards. In fact, they were fraught with danger: oil lamps or naked gas-flames burnt beneath the tanks to heat the water! By the early twentieth century, the advent of electricity provided a safe, sure way to maintain the correct water temperature. As technology improved, the interest expanded so that by the 1930s fish societies began to spring up in Britain and Europe, and by the late 1940s the first fish shows were held.

## The advantages of fishes as pets

Anyone can keep fishes, no matter how small their home, and they are particularly suitable for apartment-dwellers, especially those in high-rise blocks where the keeping of other pets may be impractical or even prohibited. There are other advantages too: fishes won't need exercising, they don't disrupt your home with messy fur or feathers, they can't escape (they can't even exist outside their own environment), and they don't make any noise. All you require is a space that can accommodate an aquarium 60 cm long, 38 cm high and 30 cm front-to-back, with adequate room around it to allow you access for maintenance.

Caring for an aquarium isn't time-consuming as feeding and routine maintenance only require a few minutes of your attention daily,

***The tropical freshwater aquarium***
*Currently the most popular branch of fish-keeping, the tropical freshwater system is also the easiest for the beginner, despite the fact that technical equipment is required. The system covers a large range of fishes, most of which are small, relatively hardy and highly coloured, as this collection of Angelfishes, Guppies, Platies and Characins shows. They require less space than fishes from other systems and many, particularly the livebearers, will breed readily in the aquarium.*

**The coldwater freshwater aquarium (above)**
*Unlike their tropical counterparts, coldwater species such as these bright, metallic Fantail Goldfishes don't need a heated tank. But they do require more space.*

**The tropical marine aquarium (above)**
*Tropical marine fishes such as these Moorish Idols, Angelfishes and Butterflyfishes are expensive, and often difficult to keep.*

**The coldwater marine aquarium (right)**
*Because fishes and invertebrates, like this sea-anemone and starfish, can be collected for free and little equipment is needed, this system is very low-cost.*

supplemented by an hour or so once a week. And no special skills are required – if you have the ability to fix a fuse or change an electric plug, you are amply qualified. All the "technical" aquarium appliances are completely reliable: they are as foolproof as the manufacturers can make them, and have to conform to rigorous safety standards before they can be marketed.

## Will an aquarium cost a lot of money?

The cost of an aquarium can be divided into two distinct areas: initial capital outlay and future day-to-day running and replacement costs. In terms of initial outlay, the only difference between a coldwater and a tropical system is the extra heating equipment needed for tropical species – perhaps another 15 percent of the total price. The system with the highest initial cost is the tropical marine aquarium as the fishes are much more expensive than freshwater types. Freshwater fishes don't vary much in price, whether they are tropical or coldwater species. There are three major reasons for the high cost of tropical marine fishes: their air-freight charges, the difficulty of catching and handling them, and their rarity value (unlike freshwater fishes, they aren't bred commercially). In addition, there are a few other extra costs to be met when you first set up a tropical marine aquarium: coral sand is more expensive than ordinary aquarium gravel, corals cost more than plants (though this is balanced by the fact that there are no on-going plant costs), and you will have to buy a synthetic seawater known as "sea-mix". (Even if you have access to natural seawater, the risk of pollution rules out using it.) In contrast, a coldwater marine system can be set up on a shoestring budget (see *Setting up Case Studies*, p. 192).

Once set up and stocked, an aquarium's running costs aren't astronomically high, even with a tropical tank. Heat conservation measures

(see *Heating*, p. 135) will keep your electricity bill low, you can switch off the lighting at night, and once the correct water temperature has been reached, only a small amount of electricity is needed to maintain it at that level. Although airpumps and filters have to be run constantly, they consume only a few watts of power. In addition, you can economize on food bills by catching or culturing live foods and using leftover household scraps to supplement your fishes' diet (see *Feeding*, pp. 204–5).

## What are the benefits of fishkeeping?

As you become involved with the day-to-day running of the aquarium, you will acquire new knowledge and skills. For example, the fishes' behaviour may develop your interest in biology, and a study of their lifestyles in their original habitats may expand your geographical knowledge. Even working out the number of fishes that are suitable for your aquarium should exercise your mathematical skills! Furthermore, responsibility for the welfare of other living things is an important part of a child's learning process, and fishkeeping is an ideal way to introduce a young person to animal care.

***Choosing fishes from a specialist dealer (left)***
*For the serious fishkeeper, a dealer who specializes in aquarium fishes is the best supplier. This type of store will have a much wider choice of species than a local pet shop, and the fishes will often be in better condition.*

***Siting an aquarium in the home (right)***
*The exterior of your tank will be a feature of the room that it is installed in, and it is therefore important to make it blend with your home decor. The tank in this picture has been set up as the focal point of the living room. It is sited above a built-in cupboard, which makes an ideal storage area for the aquarium equipment. Alternative designs include tanks built into false chimney breasts, and aquariums used as room dividers.*

# 1

# FISH ANATOMY

Although fishes are vertebrates like us, in most other respects they are a completely alien animal. They are designed for an aquatic environment; they balance in and move through water with the aid of their fins, and they breathe by obtaining oxygen from the water via their gills. Their body temperature is usually the same as the surrounding medium (this is what is meant by cold-blooded), and unlike warm-blooded creatures such as ourselves, they can't maintain their body temperature at a chosen, fixed level. So whilst we can survive in the fishes' environment (albeit supported by artificial aids), fishes can't exist in ours.

# The basic fish

The traditional torpedo-shaped image of a fish doesn't reflect the many variations of fish shape that exist. Fishes come from a variety of locations, and their bodies have become adapted to suit these different environments. A fish's shape will tell you a good deal about its living style: what type of system it inhabits, how it feeds, and what kind of swimmer it is.

**The mouth**
The structure of a fish's mouth can reveal its feeding habits. Fishes can be divided into three feeding groups: top-, midwater- and bottom-feeders.

*Top-swimmers' mouths*
This type of fish has a straight dorsal surface, and an upturned, scooplike mouth for gathering floating insects.

*Midwater-swimmers' mouths*
Species that swim in midwater have mouths at the very tip of their snouts, and generally snatch their food as it falls through the water. A few have underslung mouths fringed with rasplike folds, enabling them to graze on algae.

*Bottom-dwellers' mouths*
These fishes have underslung mouths with flattened ventral surfaces which can be brought into close contact with the riverbed where much of their food lies.

**The gills**
The fish's equivalent of lungs, these delicate membrane layers diffuse oxygen into the fish's bloodstream. They are protected from damaging particles in the water by an arch of gill-rakers.

## BODY SHAPE AND PURPOSE

The basic "fusiform" fish shape has evolved to suit different living conditions such as the rate of waterflow in the habitat and the location of food.

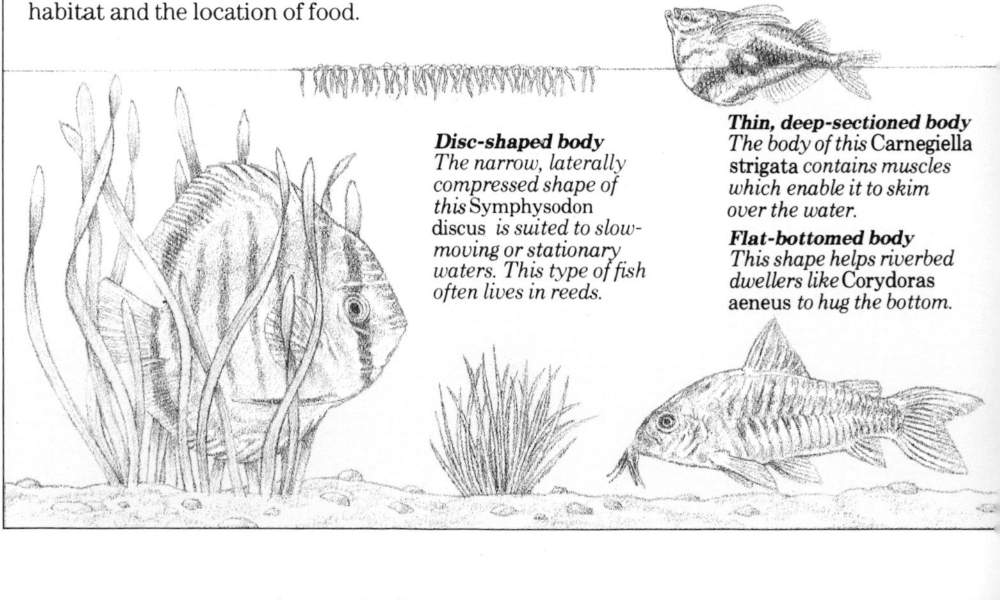

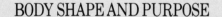

***Disc-shaped body***
*The narrow, laterally compressed shape of this* Symphysodon discus *is suited to slow-moving or stationary waters. This type of fish often lives in reeds.*

***Thin, deep-sectioned body***
*The body of this* Carnegiella strigata *contains muscles which enable it to skim over the water.*

***Flat-bottomed body***
*This shape helps riverbed dwellers like* Corydoras aeneus *to hug the bottom.*

## THE ANATOMY OF A FISH

This cutaway diagram shows the principle organs and structures found in the fish.

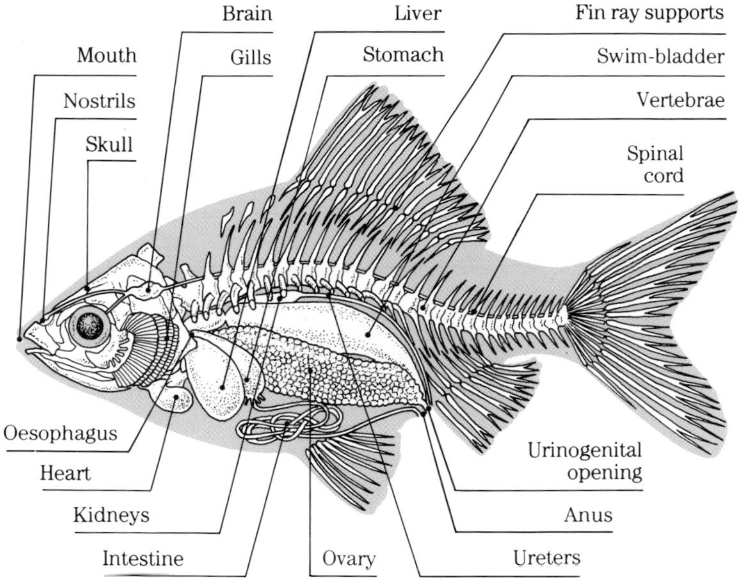


***Mouth position***
*Superior, terminal and inferior mouths indicate surface-, midwater- and bottom-feeders respectively.*

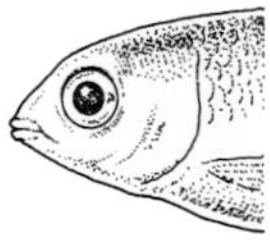

Terminal mouth

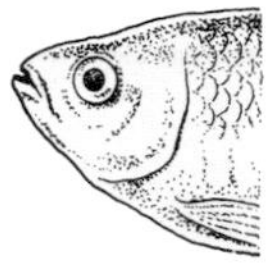

Superior mouth

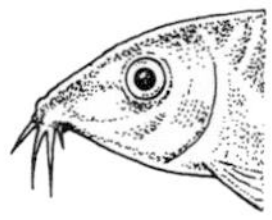

Inferior mouth

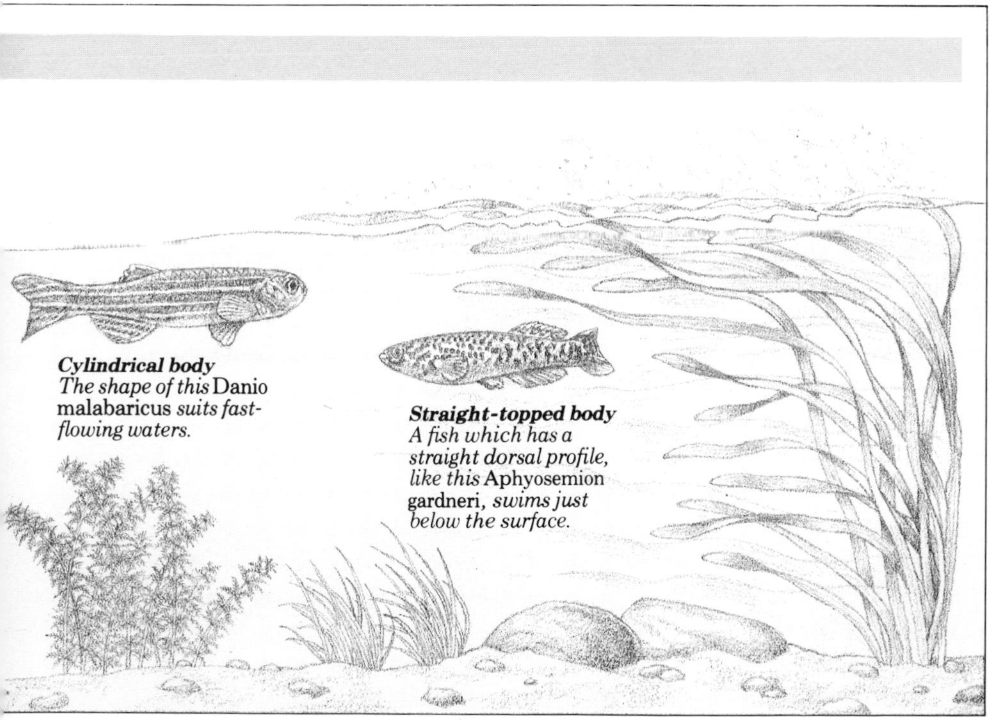

***Cylindrical body***
*The shape of this* Danio malabaricus *suits fast-flowing waters.*

***Straight-topped body***
*A fish which has a straight dorsal profile, like this* Aphyosemion gardneri, *swims just below the surface.*

# The outer covering

Fishes have two layers of skin: a thin outer layer known as the *epidermis*, and a thicker inner layer known as the *dermis*. In most cases, the fish's body is encased in a scaly exterior. These overlapping plates, which grow out of the skin, provide streamlining and protection against injury. The scales are covered in a thin mucus layer that protects against parasites and gives "slip-ability".

### Colour as camouflage

Fishes are colour-shaded from top to bottom, with a dark top and a light underside, camouflaging them against the riverbed from predators above. Other colour patterns serve as species and sex recognition, offer camouflage within the fish's natural habitat, and/or give visual warnings to other species that the fish may be poisonous. Some marine fishes have colour patterns that mimic those of species that they prey on. In some species the eye is hidden in a black area, and a false "eye" is featured elsewhere in the colour scheme to give diversionary protection if the fish is attacked.

## SCALE TYPES

Fishes' scales are either *placoid* or*elasmoid*. Placoid scales are found in Sharks and Rays, and resemble small teeth-like projections from the skin. Aquarium fishes have elasmoid scales, which rise directly from the dermis, and these may be of two types: *ctenoid* or *cycloid*. Some fishes have one sort only, others have both types.

**Ctenoid scales**
These scales have comb or teeth-like extensions to the rear edge.

**Cycloid scales**
This type of scale is round and smooth.

**Scutes**
Armoured Catfishes don't have scales; instead, their bodies are covered with two (or three) rows of overlapping bony plates known as scutes.

**"Naked" fishes**
African Catfishes have neither scales nor scutes, merely skin, and for this reason are often known as "Naked Catfishes".

### How is the fish's colour formed?

Colour is produced in two ways – by light reflection and by pigmentation. Iridescent fishes owe their "sparkle" to light reflecting back from a layer of guanin just beneath the skin.

Colour changes are made in the pigmentation cells, and can be brought about by excitement, fear or hormonal activity. Pencilfishes have a nocturnal colour pattern, and change back to their "normal" coloration at daylight.

### Juvenile coloration

The majority of juvenile fishes look like smaller versions of their parents. However, the colouring and patterning of juvenile marine Angelfishes change radically as the fishes mature.

***A juvenile Angelfish***
*In colour and patterning this young* Pomacanthus imperator *differs from the adult (p.99).*

# The fins

The fins consist of rays which are webbed with tissue; these rays may be "hard" (non-articulated and quite rigid) or "soft" (having many articulations or branches). With the help of small muscles, fins can be folded or extended.

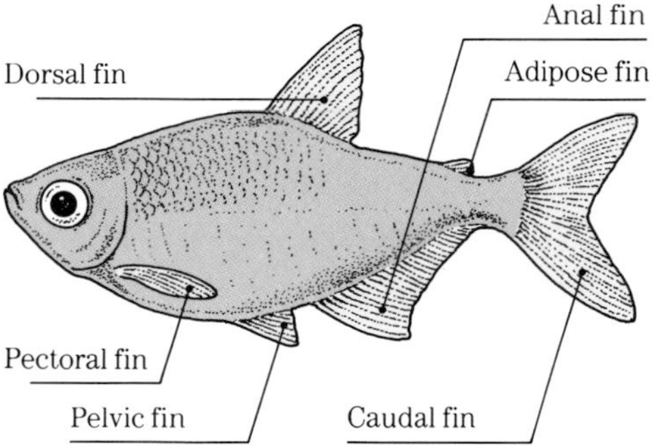


***Fin types***
*A fish usually has seven fins: three singles — dorsal, caudal and anal — and two sets of paired — pelvic and pectoral.*

## The dorsal and anal fins

The dorsal and anal fins are used to keep the fish upright, acting like keels.

In some species, the anal fin has become modified as a spawning aid. In male livebearers (see p. 245) the anal fin has developed into a rod-like tube that enables the sperm to be directed at the female's vent during spawning. The anal fins of some Characins have tiny hooks on them, and it is believed that these help keep the male and female close together when spawning.

## The caudal fin

Better known as the tail, this fin provides the "final drive" to push the fish through the water. Power is generated by muscles in a series of strong, wave-like motions along the length of the body.

## The adipose fin

This small, extra single fin is carried by some fishes on their dorsal surface, between the proper dorsal fin and the caudal fin. It has no rays within its fatty tissue structure, and seems to serve no apparent purpose.

## The paired fins

Fishes manoeuvre by means of the paired fins. These are the pectoral fins which are situated just behind the gill cover, and the pelvic or ventral fins which emerge just in front of the anal fin. These fins correspond roughly to the limbs of mammals, and can be used in a wide variety of ways.

### CAUDAL FIN SHAPE

The shape of the tail often indicates the swimming habits of the fish.

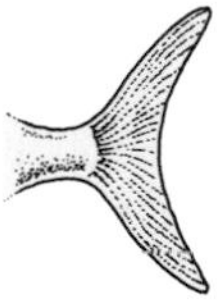

***Crescentic*** *Found in some continuous, high-speed swimmers.*

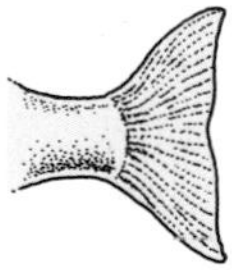

***Emarginate*** *Found in slow-movers capable of fast dashes.*

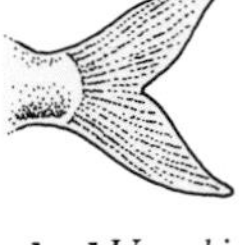

***Forked*** *Usual in continuous, high-speed swimmers.*

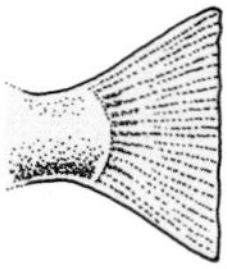

***Truncate*** *Usual in slow-movers capable of few fast dashes.*

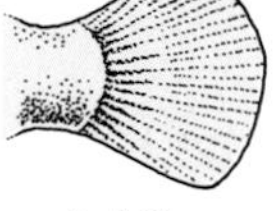

***Rounded*** *Common in very slow-moving and cultivated varieties.*

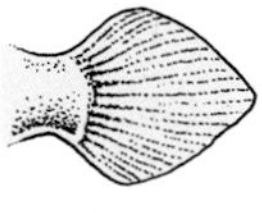

***Pointed*** *Found in some slow-moving and cultivated varieties.*

*Uses of pectoral fins*

- For the fish to spin around on its axis (achieved by moving one pectoral fin in the opposite direction to the other).
- To act as brakes (the fish sticks out two pectorals from its body at the same time).
- To gently fan water over the fish's eggs.
- To dislodge food from the bottom of the aquarium.

*Uses of pelvic fins*

- To act as hydroplanes.
- Female *Corydoras* Catfishes carry their fertilized eggs to the hatching site between their pelvic fins.
- Freshwater Angelfishes have very hard, filamentous pelvic fins which they often use to threaten members of the same species, particularly any trespassing rivals.
- Gouramis use their pelvic fins to search around the aquarium; they have taste cells at the tip and can locate food with them.

**"FANCY" CAUDAL FINS**

Many Fancy Goldfishes and aquarium-developed strains of tropical fishes have exaggerated caudal fins known as Lace- or Veil-tails. Fishes with such fins are generally slow swimmers and aren't found in nature.

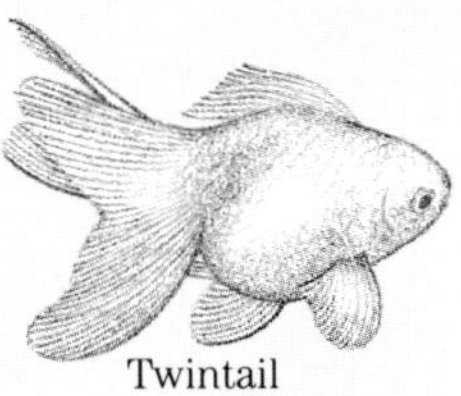

Twintail
see p. 86

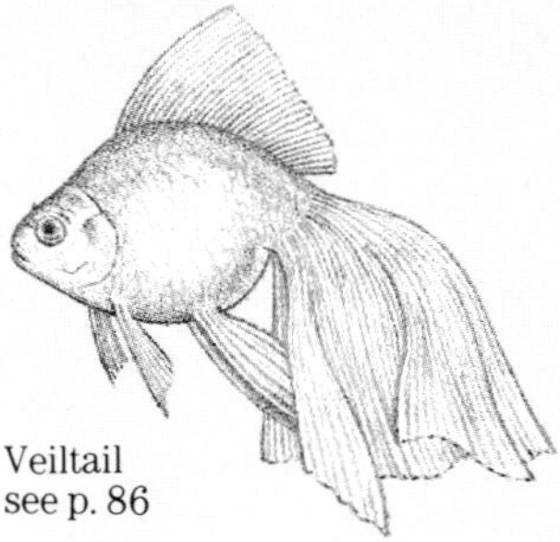

Veiltail
see p. 86

**CAUDAL FIN EXTENSIONS**

The presence of extended rays on the caudal fin of a fish such as a Swordtail or an Emperor Tetra is often a reliable indication that the fish is male. These extensions, which are found in wild fishes, serve no apparent purpose.

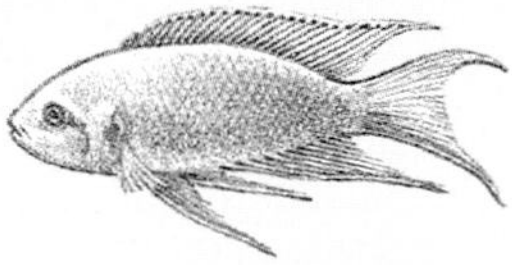

*Lamprologus brichardi*
Lyretail Lamprologus
see p. 58

*Nematobrycon palmeri*
Emperor Tetra
see p. 51

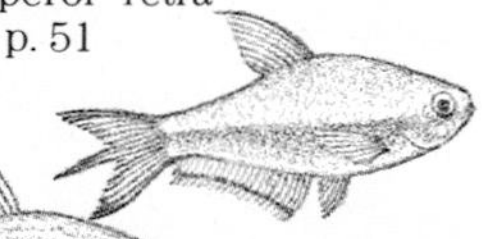

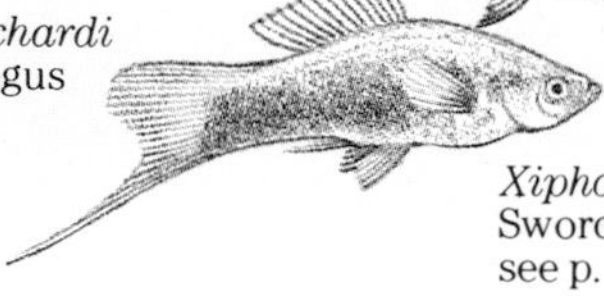

*Xiphophorus helleri*
Swordtail
see p. 68

**DEFENSIVE FINS**

The long, hollow spines of the Lionfish are poisonous, and are used in self-defence. Triggerfishes have a fin which can be locked in an erect position to prevent capture. This fin normally lies just in front of the dorsal fin.

*Pterois volitans*
Lionfish
see p. 107

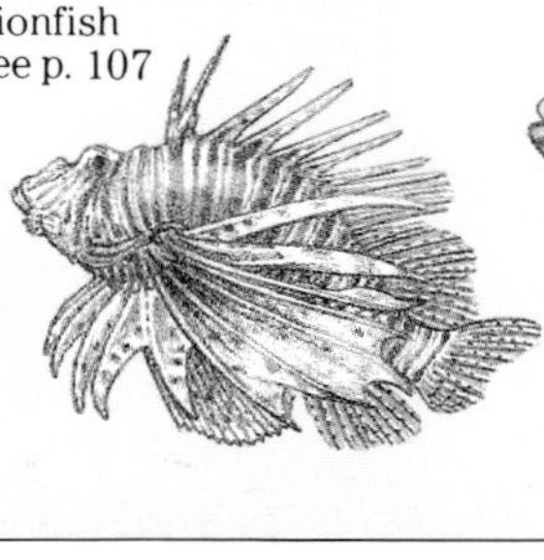

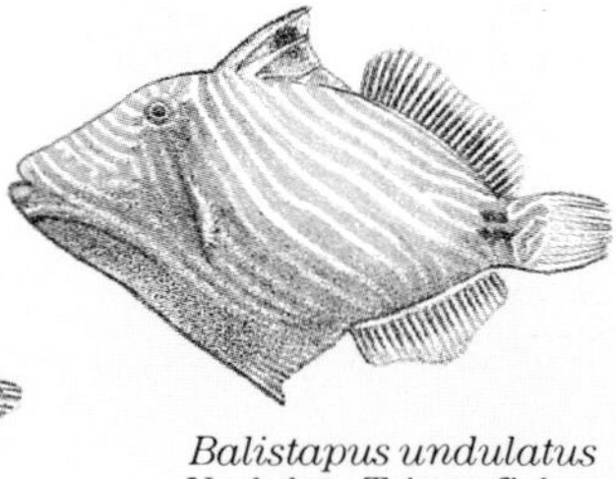

*Balistapus undulatus*
Undulate Triggerfish
see p. 104

# Bodily functions

A basic knowledge of how your fishes function will help you to understand their needs and habits.

**Respiration**
Fishes breathe oxygen which is dissolved in the surrounding water. They do so by taking in water through the mouth and expelling it through the gills. As it passes across the delicate gill membranes, oxygen is absorbed into the blood and carbon dioxide expelled. A certain amount of ammonia may also be released via the gills, and in freshwater fishes some water is released too.

Some fishes, notably Anabantoids (see p. 60), can breathe atmospheric air via a special labyrinthine chamber in the head behind the gills; others such as *Corydoras* Catfishes (see p. 74) can process it in the hind part of the gut.

Fishes with suckermouths – such as the Suckermouth Catfish, *Hypostomus* (see p. 76) – breathe through extra slits behind the head, thus releasing their special mouth to carry out the more important duties it evolved for. These include feeding and maintaining position in fast-moving water by clinging onto rock surfaces.

### THE LABYRINTH ORGAN

Found in Anabantoid fishes (see p.60), this organ consists of rosette-shaped plates which carry hundreds of blood vessels that absorb oxygen from inhaled atmospheric air.

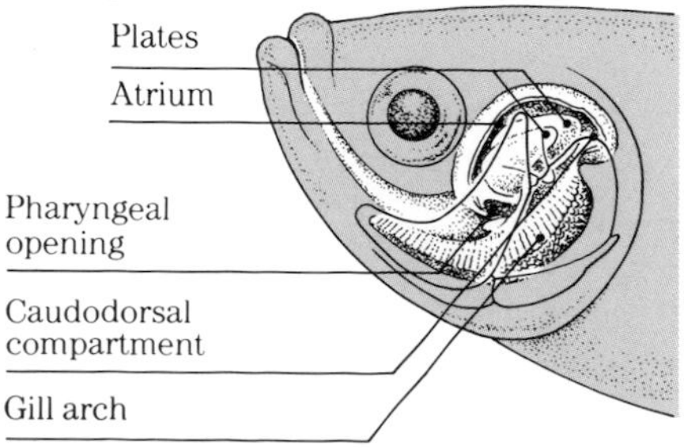


**Do fishes sleep?**
Because they don't have eyelids, and therefore can't close their eyes, it is sometimes assumed that fishes don't sleep. However, they need rest, and this takes the form of suspended animation, where the fish lies motionless for several hours. Some marine species such as the Wrasses (*Labridae*) may bury themselves in the coral sand or spin "sleeping bags"—cocoons of mucus – each night to sleep in.

**Excretion**
In addition to the usual disposal of waste products from digestive processes, fishes excrete ammonia from the gills. Freshwater species also excrete water from the gills. Moreover, fishes deposit waste products such as guanin within their own bodies (usually just under the skin). It is these guanin deposits which contribute iridescence to fishes' colourings.

**Body fluid levels**
Strange as it may seem, fishes have a drinking problem, despite being surrounded by water.

*Saltwater fishes*
The concentration of blood salts in marine species is lower than the salt concentration of the surrounding water. Due to a phenomenon known as osmosis, water is continuously lost from fishes' bodies, and to make up for this loss they must drink. By drinking copiously, passing very small quantities of urine and excreting excess salt they maintain their body-fluid levels.

*Freshwater fishes*
The situation is reversed in freshwater, where the blood concentration is higher than the surrounding water. In this instance, water is always being absorbed into the body. So, to avoid bursting, freshwater fishes have to excrete large amounts of water (up to 10 times their body weight daily!). They do this in two ways: as urine, and through the gills.

**The sense of smell**
Fishes smell through their nostrils, which, unlike ours, aren't used for breathing at all. They consist of two or four openings on the front of the snout, connected directly to the olfactory system. The piscine sense of smell assists in the detection of pheromones – for example, creating a "fear reaction" when one fish in a shoal releases the fear pheromone into the water – and helps to locate food or spawning areas.

**The sense of taste**
In fishes, the taste buds are primarily concentrated in the mouth, tongue and lips. However, they may also occur over other parts of the body, the pelvic fins and, of course, on the barbels of bottom-dwelling species.

**Sight**
In most species, vision is monocular – they can see in two directions, but can't focus both eyes on the same object at the same time. However, where the eyes are located high on the tip of the head some degree of binocular vision may occur – they can focus both eyes on the same object at once, giving a stereoscopic effect. Fishes can only focus up to 45 cms, but they can detect things much further off via their "lateral line system". Fishes are able to respond to colours, but may be confused by varying brightnesses.

## EYE STRUCTURE

Unlike our eyes, where the curvature of the lens is altered in order to focus the image on the retina, in a fish's eye the lens shape isn't altered; instead, the lens itself is moved backwards or forwards.

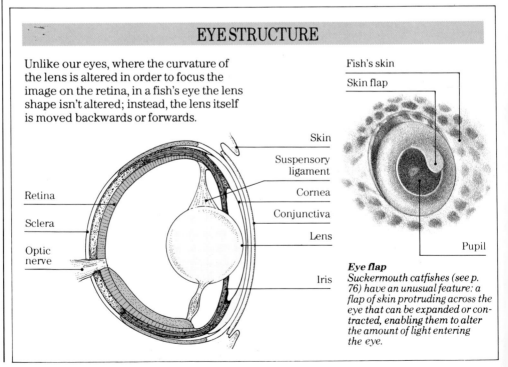

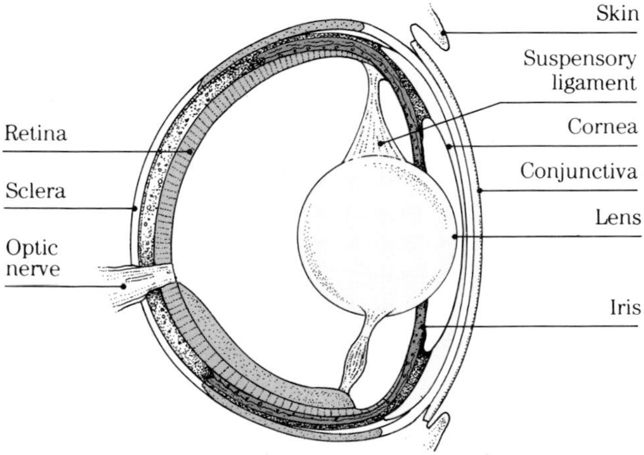


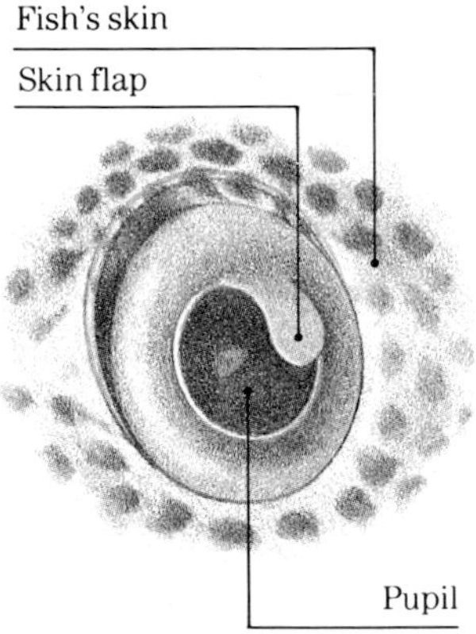


***Eye flap***
*Suckermouth catfishes (see p. 76) have an unusual feature: a flap of skin protruding across the eye that can be expanded or contracted, enabling them to alter the amount of light entering the eye.*

### The eyelid
Unlike terrestrial animals, fishes have no eyelids because they have no need to keep their eyes moistened.

### Hearing
Fishes' "ears" are much less complicated than ours, being made up of the equivalent of our inner ear only, since the connecting apparatus of the human middle and outer ear is unnecessary. The reason for this is that water is a very dense medium, and therefore sounds or vibrations — which travel five times faster than through air — are more easily detected.

Although fishes can hear sounds over much the same frequency range as we can, different species are sensitive to only parts of that range. Fishes' hearing is supported by their ability to sense vibrations. These may be detected by the swim-bladder and passed on to the ear by means of the Weberian Ossicles (a series of interconnecting bones), or by a tubular structure, resembling a doctor's stethoscope, that connects the swim-bladder with the ear.

### The lateral line system
This system is the fish's "sixth sense", enabling it to detect vibrations and currents. It gets its name because the inputs to the vibration-sensing nerves are visible as a line of tiny perforations or "portholes" through a row of scales along the lateral line of a fish. This line is often incomplete, or may be extended right over the top of the gills to the nostrils.

### The swim-bladder
With the exception of a few bottom-dwellers such as Gobies, most aquarium fishes have a swim-bladder; this organ enables them to maintain their position at any level in the water. In some fishes it serves other purposes: it may act as an amplifier for any sound that the fish is able to produce, or it may supplement the ear.

#### NAVIGATIONAL AIDS

- The lateral line system helps the fish to find its way around unimpeded
- Migrating fishes may navigate by the position of the sun or stars
- Locally, fishes use electricity to navigate. Weak electricity-generating muscles set up an electric field around the fish, which is affected by any obstacle in the vicinity. Detecting differences in the strength of this field helps the fish to recognize its surroundings or the presence of other fishes

***Lateral line navigation*** *Fishes that live in dark, muddy waters or that lack eyes — for example, this* Astyanax mexicanus, *the Blind Cave Fish (**see p. 46**) — navigate perfectly well using the lateral line system to inform them of obstacles, other fishes or moving food in their vicinity.*

# 2

# CHOOSING FISHES

To help you in your selection, this chapter outlines the four basic systems of fishkeeping: tropical freshwater, coldwater freshwater, tropical marine and coldwater marine. You have two choices to make: the first is to choose one of the four systems, and the second is to select particular species from the wide range of fishes that can live in your chosen system. Although you may have already made up your mind what sort of fishes you want to keep, it is worth examining all the choices before you actually take the plunge. Once you have made your choice of system, your next step is to read the information on fishes in the *Species Guide* (see pp. 32–111) so that you can make a selection. Before you set off to buy your chosen fishes read the information given here to help you select healthy, compatible stock.

# Choosing the system

Before you choose your fishes, you must decide whether you want a freshwater or a saltwater system, and within these two categories whether you would prefer a coldwater or a tropical aquarium.

**Freshwater aquariums**
The vast majority of aquarium-kept fishes are freshwater species. For fish-keeper's purposes they are divided into two groups – coldwater and tropical. The table opposite sets out the main differences between the groups. In historical terms, the coldwater aquarium is the oldest system, but nowadays the tropical type has the most devotees. The reason for this is quite simple – tropical fishes have two advantages: you can keep more of them in a given aquarium size, and there is a far wider choice of brilliant colours. Coldwater fishkeeping has, until fairly recently, evolved around a single species, the Goldfish, and its aquarium-developed fancy varieties.

**Saltwater aquariums**
Marine or saltwater fishkeeping is the youngest branch of the hobby, and is also divided into coldwater and tropical groups. Of these, the most popular is the tropical. This is because there is an almost infinite variety of highly coloured specimens to choose from. And, obviously, these attractive fishes have a greater appeal than the coldwater types, which are much fewer in number and plainer in appearance. Fishes are collected from the tropical seas of the world (see p. 95).

The main difference between looking after marine fishes and caring for fresh-water types is that marine fishes require saltwater to live in. The best way of providing this is by adding a synthetic commercial "sea-mix" (available from your aquatic dealer) to domestic tap water. Another difference is that you can't put aquarium plants in a marine tank, but this aesthetic disappointment is offset by the fact that you can use the colourful, decorative corals.

If the cost of tropical marine specimens seems prohibitive, but you are determined to keep marine fishes, you could keep an aquarium stocked with small native, coldwater fishes collected from the local seashore's rockpools. This is a very good system for fishkeepers on a budget, as stocks can be replenished free of charge and, provided care is taken to collect clean supplies, natural seawater can be used.

If you have a pioneering spirit, and can accept the occasional setback, then marine fishkeeping may be just the thing for you. However, you should gain practical experience with freshwater fishes first before tackling the more demanding marine aquarium.

## THE FISHLESS AQUARIUM

Although it may seem a contradiction in terms, some enthusiasts keep a marine aquarium that, instead of housing fishes, contains invertebrates such as sea-anemones, tubeworms, shrimps and living coral. This type of aquarium may be stocked either with locally captured coldwater species, or commercially obtained, imported specimens.

An invertebrate aquarium actually *needs* to be kept free of fishes because they will eat the invertebrates. Many invertebrates are filter-feeders, drawing nourishment directly from miniscule plankton in the water and therefore the filtration should be switched off when you add the food and turned back on once they have fed.

## SELECTING A SALTWATER SYSTEM

| Factor | Coldwater (Salt) | Tropical (Salt) |
|---|---|---|
| **Tank size and type** | Fairly large space-to-fish ratio, must be all glass | Fairly large space-to-fish ratio, must be all glass |
| **Water** | Collected seawater or synthetic mix | Synthetic mix only |
| **Heating and temperature** | None required, ambient temperature | A heater and thermostat without exposed metal parts. Temperature around 24°C. |
| **Lighting** | Fluorescent or tungsten | Fluorescent or tungsten |
| **Filtration** | Powerful | Standard |
| **Choice of species** | What you can find | Very high |
| **Feeding** | Manufactured or live | Manufactured or live |
| **Care** | May be liable to disease if from polluted sources | Reasonably hardy |
| **Cost of fishes** | Travel between seashore and aquarium | Quite high |
| **Breeding** | Very rare in aquarium, sexes difficult to distinguish | Very rare in aquarium, sexes difficult to distinguish |

## SELECTING A FRESHWATER SYSTEM

| Factor | Coldwater (Fresh) | Tropical (Fresh) |
|---|---|---|
| **Tank size and type** | Fairly large space-to-fish ratio, any type | Small space-to-fish ratio, any type |
| **Water** | Domestic supply | Domestic supply |
| **Heating and temperature** | None required, ambient temperature | Any type, around 24° C |
| **Lighting** | Fluorescent or tungsten, Grolux bulbs aid plants | Fluorescent or tungsten, Grolux bulbs aid plants |
| **Filtration** | Powerful | Standard |
| **Choice of species** | Limited number of ornamentals | Hundreds of species |
| **Feeding** | Manufactured or live | Manufactured or live |
| **Care** | Most are hardy, a few require special care | Most are hardy, a few require special care |
| **Cost of fishes** | Inexpensive | Medium to expensive |
| **Breeding** | Egglayers – possible in aquarium, sexes sometimes recognizable | Egglayers and livebearers – possible in aquarium, sexes often recognizable |

# Buying fishes

To make sure that you get off to a good start, you must select the best stock. This is particularly important with your very first aquarium as it will take you (and the fishes) a little time to settle down into a routine, and your fishes must have enough inbuilt stamina to get through any troubles that arise as a result of your inexperience. Don't be tempted by the more exotic species at first; they will be more expensive and, unless you have enough experience to see to their exacting requirements, you could be disappointed.

### Selecting healthy fishes

If you are a regular visitor to your local aquatic shop, you will soon spot any

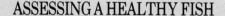

## ASSESSING A HEALTHY FISH

The key to selecting a healthy specimen is to pick a fish that looks fit and active. So when you choose a new specimen for your aquarium, you should look for one that displays all the points shown in the illustration below.

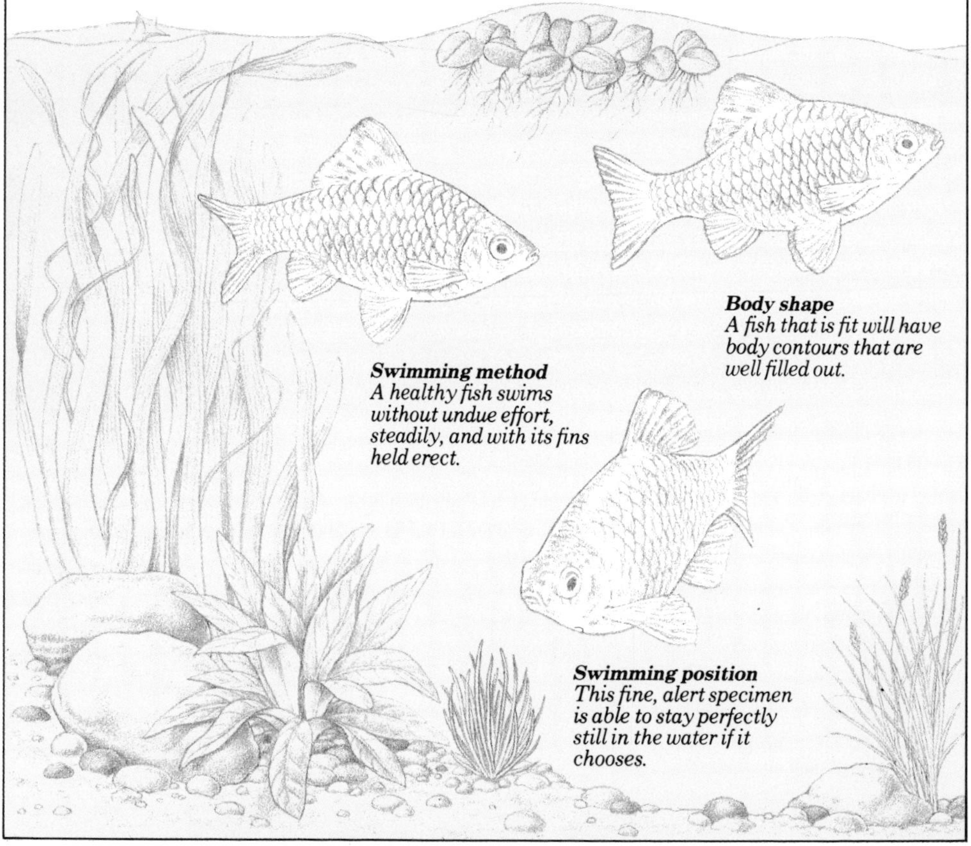
***Body shape***
*A fish that is fit will have body contours that are well filled out.*
***Swimming method***
*A healthy fish swims without undue effort, steadily, and with its fins held erect.*
***Swimming position***
*This fine, alert specimen is able to stay perfectly still in the water if it chooses.*

"new arrivals". If you want to buy one of these it is always a good idea to find out if the fish has been in stock for quarantining for two or three weeks, prior to being offered for sale. You should avoid fishes that are sold "straight off the plane" as they may die from disease or stress. A good dealer will often recommend that you don't buy new arrivals straight away, but wait until they have settled down. The dealer will reserve the fishes for you meanwhile.

**What to look for in a fish**

There are a number of visual "clues" that will help you select good quality stock. The colour of the fish should be dense, and where a pattern is formed there should be no blurring of adjacent colours. In species where the colour patterns are a special feature, the patterns should conform to the standard expected. Also, the body shape should conform to the recognized show "standards".

## ASSESSING AN UNHEALTHY FISH

The best way to avoid choosing an unhealthy fish is by careful observation; get to know the outward signs of illness.

The problems shown below are the most common. For further information on diseases see pp. 220—235.

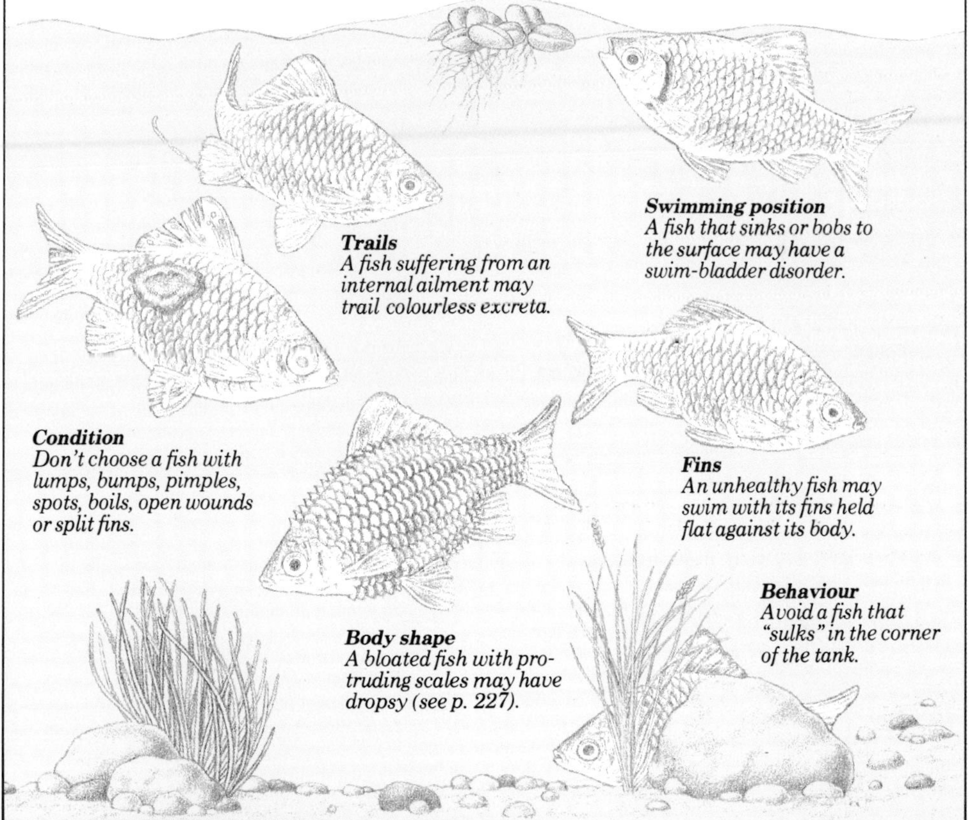
**Swimming position**
*A fish that sinks or bobs to the surface may have a swim-bladder disorder.*
**Trails**
*A fish suffering from an internal ailment may trail colourless excreta.*
**Condition**
*Don't choose a fish with lumps, bumps, pimples, spots, boils, open wounds or split fins.*
**Fins**
*An unhealthy fish may swim with its fins held flat against its body.*
**Behaviour**
*Avoid a fish that "sulks" in the corner of the tank.*
**Body shape**
*A bloated fish with protruding scales may have dropsy (see p. 227).*

Be careful not to interpret normal behaviour as a sign of illness. For example, although flat fins are usually a sign of sickness, some healthy marine species keep their fins folded when swimming. And there is an important difference between an "off-colour" fish and one like the Pyjama Cardinalfish (*Apogon nematopterus*) that lays up during the day because it is nocturnal.

**Considering fish size**

Once you have decided that your chosen fish looks healthy, you should consider whether its size is right for your aquarium. The fish may look just the size you want in the dealer's tank, but bear in mind that all the fishes are juveniles, and their eventual size will be at least double their present size. It is possible that you could buy a number of juveniles of mixed species that are all of the same size, but that in time some will grow big enough to eat the others. To avoid this, always try to find out what adult, or maximum aquarium, size a fish will attain before you buy it. A useful "sizing" tip to bear in mind is that fishes with big eyes or scales usually turn out to be large.

**Sources for fishes**

Unless you are a member of a local aquatic society, your only source of fishes will be the aquarium dealer. There are differences between dealers: some may be general pet stores or garden centres, with a section for pond or aquarium fishes, whilst others may be specialist aquarium shops. If you are looking for equipment or "dry goods" such as food, then the nearest stockist will do. But if you are looking for fishes, you will find that general pet shops don't carry a comprehensive range – they usually limit their stock to the most popular species. If you want to find rarer species you will have to go to a specialized dealer. Dealers obtain their fishes through a central importing wholesaler; a few may import their fishes directly. A good dealer should be able to get you any fish (albeit at a price), no matter where or how his or her stock is obtained.

**PURCHASING DON'TS**

- Don't buy an ailing fish in the mistaken belief that you can nurse it back to health
- NEVER buy an apparently healthy fish from a tank in which there are dead ones
- Don't be tempted to buy rare species if you are a beginner
- Don't buy new arrivals immediately – let them get accustomed to local water conditions at the dealer's
- Don't buy a fish with a "humped back" – this is usually a sign of age
- Don't buy young fishes without taking into account their adult aquarium size
- Don't buy a single fish of a gregarious species
- Don't put new fishes straight into your tank – use the equalization method (opposite) to acclimatize them to the new water

**How are fishes collected?**

In temperate countries native marine fishes can be caught relatively easily. The best place to collect them for yourself is in a rockpool, using a net. Freshwater fishes aren't so easy to obtain yourself as often there are legal restrictions on taking fishes from rivers and lakes. However, coldwater freshwater fishes are inexpensive and readily available from aquatic dealers.

Tropical freshwater fishes are captured with a net, usually from a boat. Marine fishes are caught by divers, who herd them into a fish trap by hand.

## TRANSPORTING FISHES SAFELY

Once you have chosen and paid for your fishes, your next task is to get them home. Most shops put fishes in a plastic bag filled with water. This is a perfectly acceptable way of moving fishes in the summer, or for short journeys. But in colder weather it is a good idea to obtain a suitable insulated carrying container.

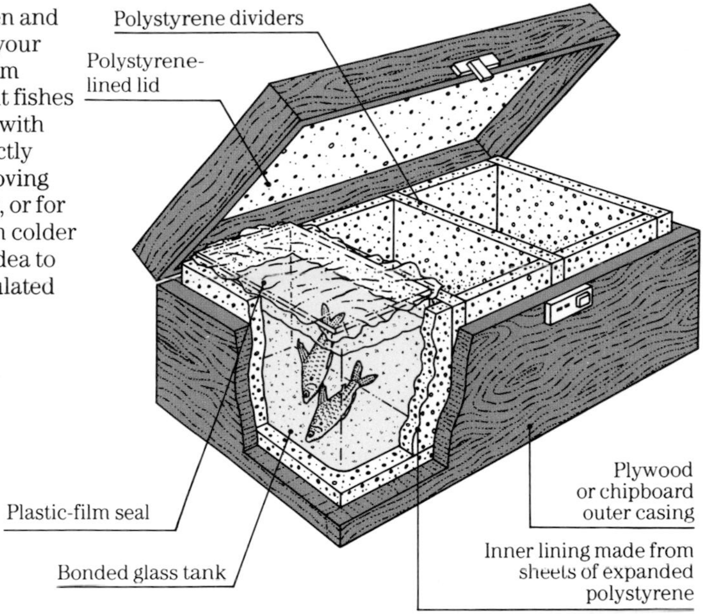


***Transportation box***
*Make an outer casing from plywood or chipboard sealed with polyurethane. Line the inside with polystyrene, and slot all-glass tanks into the lined compartments.*

### Introducing new specimens into an aquarium

When you get your newly bought fishes home you must add them to your tank very carefully. There will probably be a difference in temperature between the water in the fishes' bag and that in the tank, and any such sudden change may stress the fishes.

The best way to equalize the two temperatures is to float the bag containing your new fishes in the aquarium for 10–15 minutes before you release them. During this equalization process introduce a little of the aquarium water into the fishes' bag in case the two bodies of water are of slightly different qualities. And if you are introducing marine fishes into a new tank, you should switch the aquarium lights off, but leave the room lights on.

**Releasing new fishes**
**1** *Float the bag for 10–15 minutes to equalize the water temperatures.*

**2** *Give existing fishes food to distract them, so they don't fuss the newcomers. Then release the new ones.*

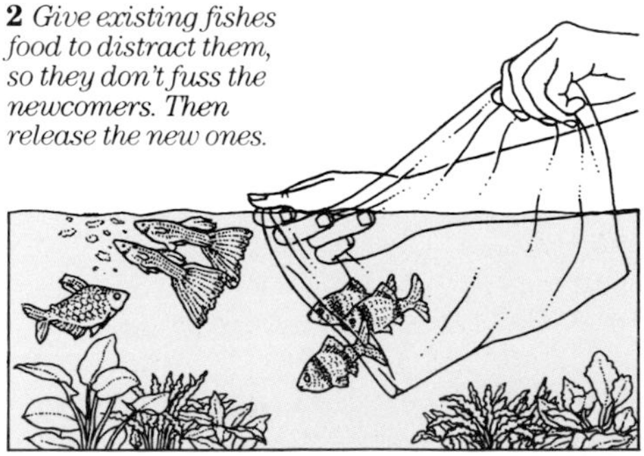

# Grouping fishes for compatibility

Fishes vary in their need for companionship. Some need company, others can be anti-social. It is very important to select a compatible group of fishes for your aquarium. Information on the sociability of individual species is given in the *Species Guide* chapter (see pp. 32–111).

### Sociable fishes

Many fishes are gregarious, and need the companionship of their own species. If you keep a single specimen of any such species you may run into problems, even if it is in an otherwise crowded tank. It won't show off its full colours, it may hide or sulk in a corner, or, worse still, it may develop an aggressive habit like fin-nipping (this may be a sign not of viciousness but of boredom). Such fishes need to be kept in a group of their own species to give them a feeling of security. Moreover, it is a simple fact that a group or "shoal" of fishes of the same species often looks far better than one or two mixed in with other fishes. You will get a very attractive display if you take advantage of this fact and set these fishes up in a "one-species-only" aquarium.

### Anti-social fishes

Fishes can be very territorially minded in their habits, even with members of their own species. Some fishes may fight continuously, especially if the tank isn't large enough for each fish to have its own area. These battles may be more frequent at breeding times. Quarrelling amongst males is quite common, and with species like the Siamese Fighting Fish you should impose a rule of "one male only in each tank". And some marine fishes, such as *Abudefduf oxyodon* (Blue-velvet Damselfish), are a complete contrast to the gregarious fishes; they won't tolerate the presence of another member of their own species in the aquarium.

Feeding habits can also cause problems. Carnivores may attack other fishes, and species that like green stuff may denude the tank of plants. Keep carnivores with same-sized or larger fishes, and don't plant display specimens in a tank of vegetarian fishes.

### Breeding potential

When you buy fishes that you plan to use for breeding purposes, quality considerations become even more important, as any defects in the adults will be passed on to the young.

Although the majority of freshwater fishes can be sexed at breeding time, this isn't always possible with juveniles. If this is the case, buy six or so of each species; this should ensure that you have some of each sex.

### Breeding compatibility problems

Compatibility can, paradoxically, be taken too far. With livebearing fishes, you should avoid mixing different colour strains if you want to keep them pure, as these fishes will mate with each other regardless of colour pattern.

### LONGEVITY

Generally speaking, the larger the fish, the longer it will live. In the aquarium, fishes are protected from predators, and so may live longer than in the wild. Lifespan varies greatly from species to species – from as little as one year for some "seasonal" types, up to ten years or more for larger freshwater tropical and coldwater fishes. Marine fishkeeping is too new for an accurate prediction of lifespans to be made.

## UTILIZATION OF AQUARIUM SPACE

Not all fishes have the same habits; some are better adapted for taking food from the water surface, others cruise about in mid-water, whilst a third group hardly ever leave the aquarium floor. By selecting fishes from each of these levels you can make full use of every bit of water depth, and swimming space, in your aquarium. Shown here is a selection of very popular species chosen to fill all the tropical freshwater levels in a tank.

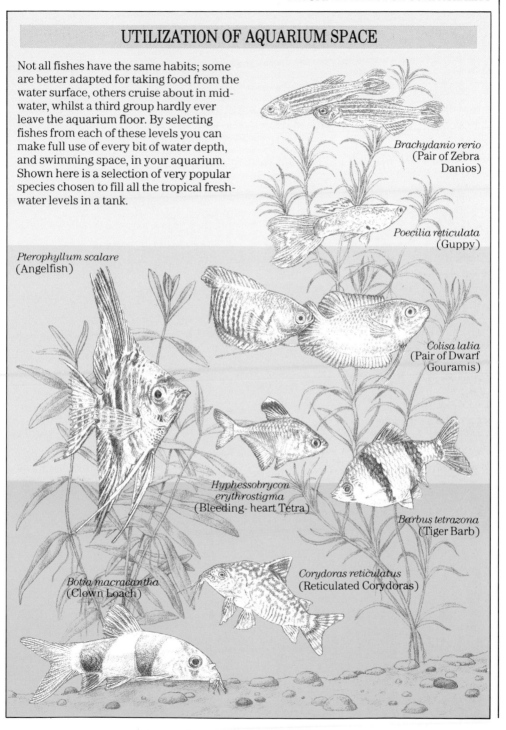

3

# SPECIES GUIDE

An enormous number of fish species are now commonly available as aquarium specimens: there are fishes of every shape, size, colour and habit from both coldwater and tropical, freshwater and marine environments. When choosing suitable species for your tank, you will first need to study their individual requirements, and equally importantly, their compatibility with other fishes. This chapter will help you to do that as it introduces 121 of the most interesting aquarium-suitable species, and gives all the relevant information for maintaining them in a healthy condition. In addition, each fish is rated for ease of keeping as some are ideal for the novice fishkeeper, whilst others demand some experience of aquarium management.

# How to use the species guide

The 121 aquarium-suitable fishes illustrated in this section are divided according to the four fishkeeping systems: tropical freshwater, coldwater freshwater, tropical marine and coldwater marine. Within each system they are further subdivided into the groups by which they are known for showing. Each group is identified by a different coloured band (see right).

Following a general introduction to each group of fishes, a selection of popular individual species are covered, usually in alphabetical order by scientific name (see p. 279), each with a chart outlining details of the species and the necessary aquarium conditions.

**Basic shape identification**

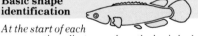

*At the start of each group section, diagrams show the basic body shape of the main types of fishes in the group.*

**Definition of terms**

Most of the terms used in the "Species details" and "Aquarium conditions" charts are self-explanatory, but a few require additional definition:

- **Size** – the length of fish from snout to caudal peduncle (i.e., excluding finnage).
- **Water** Standard mix – refers to synthetic sea-water which you make up (see p. 95).
- **Temperature** – the level at which the species should be kept, but higher temperatures may be required for breeding (see p. 249).
- **Tank type** Species – only that species should be kept in one aquarium, as it isn't compatible with other fishes.
- **Tank type** Community – this species can be kept in a tank housing a variety of compatible fishes.

**TROPICAL FRESHWATER SPECIES**

Cyprinids

Characins

Cichlids

Anabantoids

Livebearers

Killifishes

Catfishes

Loaches

Other tropical egglaying species

**COLDWATER FRESHWATER SPECIES**

Goldfishes

Koi

Other coldwater species

**TROPICAL MARINE SPECIES**

Anemonefishes and Damselfishes

Angelfishes and Butterflyfishes

Other tropical marine species

**COLDWATER MARINE SPECIES**

# TROPICAL FRESHWATER SPECIES

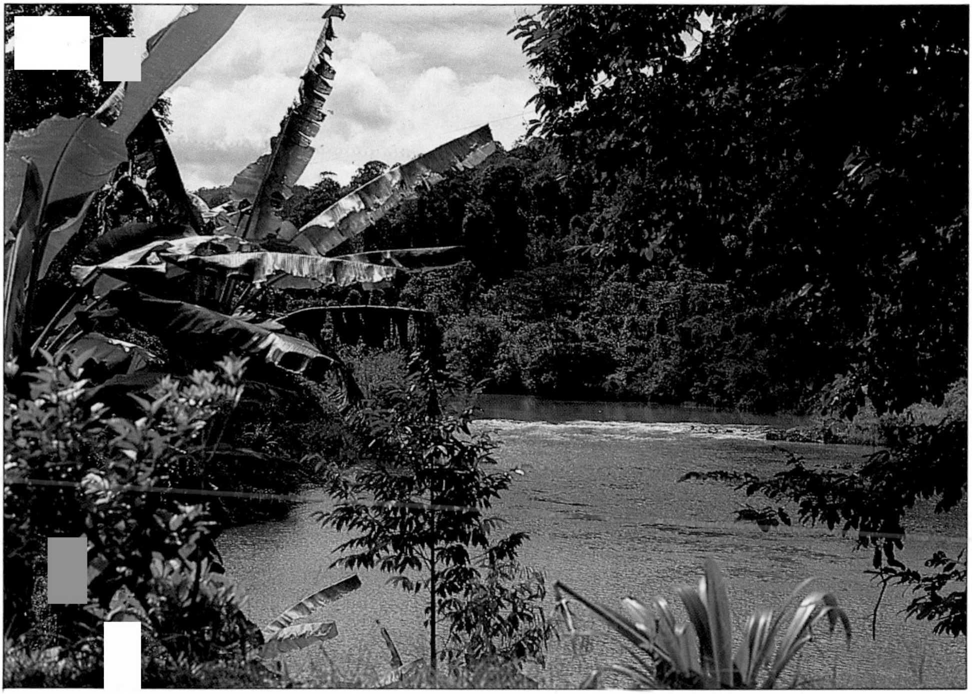

Mainly native to the lakes and rivers of tropical Africa, Asia and America, most of the fishes in this system need to be kept in a heated aquarium at a constant temperature – this figure is generally between 21–7°C. However, apart from a heating system (see pp.130–5), the technical equipment is identical to that required for coldwater freshwater fishkeeping.

The tropical freshwater system is the most popular aquarium type today. There are several reasons for this: the fishes are reasonably small, most are easy to care for, and you can generally keep more of them in any given tank size than coldwater freshwater or tropical marine species. All are highly coloured, and they will eat widely available, commercially produced foods. Many will reproduce readily in the aquarium, displaying a broad range of breeding techniques. With the obvious exception of extremes in size between species, most will tolerate each other's company in captivity.

Thanks to reliable, high-speed air transport and good fish handling, there is a very large selection of good-quality fishes available. However, they must all be put in quarantine for a suitable period before you add them to your tank (see p. 214).

# Cyprinids

Originating in Eurasia, Africa, Asia and North America, the Cyprinidae family includes a wide choice of moderately sized, hardy, undemanding, colourful fishes. Cyprinids have a unique feature: they have no teeth in their jaws; instead they grind up their food using teeth in their throats (known as "pharyngeal" teeth). There are three aquarium-suitable groups in the family – Barbs, Danios and Rasboras. However, there are also many Cyprinid species which don't fall into any one of these categories.

In general, Barbs are very active, midwater and bottom-dwelling fishes. The common name is taken from the generic name *Barbus*, the Latin for bearded. This refers to the "barbels" – whisker-like growths which many Barbs carry at the side of the mouth. These help the fish to search for food on the riverbed or aquarium floor.

Most Danios are active fishes, swim in the upper levels of the aquarium and feed at the water surface. Whereas Barbs are quite happy in still water, Danios are native to faster-flowing, slightly cooler waters, and so they appreciate well-filtered, well-oxygenated water in their tank. Being naturally gregarious, they thrive better when kept in shoals.

Rasboras combine both Barb and Danio characteristics: they inhabit top and midwater levels of the water, and are very active. Like Danios they are happier in shoals.

## Barbus conchonius

*Rosy Barb*

An aquarist's dream, the Rosy Barb is hardy, active, colourful, and a very ready breeder. Mature males change to a deep pink colour when they are ready to breed (hence their common name). Recently, long-finned varieties have been developed.

*Pair of Rosy Barbs (female below)*

SPECIES DETAILS

**Size** 100 mm long
**Origin** N. India
**Sexing** Males have black fins, and a deep copper-pink body at breeding time
**Ease of keeping** Easy
**Food** All foods
**Breeding method** Egg-scatterer, will guard eggs
**Breeding potential** Easy; very prolific

AQUARIUM CONDITIONS

**Water** Soft, medium-hard
**Temperature** 24°C
**Tank type** Community
**Swimming level** Midwater and lower levels

# Barbus cumingi

*Cuming's Barb*

This small, neat-looking Barb is golden-brown in colour, with two distinct black blotches on each side. The leading edges of its scales are dark, creating a net-like pattern all over its body. This species lacks barbels. There are two local variants of Cuming's Barb: one with red fins, the other with yellow.

*Male Cuming's Barb*

SPECIES DETAILS

**Size** 50 mm long
**Origin** Sri Lanka
**Sexing** Females are duller and, at breeding time, plumper
**Ease of keeping** Easy
**Food** All foods, including vegetable matter
**Breeding method** Egg-scatterer
**Breeding potential** Moderately easy

AQUARIUM CONDITIONS

**Water** Soft, medium-hard
**Temperature** 24°C
**Tank type** Community
**Swimming level** Midwater and lower levels

# Barbus nigrofasciatus

*Black Ruby Barb, Purple-Headed Barb*

Like the Rosy Barb, the Black Ruby male undergoes a dramatic colour change at breeding time. His body turns a deep black-red colour, and his head becomes deep purple. When out of breeding colours, the male is a similar colour to the female – pale yellow with dark stripes – but he has more traces of black on his body, and his fins are black, whereas the female's are clear.

*Male Black Ruby Barb*

SPECIES DETAILS

**Size** 65 mm long
**Origin** Sri Lanka
**Sexing** Males are black-red at breeding time; females plumper
**Ease of keeping** Easy
**Food** All foods
**Breeding method** Egg-scatterer
**Breeding potential** Easy; may need higher temperature

AQUARIUM CONDITIONS

**Water** Soft, medium-hard
**Temperature** 24°C
**Tank type** Community
**Swimming level** Midwater and lower levels

# Barbus oligolepis

*Checker Barb, Island Barb*

The common name describes this small, reddish Barb exactly – each scale has a dark edge which gives the fish a checkerboard pattern. It differs from the usual shape of the Barb group (see p.36) by having a more cylindrical body. Wild-caught specimens are better coloured than those bred commercially. And in both wild and cultivated types, the male is usually larger and redder in colour than his mate.

*Female Checker Barb*

SPECIES DETAILS

**Size** 50 mm long
**Origin** Indonesia, Sumatra
**Sexing** Females are duller and smaller
**Ease of keeping** Easy
**Food** All foods
**Breeding method** Egg-scatterer
**Breeding potential** Easy; prolific

AQUARIUM CONDITIONS

**Water** Soft, medium-hard
**Temperature** 24°C
**Tank type** Community
**Swimming level** Midwater and lower levels

# Barbus schwanenfeldi

*Tinfoil Barb, Goldfoil Barb*

This very large fish has a "chromium-plated" appearance to its scales, hence its common name. The metallic colouring covers the whole fish, but for its red dorsal fin. Although the Tinfoil Barb is a peaceful species, its sheer bulk may intimidate or disturb smaller fishes. It has a voracious appetite for vegetation.

AQUARIUM CONDITIONS

**Water** Soft, medium-hard
**Temperature** 24°C
**Tank type** Community, but large fishes only
**Swimming level** Midwater and lower levels
**Special needs** Space

SPECIES DETAILS

**Size** 300 mm long
**Origin** Far East
**Sexing** No visible differences
**Ease of keeping** Easy
**Food** All foods
**Breeding method** Egg-scatterer
**Breeding potential** Moderately difficult

*Group of Tinfoil Barbs*

# Barbus tetrazona

*Tiger Barb*

With its red-brown body fading to silver on the underside and its four distinctive black stripes, it is obvious how the Tiger Barb gained its name. Recently "albino" and "green" varieties have been developed. The reputation of these fishes as fin-nippers may be ill-deserved; keeping them in a group may distract their attention from other fishes' fins.

*Group of Tiger Barbs*

SPECIES DETAILS

**Size** 57 mm long
**Origin** Sumatra
**Sexing** Males have more red in the fins and usually have a bright red nose; females are plumper
**Ease of keeping** Easy
**Food** All foods
**Breeding method** Egg-scatterer
**Breeding potential** Moderately easy

AQUARIUM CONDITIONS

**Water** Soft, medium-hard
**Temperature** 24°C
**Tank type** Community
**Swimming level** Midwater and lower levels
**Special needs** Company

# Barbus titteya

*Cherry Barb*

Similar in build to the Checker Barb (see opposite top), the Cherry has a silver-coloured body with a dark horizontal stripe topped by a golden one. During the breeding period, the male becomes bright red. Imported wild-caught fishes are more intensely coloured than aquarium-bred types. This Barb tends to be shy.

*Pair of Cherry Barbs (female above)*

SPECIES DETAILS

**Size** 50 mm long
**Origin** Sri Lanka
**Sexing** Females are duller and plumper
**Ease of keeping** Easy
**Food** All foods, including green algae scrapings
**Breeding method** Egg-scatterer
**Breeding potential** Moderately easy

AQUARIUM CONDITIONS

**Water** Soft, medium-hard
**Temperature** 24°C
**Tank type** Community, but small fishes only
**Swimming level** Midwater and lower levels

# Brachydanio albolineatus

*Pearl Danio*

The Pearl's delicate colours of pastel-blue marked with gold lines are best seen in natural sunlight, particularly if the light enters through the front glass of the aquarium. The fast-moving Pearl Danio usually keeps to the upper levels, where it is constantly active, looking for food. It is therefore a good idea to fit a lid to prevent jumping.

*Pair of Pearl Danios (female above)*

### SPECIES DETAILS

**Size** 57 mm long
**Origin** S.E. Asia
**Sexing** Females are duller and plumper
**Ease of keeping** Easy
**Food** All foods
**Breeding method** Egg-scatterer; will spawn either in a group or as a pair
**Breeding potential** Easy

### AQUARIUM CONDITIONS

**Water** Soft, medium-hard
**Temperature** 24°C
**Tank type** Community
**Swimming level** Upper levels and midwater
**Special needs** Sunlight; space; lid

# Brachydanio frankei

*Leopard Danio*

The golden-coloured body of this fish is covered with tiny leopardskin-like spots, hence the common name. A slim, fast-swimming species, the Leopard Danio is very similar in shape to the Zebra Danio (see opposite top). And like the Zebra, its dorsal fin is set a long way back on its body. There is some confusion about the origins of the fish, simply because no-one is certain if it is a natural species or aquarium-developed. In either case this Danio's origins lie in the Far East – in mountain streams or in a fishbreeder's establishment.

*Male Leopard Danio*

### SPECIES DETAILS

**Size** 57 mm long
**Origin** Far East
**Sexing** Females are plumper
**Ease of keeping** Easy
**Food** All foods
**Breeding method** Egg-scatterer
**Breeding potential** Easy

### AQUARIUM CONDITIONS

**Water** Soft, medium-hard
**Temperature** 24°C
**Tank type** Community
**Swimming level** Upper levels and midwater

# Brachydanio rerio

*Zebra Danio*

The slender body of the Zebra Danio is covered in alternating horizontal stripes of dark blue and silver or gold (depending on the light). The pattern continues on the anal fin, but the remaining fins are practically colourless. Being easy to breed, it is often recommended as the ideal beginner's breeding fish. This Danio will spawn as a large group as well as simple pairs, but it is an avid egg-eater so protect the eggs (see p. 248). A group of these active fishes will add movement and excitement to your aquarium.

*Female Zebra Danio*

SPECIES DETAILS

**Size** 45 mm long

**Origin** Eastern India

**Sexing** Females are duller and plumper

**Ease of keeping** Easy

**Food** All foods

**Breeding method** Egg-scatterer

**Breeding potential** Easy; but protect eggs from hungry parents

AQUARIUM CONDITIONS

**Water** Soft, medium-hard

**Temperature** 24°C

**Tank type** Community

**Swimming level** Upper levels and midwater

# Danio malabaricus

*Giant Danio*

As its name implies, the Giant Danio is the largest of the Danio group, and requires plenty of swimming room. The fish has a grey-blue back, but the colour fades to a pale pink on the belly. The sides are marked with narrow horizontal bands of blue and yellow, reminiscent of those of the Pearl Danio (see opposite top). This fish is best kept in a group.

*Male Giant Danio*

SPECIES DETAILS

**Size** 100 mm long

**Origin** Sri Lanka, India

**Sexing** Females are deeper-bodied; central stripe runs horizontally on males, but turns up at base of caudal fin on females

**Ease of keeping** Easy

**Food** All foods

**Breeding method** Egg-scatterer

**Breeding potential** Easy

AQUARIUM CONDITIONS

**Water** Soft, medium-hard

**Temperature** 24°C

**Tank type** Community

**Swimming level** Upper levels

# Tanichthys albonubes

*White Cloud Mountain Minnow*

A very hardy, colourful species, the White Cloud Mountain Minnow has an olive-brown body with a distinctive series of iridescent blue and red-brown bands along the flanks. It has a slightly upturned mouth with no barbels. It tolerates cooler temperatures than other Cyprinids, and can be kept outdoors during summer months (see p. 279).

*Pair of White Cloud Mountain Minnows (female above)*

SPECIES DETAILS

**Size** 45 mm long
**Origin** China
**Sexing** Females are plumper
**Ease of keeping** Easy; may be kept outdoors in summer
**Food** All foods
**Breeding method** Egg-scatterer
**Breeding potential** Easy; but protect eggs from hungry parents

AQUARIUM CONDITIONS

**Water** Soft, medium-hard
**Temperature** 24°C in summer, lower in winter
**Tank type** Community
**Swimming level** Upper levels and midwater

# Rasbora heteromorpha

*Harlequin Fish*

Basically silver in colour, the Harlequin Fish has gained its common name from the blue wedge-shaped marking which starts beneath its dorsal fin and runs to its tail. Its method of breeding differs from that of many other Cyprinids as it deposits its eggs rather than merely scattering them in the water. The usual site for this is the underside of a broad-leaved aquatic plant. It is therefore essential to stock the tank with suitable foliage. The Harlequin Fish thrives in a group.

*Female Harlequin fish*

SPECIES DETAILS

**Size** 40 mm long
**Origin** Thailand, Malaya, Indonesia
**Sexing** Females have less clearly defined "triangle"
**Ease of keeping** Easy
**Food** All foods
**Breeding method** Egg-depositor
**Breeding potential** Moderately difficult

AQUARIUM CONDITIONS

**Water** Soft, medium-hard
**Temperature** 24°C
**Tank type** Community
**Swimming level** All levels
**Special needs** Plants

# Rasbora maculata

*Pigmy Rasbora, Spotted Rasbora*

The smallest of the Rasboras, this fish has a red-brown body and a lighter orange underside. It is marked with one or two dark spots, which vary in size. Because it is so small, it should only be kept with aquarium companions of a similar size. However, it flourishes best in a group.

The Pigmy Rasbora's tank should be furnished with driftwood and dark-coloured gravel to contrast well with the fish's colours. Considering the Pigmy Rasbora's size, the female lays a surprisingly large number of eggs – sometimes as many as 200 at a time.

*Pigmy Rasbora*

SPECIES DETAILS

**Size** 25 mm long
**Origin** Malaya, Sumatra
**Sexing** No visible differences
**Ease of keeping** Easy
**Food** All small foods
**Breeding method** Egg-scatterer
**Breeding potential** Moderately difficult; protect eggs from hungry parents

AQUARIUM CONDITIONS

**Water** Soft, slightly acid
**Temperature** 24°C
**Tank type** Community, but small fishes only
**Swimming level** Upper levels and midwater

# Rasbora trilineata

*Scissortail*

This Rasbora is mostly silver in colour, with a black blotch in each lobe of the caudal fin. It is considerably larger than other fishes in the same group, but an aquarium-kept specimen will never grow as large as its wild counterparts. As an active, shoaling fish, the Scissortail requires plenty of swimming space. When standing still, it lives up to its common name by constantly twitching its forked caudal fin open and shut like a pair of scissors.

*Male Scissortail*

SPECIES DETAILS

**Size** 110 mm long
**Origin** Malaya, Sumatra, Borneo
**Sexing** Females are deeper-bodied
**Ease of keeping** Easy
**Food** Live foods and insects preferred
**Breeding method** Egg-scatterer
**Breeding potential** Moderately easy

AQUARIUM CONDITIONS

**Water** Soft, medium-hard
**Temperature** 20–5°C
**Tank type** Community
**Swimming level** Upper levels
**Special needs** Space

# Epalzeorhynchus kallopterus

*Flying Fox*

This torpedo-shaped fish has two stripes running along its flanks – one bright gold-yellow and one black. The flanks themselves are brown at the top and white at the bottom. And its brown dorsal, anal and pelvic fins are tipped with white. When stationary, the Flying Fox rests on its pelvic fins at the bottom of the aquarium. At other times, it will dash off after some titbit. It has a tendency to be quarrelsome, particularly with members of its own species. It spends a lot of time in its own territory at the bottom of the aquarium, where it scrapes algae off the rocks and plants.

SPECIES DETAILS

**Size** 140 mm long
**Origin** Sumatra, Borneo
**Sexing** No visible differences
**Ease of keeping** Easy, but quarrels with own species
**Food** All foods, algae
**Breeding method** Unknown
**Breeding potential** Not known to have bred in an aquarium

AQUARIUM CONDITIONS

**Water** Soft, medium-hard
**Temperature** 24°C
**Tank type** Community
**Swimming level** Midwater and lower levels

*Flying Fox*

# Labeo bicolor

*Red-tailed Black Shark*

This long-bodied fish is very popular, probably because of the striking colour contrast between its jet-black body and scarlet caudal fin. Despite its common name it isn't from the shark family, although the dorsal fin shape is reminiscent of that of the marine species. It may quarrel with its own species.

*Red-tailed Black Shark*

SPECIES DETAILS

**Size** 120 mm long
**Origin** Thailand
**Sexing** No visible differences
**Ease of keeping** Easy; sometimes quarrelsome
**Food** All foods
**Breeding method** Egg-scatterer
**Breeding potential** Difficult

AQUARIUM CONDITIONS

**Water** Soft, medium-hard
**Temperature** 24°C
**Tank type** Community
**Swimming level** Midwater and lower levels
**Special needs** Plants

# Characins

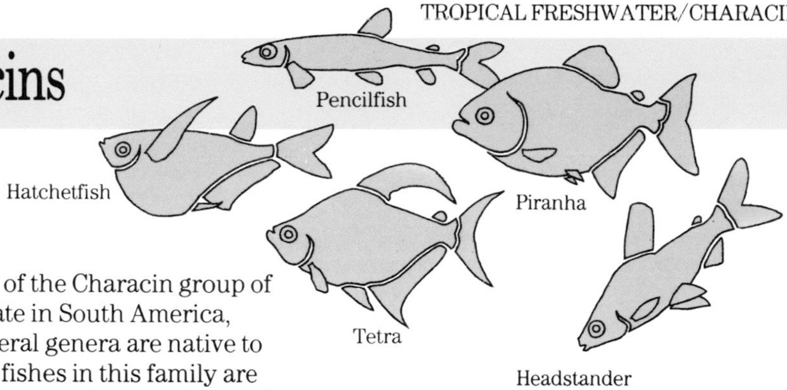


The majority of the Characin group of fishes originate in South America, although several genera are native to Africa. Many fishes in this family are suitable for the aquarium, and range in size from the diminutive Neon Tetra to the Piranha, whose size almost equals its reputation for bloodthirstiness.

Most Characins are equipped with a small extra fin on the dorsal surface. This is known as the *adipose* fin and as yet scientists are unsure of its purpose. The males of some species also have a tiny hook on the anal fin.

This group gives you a choice of species for all water levels (see p. 31).

Hatchetfishes hang just below the surface waiting for unsuspecting insects, and also break the surface to catch food, "flying" through the air by flapping their large pectoral fins. Torpedo-shaped Pencilfishes also prefer upper levels. In contrast, most of the Tetras are midwater swimmers, whilst the head-down attitude of the Headstanders indicates that they feed at lower levels (although they may swim at a variety of levels).

## Anostomus anostomus

*Striped Headstander*

This large Characin has two gold bands running the length of its dark brown body and bright red patches in the otherwise colourless dorsal and caudal fins. Its mouth is relatively small. The Striped Headstander gets its name from the attitude it adopts when at rest. It will remain stationary for quite long periods in a "head-down" position.

AQUARIUM CONDITIONS

**Water** Soft, medium-hard
**Temperature** 24°C
**Tank type** Community, but large fishes only
**Swimming level** All levels
**Special needs** Retreats

SPECIES DETAILS

**Size** 140 mm long
**Origin** S. America
**Sexing** No visible differences
**Ease of keeping** Easy
**Food** All foods
**Breeding method** Egg-scatterer
**Breeding potential** Difficult

*Striped Headstander*

# Astyanax mexicanus

*Blind Cave Fish*

Pale pink in colour (almost albino), this Characin (a variety of the normal-looking Mexican Tetra) may have either colourless or reddish fins. An aquarium curiosity, the Blind Cave Fish has no eyes. It navigates around the crowded aquarium by means of its lateral line system (see p. 21), which detects vibrations from obstacles in the fish's path.

*Female Blind Cave Fish*

SPECIES DETAILS

**Size** 90 mm long
**Origin** Central America
**Sexing** Females plumper
**Ease of keeping** Easy
**Food** All foods
**Breeding method** Egg-scatterer
**Breeding potential** Moderately difficult

AQUARIUM CONDITIONS

**Water** Soft, medium-hard
**Temperature** 24°C
**Tank type** Community or species
**Swimming level** Midwater and lower levels

# Carnegiella strigata

*Marbled Hatchetfish*

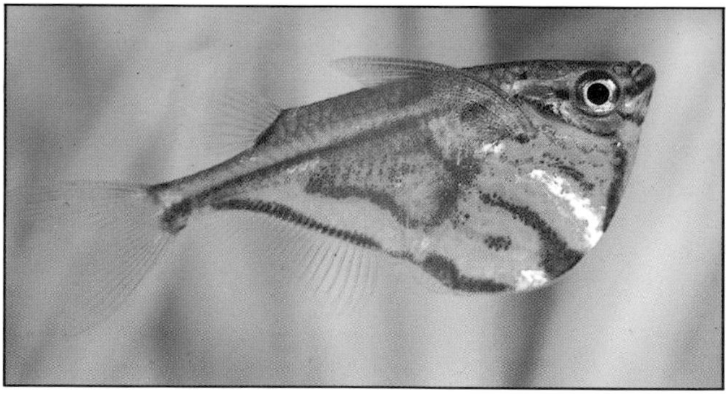

*Marbled Hatchetfish*

This unusually shaped fish is basically white. A gold-tipped line runs from its eye to its caudal fin, and the area below this is attractively marked with a brown marbled patterning (hence its common name). As it can "fly" above the water surface by flapping its pectoral fins, you should fit a cover on the tank. The Marbled Hatchetfish also requires long, trailing-leaved plants in its tank to create shade.

SPECIES DETAILS

**Size** 35 mm long
**Origin** S. America
**Sexing** No visible differences
**Ease of keeping** Easy
**Food** Floating foods
**Breeding method** Egg-scatterer
**Breeding potential** Moderately difficult

AQUARIUM CONDITIONS

**Water** Soft, medium-hard
**Temperature** 24°C
**Tank type** Community
**Swimming level** Upper levels
**Special needs** Plants; lid

# Cheirodon axelrodi

*Cardinal Tetra*

Now reclassified as *Paracheirodon axelrodi*, the Cardinal Tetra has a vivid red coloured lower half, surmounted by an electric-blue band, and a red-brown upper body. The fins are colourless, although the red may extend slightly into the caudal fin. The female fish is usually a little larger than the male. This colourful Characin looks very spectacular when kept in a group. The scientific name of the Cardinal Tetra is a dedication to its discoverer, Dr Herbert Axelrod, the American fish expert.

*Female Cardinal Tetra*

SPECIES DETAILS

**Size** 40 mm long
**Origin** S. America
**Sexing** Females are larger and plumper; males may have tiny hooks on anal fin
**Ease of keeping** Easy
**Food** All foods
**Breeding method** Egg-scatterer
**Breeding potential** Moderately difficult

AQUARIUM CONDITIONS

**Water** Soft, medium-hard
**Temperature** 24°C
**Tank type** Community
**Swimming level** All levels

# Chilodus punctatus

*Spotted Headstander*

This dark-spotted, silver-coloured fish has a dark line running along its flanks from snout to caudal fin, and red eyes. It also has two unmistakable features: a "head-down" resting position, from which it gets its name, and a square-shaped dorsal fin. Although generally regarded by fish-keepers as a Characin, scientifically the Spotted Headstander belongs to the Curimatidae family.

*Spotted Headstander*

SPECIES DETAILS

**Size** 75 mm long
**Origin** S. America
**Sexing** No visible differences
**Ease of keeping** Moderately easy
**Food** All foods, including greenstuffs
**Breeding method** Egg-scatterer
**Breeding potential** Moderately difficult

AQUARIUM CONDITIONS

**Water** Soft, medium-hard
**Temperature** 24°C
**Tank type** Community
**Swimming level** Midwater and lower levels

# Copella arnoldi

*Splashing Tetra*

Formerly known as *Copeina arnoldi*, the plain yellow Splashing Tetra is a fish with a most unusual method of breeding. To safeguard its eggs, the fish deserts its natural habitat and lays them out of water on the underside of an over-hanging plant leaf. The male then takes on the thankless task of splashing them repeatedly with water to prevent them drying out until hatching occurs and the fry drop into the water. In the aquarium, the eggs may be laid on the cover-glass (which makes a good spawning site substitute).

*Male Splashing Tetra*

SPECIES DETAILS

**Size** 80 mm long
**Origin** S. America
**Sexing** Females smaller with less pointed fins
**Ease of keeping** Moderately easy
**Food** All foods
**Breeding method** Egg-depositor
**Breeding potential** Moderately difficult

AQUARIUM CONDITIONS

**Water** Soft, medium-hard
**Temperature** 24°C
**Tank type** Community
**Swimming level** Upper levels
**Special needs** A cover-glass

# Hemigrammus rhodostomus

*Rummy-nosed Tetra, Red-nosed Tetra*

A medium-sized Characin, this fish has a silver-coloured body, a black and white caudal fin and a deep red "cap" on its head, from which it gets its name. A central black line on the flank runs from just behind the dorsal fin into the caudal fin. The male of this species has a tiny hook on the anal fin. The Rummy-nosed Tetra is a shy fish, and is extremely sensitive to changes in water conditions.

*Group of Rummy-nosed Tetras*

SPECIES DETAILS

**Size** 55 mm long
**Origin** S. America
**Sexing** Females are slimmer
**Ease of keeping** Difficult; needs peaceful companions
**Food** All foods
**Breeding method** Egg-scatterer
**Breeding potential** Difficult

AQUARIUM CONDITIONS

**Water** Soft
**Temperature** 24°C
**Tank type** Community
**Swimming level** All levels

# Hyphessobrycon erythrostigma

*Bleeding Heart Tetra*

This large shoaling Tetra is very aptly named, for it has a blood-red mark on its flanks just to the rear of the gill-covers. The body is brown on top, lightening to orange at the belly. There is also a blue-mauve longitudinal band running from head to caudal fin. The well-developed dorsal and anal fins are coloured red and black, and blue and white respectively. Adult fishes are nervous, and may dash about frantically when put in a new aquarium or bare show tank.

*Male Bleeding Heart Tetra*

SPECIES DETAILS

**Size** 70 mm long
**Origin** S. America
**Sexing** Males have longer dorsal fins; females are plumper
**Ease of keeping** Moderately easy
**Food** All foods
**Breeding method** Egg-scatterer
**Breeding potential** Difficult

AQUARIUM CONDITIONS

**Water** Soft, medium-hard
**Temperature** 24°C
**Tank type** Community
**Swimming level** All levels
**Special needs** Plants

# Hyphessobrycon pulchripinnis

*Lemon Tetra*

The body colouring of this fish is a very delicate pale yellow, although the leading edge of the anal fin is bright yellow. The flanks are silver-coloured. The upper part of the eye is bright red, and there is a yellowish adipose fin. Although an "albino" form now exists, most fish-keepers prefer the original colouring. The fish's colours are best seen in a well-planted aquarium against a dark background such as driftwood.

*Male Lemon Tetra*

SPECIES DETAILS

**Size** 50 mm long
**Origin** S. America
**Sexing** Females are plumper; males have tiny hooks on the anal fin
**Ease of keeping** Easy
**Food** All foods
**Breeding method** Egg-scatterer
**Breeding potential** Moderately easy

AQUARIUM CONDITIONS

**Water** Soft, medium-hard
**Temperature** 24°C
**Tank type** Community
**Swimming level** All levels

# Micralestes interruptus

*Congo Tetra*

A relatively large, brown Characin, the Congo Tetra has highly reflective scales, and the male fish shows gold and turquoise iridescences when seen in sidelighting (see p. 280). The female isn't so spectacularly coloured. The dorsal fin of a mature male is very elongated, often over-hanging the caudal fin (which also has central extensions). Irrespective of the fish's sex, the fins are grey-brown with white edges, and are seen to their best advantage where the tank is furnished with dark bogwood. The Congo Tetra should be kept in a sizeable group in a large tank.

*Male Congo Tetra*

SPECIES DETAILS

**Size** 90 mm long
**Origin** Zaire
**Sexing** Males have longer fins
**Ease of keeping** Moderately easy
**Food** All foods
**Breeding method** Egg-scatterer
**Breeding potential** Moderately difficult

AQUARIUM CONDITIONS

**Water** Soft, acidic
**Temperature** 24°C
**Tank type** Community
**Swimming level** Upper levels and midwater

# Moenkhausia pittieri

*Diamond Tetra*

This Tetra has attractive, violet scales with green and gold iridescences, which give it a sparkling appearance. Apart from the colourless pectorals, the fins are grey-violet with white edges. The male has a large, sickle-shaped dorsal fin.

*Male Diamond Tetra*

SPECIES DETAILS

**Size** 60 mm long
**Origin** S. America
**Sexing** Males have larger fins
**Ease of keeping** Moderately easy
**Food** All foods
**Breeding method** Egg-scatterer
**Breeding potential** Moderately easy

AQUARIUM CONDITIONS

**Water** Soft, medium-hard
**Temperature** 24°C
**Tank type** Community
**Swimming level** All levels

# Nannostomus unifasciatus

*One-lined Pencilfish*

This fish has a slender, light brown body, a silver-coloured belly and a single black stripe. Like all Pencil-fish, it takes on a nocturnal pattern of dark blotches, before reverting to its striped marking at dawn.

*Male One-lined Pencilfish*

SPECIES DETAILS

**Size** 70 mm long
**Origin** S. America
**Sexing** Males have rounded anal fin, females' is straight
**Ease of keeping** Easy
**Food** All foods, of small proportions
**Breeding method** Egg-scatterer
**Breeding potential** Moderately easy

AQUARIUM CONDITIONS

**Water** Soft, medium-hard
**Temperature** 26°C
**Tank type** Community of small fishes or species
**Swimming level** Upper levels

# Nematobrycon palmeri

*Emperor Tetra*

The brilliant blue eye of this species is very noticeable, although the bright yellow anal fin, sickle-shaped dorsal fin and centre-spiked caudal fin are all aids to identification. The body is olive-brown with a dark band running along the flanks. The female's dorsal and caudal fins are less developed. The Emperor Tetra can be bred quite successfully in hard water after proper aquarium acclimatization.

AQUARIUM CONDITIONS

**Water** Soft, medium-hard
**Temperature** 24°C
**Tank type** Community
**Swimming level** All levels

SPECIES DETAILS

**Size** 60 mm long
**Origin** S. America
**Sexing** Males have longer caudal fins
**Ease of keeping** Easy
**Food** All foods
**Breeding method** Egg-scatterer
**Breeding potential** Moderately easy

*Juvenile male Emperor Tetra*

# Paracheirodon innesi

*Neon Tetra*

With a bright red coloration topped by an electric-blue flash, this well-known Characin caused quite a stir when first introduced to the fishkeeping world in the 1930s, but it has since been overshadowed by the discovery of gaudier species like the Cardinal Tetra (see p. 47). A small fish, it should be kept in a group. Occasionally, it will fall victim to the incurable Neon Disease (*Plistophora*), characterized by a spreading pale area beneath the dorsal fin.

*Group of Neon Tetras*

### SPECIES DETAILS

**Size** 45 mm long
**Origin** S. America
**Sexing** Females are plumper, deeper-bodied
**Ease of keeping** Easy
**Food** All foods
**Breeding method** Egg-scatterer
**Breeding potential** Moderately easy; not a prolific spawner so can be bred in small tanks

### AQUARIUM CONDITIONS

**Water** Soft, medium-hard
**Temperature** 24°C
**Tank type** Community, but small fishes only
**Swimming level** All levels

# Serrasalmus nattereri

*Red Piranha*

A chunky fish whose pale body is marked with darker spots and a bright red flush to the lower parts, the Red Piranha's most striking physical characteristic is its teeth-filled mouth. Its aggressive nature has been over-emphasized, but it will attack if it smells blood. Only young specimens are suitable for domestic tanks. The fish is also known as *Rooseveltiella nattereri*.

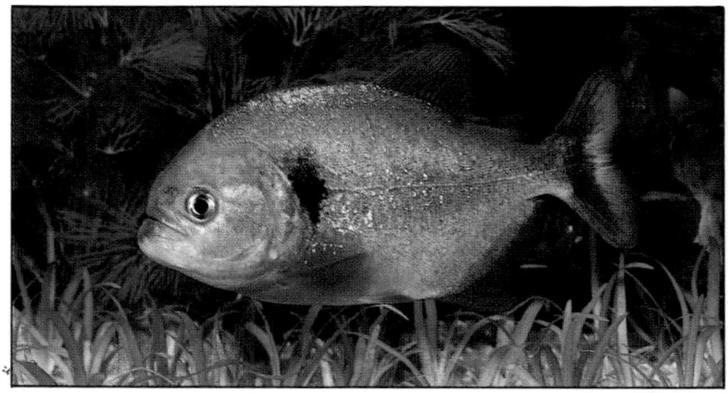

*Red Piranha*

### SPECIES DETAILS

**Size** 300 mm long
**Origin** S. America
**Sexing** No visible differences
**Ease of keeping** Moderately easy
**Food** Meaty foods such as insects, worms, live fish
**Breeding method** Egg-scatterer
**Breeding potential** Few known to have bred in an aquarium

### AQUARIUM CONDITIONS

**Water** Soft, medium-hard
**Temperature** 24°C
**Tank type** Species
**Swimming level** All levels

# Cichlids

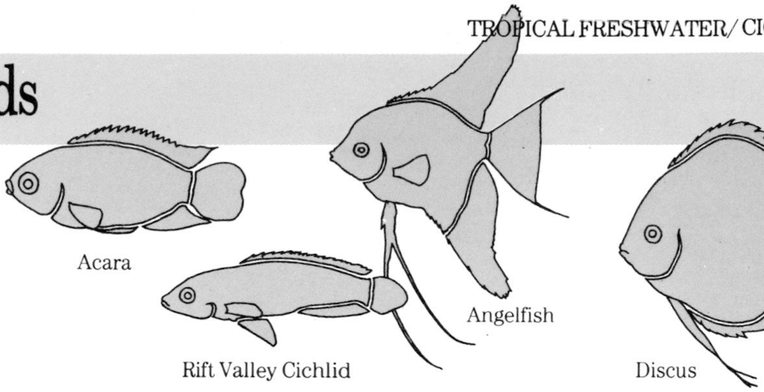


Of all the tropical fishes kept in the aquarium, no single group can offer such a diversity of form, size, swimming level and spawning behaviour as can the Cichlids.

Body shapes range from the disk-like Angelfish and Discus of the Amazon, to the cylindrical-bodied *Julidochromis* from Africa, and the more familiar, chunky proportions of the *Aequidens* and *Cichlasoma* genera. Many are extremely colourful, particularly those from the Rift Valley in Africa.

Some Cichlids grow very large and need an aquarium to themselves, but others may be kept in a community tank. However, they all defend their territory very fiercely at breeding times.

Cichlids, especially Acaras, exercise exemplary parental care. Breeding methods range from secretive breeding in caves, through the open water spawners, to the almost self-sacrificing care of the mouth-brooders.

## Aequidens curviceps

*Sheepshead Acara, Flag Cichlid*

This small, colourful Cichlid has a brown-green upper body and a speckled light blue underside. Its caudal fin is brown in colour and is speckled with blue dots. As it is an open-water spawner, it needs flat rocks.

*Male Sheepshead Acara*

### SPECIES DETAILS

**Size** 75 mm long
**Origin** S. America
**Sexing** Males have longer pointed fins
**Ease of keeping** Easy
**Food** All foods
**Breeding method** Egg-depositor
**Breeding potential** Moderately easy; improves to easy if given a separate tank, but needs flat stones

### AQUARIUM CONDITIONS

**Water** Soft, medium-hard
**Temperature** 24°C
**Tank type** Community, but small fishes only
**Swimming level** All levels

# Aequidens maronii

*Keyhole Cichlid*

A drab, beige fish, this Cichlid is easily recognized by the distinctive, keyhole-shaped dark patch on the side of its body and the dark band running downwards through its eye and across its head. As maturity is reached, the dorsal and anal fins of the male become elongated. The Keyhole Cichlid is a peaceful species, and will breed in a community.

*Male Keyhole Cichlid*

### SPECIES DETAILS

**Size** 100 mm long
**Origin** Northern S. America
**Sexing** Males have longer dorsal and anal fins when mature
**Ease of keeping** Easy
**Food** All foods
**Breeding method** Egg-depositor
**Breeding potential** Moderately easy, but needs flat stones

### AQUARIUM CONDITIONS

**Water** Soft, medium-hard
**Temperature** 24°C
**Tank type** Community
**Swimming level** All levels

# Apistogramma ramirezi

*Ram*

Also known as *Papiliochromis ramirezi*, this "dwarf" variety of Cichlid is a very colourful fish with many sparkling iridescences, especially when viewed under side-lighting (see p. 280). The body colours include yellow and purple. Unlike other egg-depositors, the Ram lays its eggs in a pit dug in the gravel, not on stones. It is sensitive to changes in the tank's water conditions.

### AQUARIUM CONDITIONS

**Water** Soft
**Temperature** 24°C
**Tank type** Community
**Swimming level** All levels

### SPECIES DETAILS

**Size** 70 mm long
**Origin** S. America
**Sexing** Males have long 2nd/3rd dorsal fin ray
**Ease of keeping** Moderately difficult, but easy once acclimatized
**Food** All foods
**Breeding method** Egg-depositor
**Breeding potential** Moderately easy

*Male Ram*

# Cichlasoma festivum

*Flag Cichlid, Festive Cichlid*

Although the body colour of this fish varies, it always has a dark diagonal stripe running from the lower half of its bright red eye to the tip of the dorsal fin. The pelvic fins are filamentous, and similar to those of the Angelfish. Shy at first, it will become tame enough to take food from its owner's hand.

*Pair of Flag Cichlids*

SPECIES DETAILS

**Size** 150 mm long
**Origin** S. America
**Sexing** Males may have more pointed fins
**Ease of keeping** Easy
**Food** All foods
**Breeding method** Egg-depositor
**Breeding potential** Easy

AQUARIUM CONDITIONS

**Water** Soft, medium-hard
**Temperature** 24°C
**Tank type** Community
**Swimming level** All levels

# Cichlasoma meeki

*Firemouth Cichlid*

The common name of this species refers to its vivid red throat and belly regions. During the breeding period this colour intensifies on male fishes. The rest of the fish is blue-grey with several black blotches. The brown fins are streaked with blue. Although the Firemouth Cichlid is a peaceful fish, it may occasionally dig in the gravel and uproot plants.

*Male Firemouth Cichlids*

SPECIES DETAILS

**Size** 150 mm long
**Origin** Central America
**Sexing** Males have redder throat, more pointed fins and extended dorsal fin
**Ease of keeping** Easy
**Food** All foods
**Breeding method** Egg-depositor
**Breeding potential** Moderately easy; may be prolific; likes plenty of retreats

AQUARIUM CONDITIONS

**Water** Soft, medium-hard
**Temperature** 24°C
**Tank type** Community of similar-sized fishes
**Swimming level** Midwater and lower levels

# Crenicara filamentosa

*Checkerboard Cichlid*

This fish has a brown, cylindrical body that is covered in a checkerboard pattern of deep blue and orange-brown. The dorsal, anal and caudal fins are red and blue. The latter develops long filaments as the male fish matures. The Checkerboard Cichlid spends much time on the aquarium floor, and likes plenty of hiding places.

AQUARIUM CONDITIONS

**Water** Soft, medium-hard

**Temperature** 24°C

**Tank type** Community

**Swimming level** Lower levels

**Special needs** Retreats; well-filtered water

SPECIES DETAILS

**Size** 75 mm long

**Origin** S. America

**Sexing** Males have longer fins and filaments to caudal fin

**Ease of keeping** Moderately difficult; can be sensitive to water conditions

**Food** All foods; relishes worm foods

**Breeding method** Egg-depositor, on or between stones

**Breeding potential** Moderately easy

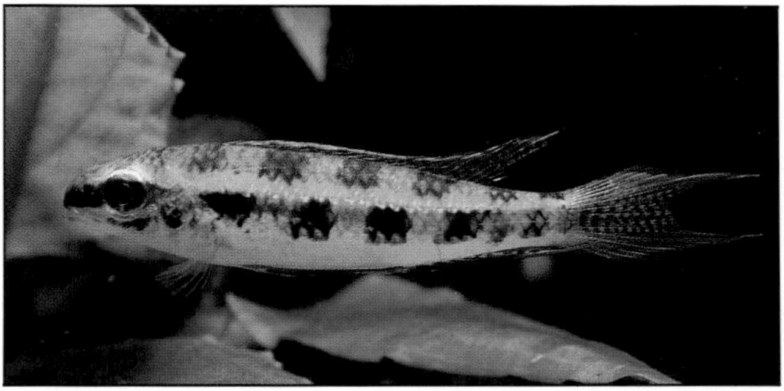

*Male Checkerboard Cichlid*

# Etroplus maculatus

*Orange Chromide*

One of only three Asian Cichlids, the Orange Chromide's oval body is gold in colour, marked with tiny red dots and three dark vertical bars. It has blue facial markings, and a red iris to the eye. This fish is susceptible to fungal disease (see p. 233), and so a little sea salt should be added to its water.

*Male Orange Chromide*

SPECIES DETAILS

**Size** 90 mm long

**Origin** India, Sri Lanka

**Sexing** No visible differences

**Ease of keeping** Moderately easy

**Food** All foods

**Breeding method** Egg-depositor

**Breeding potential** Moderately easy

AQUARIUM CONDITIONS

**Water** Hard, with sea salt

**Temperature** 24°C

**Tank type** Community

**Swimming level** Midwater and lower levels

# Julidochromis marlieri

*Marlier's Julie*

The genus *Julidochromis* is only found in Lake Tanganyika, East Africa. It has a brown body with lighter brown markings forming a checker pattern. Its cylindrical body form makes it adept at living amongst crevices, and therefore its aquarium should be furnished with plenty of rocky retreats. It is a secretive spawner, laying its eggs on a cave roof.

AQUARIUM CONDITIONS

**Water** Hard
**Temperature** 24°C
**Tank type** Species
**Swimming level** Lower levels
**Special needs** Retreats

SPECIES DETAILS

**Size** 110 mm long
**Origin** Lake Tanganyika
**Sexing** Females smaller
**Ease of keeping** Moderately easy
**Food** All types; vegetable foods
**Breeding method** Egg-depositor
**Breeding potential** Moderately difficult

*Female Marlier's Julie*

# Labeotropheus trewavasae

*Red-finned Cichlid*

This African Cichlid may vary in colour, but is usually blue with a red dorsal fin. It lives in shallow water along rocky shorelines, where it rasps algae from the rocks with its slightly underslung mouth. The Red-finned Cichlid is inclined to be aggressive. As the water in its natural habitat is hard, it will quickly acclimatize to most tap water.

AQUARIUM CONDITIONS

**Water** Hard
**Temperature** 24°C
**Tank type** Species
**Swimming level** All levels
**Special needs** Plenty of rocky retreats

SPECIES DETAILS

**Size** 150 mm long
**Origin** Lake Malawi
**Sexing** Females may be speckled yellow-brown
**Ease of keeping** Moderately easy
**Food** Most foods, including vegetable matter
**Breeding method** Mouth-brooder
**Breeding potential** Moderately easy

*Male Red-finned Cichlid*

# Lamprologus brichardi

*Lyretail Lamprologus*

The combination of a light brown body, white-tipped fins, a lyre-shaped caudal fin and brilliant blue eyes make this fish very easy to identify. It is a secretive cave-spawner, and so needs rocky retreats.

*Pair of Lyretail Lamprologus (male above)*

SPECIES DETAILS

**Size** 100 mm long
**Origin** Lake Tanganyika
**Sexing** Males have elongated dorsal and anal fins
**Ease of keeping** Moderately difficult
**Food** All foods
**Breeding method** Egg-depositor
**Breeding potential** Moderately difficult

AQUARIUM CONDITIONS

**Water** Hard
**Temperature** 24°C
**Tank type** Species
**Swimming level** Lower levels
**Special needs** Retreats

# Pelvicachromis pulcher

*Kribensis*

Formerly known as *Pelmatochromis kribensis*, the "Krib" is brown with a violet to deep purple iridescence. There is a red blotch on each flank. During spawning periods, the female darkens in colour. It is a secretive spawner – often a pair will disappear for a few days, only to emerge from beneath a rock with their newly-hatched family. The parents guard the young very aggressively.

*Male Kribensis*

SPECIES DETAILS

**Size** 100 mm long
**Origin** W. Africa
**Sexing** Males have pointed dorsal fins, spotted caudal fins; females are smaller, darker at breeding times
**Ease of keeping** Easy
**Food** All foods; meat
**Breeding method** Egg-depositor
**Breeding potential** Moderately easy

AQUARIUM CONDITIONS

**Water** Soft, medium-hard
**Temperature** 24°C
**Tank type** Community
**Swimming level** All levels
**Special needs** Retreats

# Pterophyllum scalare

*Angelfish*

A graceful, disk-shaped Cichlid, the most common Angelfish has black bars on a silver-coloured body. However, many new varieties have been developed including All-Black, Half-Black, Marbled and Blushing types. The relatively large fins may take normal, lace or veil forms (see p. 18).

*Male Angelfish*

SPECIES DETAILS

**Size** 110 mm long
**Origin** S. America
**Sexing** Females have fatter breeding tube
**Ease of keeping** Easy
**Food** All foods
**Breeding method** Egg-depositor
**Breeding potential** Moderately easy

AQUARIUM CONDITIONS

**Water** Soft, medium-hard
**Temperature** 24°C
**Tank type** Community, but medium-sized fishes only
**Swimming level** All levels
**Special needs** Plants

# Symphysodon discus

*Discus, Pompadour Fish*

An attractive brown colour, with a pale blue iridescence, this disk-shaped fish isn't recommended for the beginner as it is very delicate. The Discus will only survive if the water is constantly kept soft, acidic, well-filtered and at a temperature of 28°C.

*Discus*

SPECIES DETAILS

**Size** 150 mm long
**Origin** S. America
**Sexing** Females may have fatter breeding tube
**Ease of keeping** Difficult
**Food** Live, meaty food
**Breeding method** Egg-depositor
**Breeding potential** Moderately difficult

AQUARIUM CONDITIONS

**Water** Very soft, acid
**Temperature** 28°C
**Tank type** Species
**Swimming level** All levels
**Special needs** Excellent water conditions

# Anabantoids

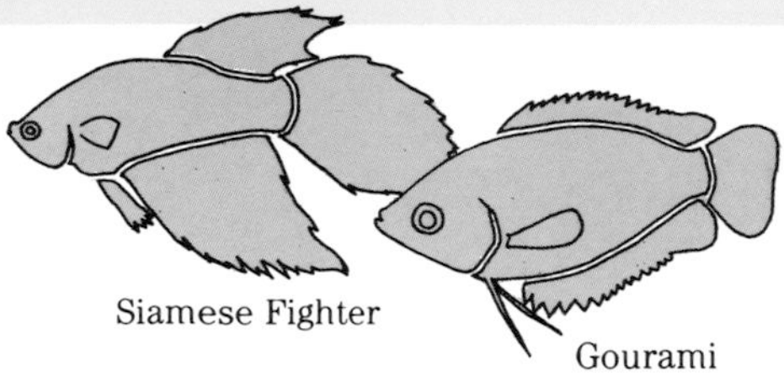


Native to parts of Africa and southern Asia, Anabantoids are medium-sized, freshwater fishes. One of them, the Siamese Fighting Fish, is known the world over for its pugnacity towards rival males, and for the wagers such human-arranged battles command. In contrast, many other fishes in this family, the Gouramis in particular, are noted for their peaceful behaviour. However, even these species can become very aggressive at breeding times.

All Anabantoids can breathe atmospheric oxygen in the event of their natural waters becoming oxygen-depleted or otherwise polluted. They do so by means of a special organ located in the head just behind the gills (see p. 19). This organ is constructed like a maze or labyrinth. Atmospheric air is trapped in the many folds and is then absorbed into the fish's bloodstream. For this reason, Anabantoids are known to some fish-keepers as Labyrinth fishes.

During the breeding period, the majority of male Anabantoid fishes build bubble-nests. They entice their mates under these and induce them to lay their eggs by giving them a spawning embrace. The eggs are immediately fertilized and placed in the nest by the male who guards them and the subsequent fry against allcomers, regardless of size. You won't find Anabantoids difficult to breed in the aquarium but, unfortunately, raising the fry is more problematical because of their tiny size.

# Belontia signata

*Combtail, Combtail Paradise Fish*

*Juvenile female Combtail*

This species has a golden-coloured body which appears to be reticulated because of the dark red-brown edge to every scale. All the fins may have some blue coloration. As the fish matures, its whole body takes on a redder appearance. The Combtail gets its common name from the extended rays of the male's caudal fin. It has a reputation for pugnaciousness, and therefore other fishes kept with it should be large types that won't be bullied.

## SPECIES DETAILS

**Size** 125 mm long
**Origin** Sri Lanka
**Sexing** Males have more pointed fins and extensions to the caudal fin
**Ease of keeping** Moderately easy
**Food** All foods
**Breeding method** Bubble-nest builder
**Breeding potential** Moderately easy; bubble-nest not so thoroughly constructed as with other genera

## AQUARIUM CONDITIONS

**Water** All types
**Temperature** 24°C
**Tank type** Community of large fishes
**Swimming level** All levels
**Special needs** Plants

# Betta splendens

*Siamese Fighter, Siamese Fighting Fish*

Although the fish illustrated here is blue and red in colour, there are now many other colour strains. The distinctive fins of aquarium-developed fishes are more ornate than those of wild ones. The Siamese Fighter is renowned for its aggressive rivalry with males of the same species, and therefore only one male should be kept in an aquarium at a time. Faced with a rival, the male erects his gill-covers and spreads his fins. A vicious battle then ensues, after which the victor is often so exhausted that he dies. Males will even display on seeing their reflection in a mirror. When breeding this fish, many small tanks are needed in which to raise the males separately.

*Male Siamese Fighter*

SPECIES DETAILS

**Size** 60 mm long
**Origin** Thailand
**Sexing** Males have long, flowing fins
**Ease of keeping** Easy
**Food** All foods
**Breeding method** Bubble-nest builder
**Breeding potential** Easy, but males must be segregated (as soon as recognizable) into separate tanks

AQUARIUM CONDITIONS

**Water** All types
**Temperature** 24°C
**Tank type** Community, but one male only
**Swimming level** All levels
**Special needs** Plants

# Colisa chuna

*Honey Gourami*

A small fish, the male Honey Gourami is a golden-brown colour with bright yellow dorsal and ventral fins. There is also a red variety with turquoise fins. The female is duller, with a dark line along her flank. When in breeding condition, her mate develops a deep turquoise area below his throat, on his ventral surface and part of his anal fin. Although the fry are small, they thrive more readily than the young of the larger *Colisa lalia* (see p. 62).

*Male Honey Gourami*

SPECIES DETAILS

**Size** 45 mm long
**Origin** India
**Sexing** Females are duller with a dark line along the side; males have a deep turquoise throat at breeding time
**Ease of keeping** Easy
**Food** All foods
**Breeding method** Bubble-nest builder
**Breeding potential** Moderately easy

AQUARIUM CONDITIONS

**Water** All types
**Temperature** 24°C
**Tank type** Community, but similar-sized fishes only
**Swimming level** All levels
**Special needs** Plants

# Colisa labiosa

*Thick-lip Gourami*

Grey-red in colour, this Gourami has pale blue markings, deep red edges to the dorsal fins, and orange pectoral fins. At breeding time, the male changes to a deep chocolate-brown and the blue markings disappear. A medium-sized fish, it has a thickened upper lip, which gives it its popular name. The Thick-lip Gourami's bubble-nest is less well constructed than that of other Gouramis. After spawning, the male is often belligerent towards his mate, and she should be moved to a separate tank for her own protection.

*Female Thick-lip Gourami*

SPECIES DETAILS

**Size** 80 mm long
**Origin** Burma
**Sexing** Males have pointed fins, and turn dark brown when breeding; females smaller
**Ease of keeping** Easy, but guard female from the male after spawning
**Food** All foods
**Breeding method** Bubble-nest builder
**Breeding potential** Easy; very prolific

AQUARIUM CONDITIONS

**Water** Most types
**Temperature** 24°C
**Tank type** Community
**Swimming level** All levels

# Colisa lalia

*Dwarf Gourami*

*Pair of Dwarf Gouramis (male above)*

One of the most brilliantly coloured aquarium fish, the Dwarf Gourami is diagonally striped in bright red and turquoise, with red-edged dorsal fins and bright orange pectoral fins. At breeding time these colours are intensified in the male. Despite his small size, he can be very pugnacious towards other fishes in the aquarium. Also, he may attack a female before spawning if he doesn't feel she is ready, and he will usually attack her afterwards. If this happens, you may need to remove the female.

SPECIES DETAILS

**Size** 60 mm long
**Origin** N.E. India
**Sexing** Females are duller and plumper
**Ease of keeping** Easy
**Food** All foods
**Breeding method** Bubble-nest breeder
**Breeding potential** Moderately easy; not difficult to spawn, but fry may be hard to raise as they are very small

AQUARIUM CONDITIONS

**Water** Most types
**Temperature** 24°C
**Tank type** Community
**Swimming level** All levels

# Ctenopoma acutirostre

*Spotted Climbing Perch*

This gold-coloured fish is attractively blotched with black. It is a predator, and it has an interesting feature: it can extend its mouth like a tubular funnel, creating a local vacuum that sucks in anything passing by. Therefore, it shouldn't be trusted in aquariums with smaller fishes. The Spotted Climbing Perch likes a shady, sheltered environment, so put it in a well-planted, not too brightly lit tank that includes floating plants and rocky hideaways.

*Spotted Climbing Perch*

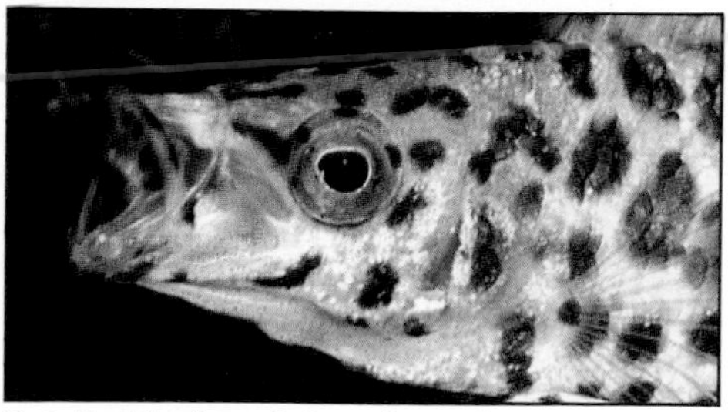

*Detail of the funnel-shaped mouth*

SPECIES DETAILS

**Size** 150 mm long
**Origin** Africa
**Sexing** No visible differences
**Ease of keeping** Easy
**Food** All meaty foods
**Breeding method** Bubble-nest builder
**Breeding potential** Moderately easy

AQUARIUM CONDITIONS

**Water** Preferably soft
**Temperature** 24°C
**Tank type** Community of large fishes
**Swimming level** All levels
**Special needs** Plants; retreats

# Ctenopoma ansorgei

*Ornate Ctenopoma*

The markings on the cylindrical body of this fish consist of alternate light and dark iridescent brown and yellow bands extending into the dorsal and anal fins. The male's fins are more pointed than the female's, and she looks much duller in colour. The Ornate Ctenopoma isn't as aggressive as other Anabantoids, but it does watch over its own interests at breeding time, when it may defend its territory more vigorously than usual. Unusually, it is said to spawn at night. This species isn't difficult to breed and although not commonly available in dealers' shops, it is well worth searching for if you are interested in Anabantoid fishes.

*Male Ornate Ctenopoma*

SPECIES DETAILS

**Size** 70 mm long
**Origin** Africa
**Sexing** Males have pointed fins; females are duller
**Ease of keeping** Easy
**Food** Meaty foods such as beef heart
**Breeding method** Bubble-nest builder
**Breeding potential** Moderately easy; the fry are very tiny at first

AQUARIUM CONDITIONS

**Water** Relatively soft
**Temperature** 28°C
**Tank type** Community, but with similar-sized fishes only
**Swimming level** Midwater and lower levels

# Helostoma temmincki

*Kissing Gourami*

There are two colour variations of the Kissing Gourami – one is olive-green and the other (as shown here) a delicate pink. The most unusual feature of this fish is its peculiar habit of "kissing" members of its own species with its protruberant lips. This isn't an expression of affection – it is probably a trial of strength between fishes. The same lips are used to rasp soft algae from the surface of rocks or the glass walls of the aquarium. Occasionally this Gourami sucks at the sides of slow-moving fishes.

*Kissing Gouramis*

*Detail of Kissing Gouramis*

SPECIES DETAILS

**Size** 200 mm long
**Origin** Far East
**Sexing** No visible differences
**Ease of keeping** Moderately easy
**Food** All foods; vegetable matter
**Breeding method** Lays floating eggs, but doesn't build a bubble-nest
**Breeding potential** Difficult

AQUARIUM CONDITIONS

**Water** All types
**Temperature** 24°C
**Tank type** Community
**Swimming level** All levels
**Special needs** Floating plants as shade

# Macropodus opercularis

*Paradise Fish*

This Anabantoid has red, blue and brown bands running across its body. An albino form has been developed in the aquarium, but doesn't occur in nature. The male has elongated dorsal, anal and caudal fins which he displays to impress females or to threaten rival males. The Paradise Fish is able to withstand lower temperatures than the usual required range for tropical fresh-water fishes, and can even be kept out of doors in a pond (see p. 279) during the summer months.

*Pair of Paradise Fishes (male below)*

SPECIES DETAILS

**Size** 75 mm long
**Origin** Far East
**Sexing** Males have longer fins
**Ease of keeping** Moderately easy; aggressive
**Food** All foods
**Breeding method** Bubble-nest builder
**Breeding potential** Moderately easy; may need a higher temperature

AQUARIUM CONDITIONS

**Water** All types
**Temperature** 16–24°C
**Tank type** Species
**Swimming level** All levels

# Trichogaster leeri

*Lace Gourami, Leeri Gourami, Mosaic Gourami*

*Pair of Lace Gouramis (male above)*

The male of this species has a brown body with orange underparts, and the female's underside is more silver-coloured. Both sexes have the same body pattern – a mosaic of silvery spots which continues into the fins, and a dark line running centrally along the flanks. The male has a long dorsal fin that may extend over the caudal fin, and very long, filamentous pelvic fins which, along with the throat and front part of the anal fin, are bright red-orange at spawning times. In good specimens, the dorsal and anal fins have extended filaments. The Lace Gourami takes three years to mature to its final adult size. It looks best with slow-moving fishes in a well-planted aquarium.

SPECIES DETAILS

**Size** 110 mm long
**Origin** Far East
**Sexing** Males have longer fins, and red throat region when spawning
**Ease of keeping** Moderately easy
**Food** All foods
**Breeding method** Bubble-nest builder
**Breeding potential** Easy; prolific

AQUARIUM CONDITIONS

**Water** All types
**Temperature** 24°C
**Tank type** Community; but slow-moving fishes only
**Swimming level** All levels

# Trichopsis pumilus

*Sparkling Gourami, Croaking Gourami*

The brown body of the Sparkling Gourami is covered with blue spots and the fins are blue with red patterning and edges. The eye is bright blue rimmed with red. This species has a similar body shape to the Siamese Fighter, but has much shorter fins. Male fishes have pointed fins and a thin red line along the body at the base of the long anal fin. When spawning, both male and female make an audible croaking sound. This species requires high temperatures.

*Female Sparkling Gourami*

SPECIES DETAILS

**Size** 40 mm long
**Origin** Far East
**Sexing** Males have more pointed fins and a red line at anal fin base; females duller
**Ease of keeping** Moderately easy
**Food** All foods
**Breeding method** Bubble-nest builder, or may lay eggs near to the bottom
**Breeding potential** Moderately easy

AQUARIUM CONDITIONS

**Water** Soft, acid preferred
**Temperature** 25°C or slightly higher
**Tank type** Community of small fishes, or species
**Swimming level** All levels

# Livebearers

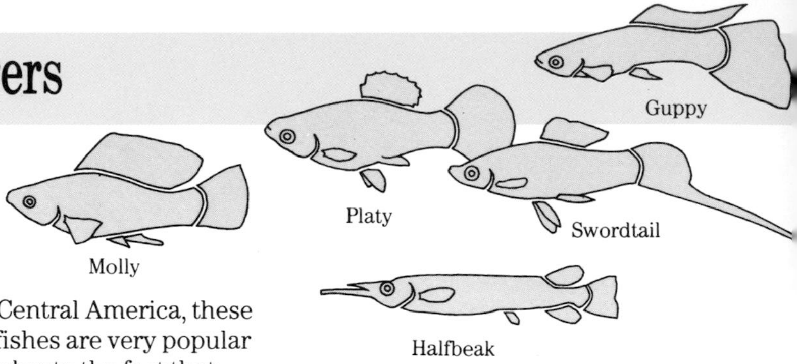


Mostly native to Central America, these small, colourful fishes are very popular with fishkeepers due to the fact that most inhabit all levels of the aquarium, are prolific breeders and produce living, free-swimming young.

Because they reproduce so readily, to keep any colour strain pure, you must isolate it from any possible contaminating "cross-breeds". You will need a number of tanks: one for males, one for females, as well as a breeding tank for each selected pair and another aquarium for the fry. Unlike other fishes, female Livebearers are fertilized internally by means of the male's modified anal fin – the gonopodium – by which sperm is transferred.

Fishes in this group include the Swordtail, so-called because the males have a long, pointed sword-like extension to the caudal fin, and the Guppy, a species which has many colour variants. Platys usually have rounded tail fins, whilst some Mollies have enlarged, sail-like dorsal fins. And the Halfbeak has an elongated lower jaw.

## Dermogenys pusillus

*Wrestling Halfbeak*

The yellow-green coloured Wrestling Halfbeak is instantly recognizable because of its unusual jaw structure – the immovable lower jaw being much longer than the upper. This makes feeding from any location other than the surface very difficult, and so floating foods and worm-feeders (see p. 208) must be used. The tank should be well-planted for two reasons: young fry need immediate sanctuary from other fishes, and there is a real risk of this fish damaging its jaw by colliding with the invisible glass walls unless the tank sides are defined by plants.

*Male Wrestling Halfbeak*

SPECIES DETAILS

**Size** 65 mm long
**Origin** Far East
**Sexing** Males have notched anal fin
**Ease of keeping** Moderately difficult
**Food** Insects, worms
**Breeding method** Livebearer
**Breeding potential** Moderately difficult

AQUARIUM CONDITIONS

**Water** All types; 1 tsp salt per 5 litres
**Temperature** 24°C
**Tank type** Species
**Swimming level** Upper levels
**Special needs** Plants

# Poecilia hybrid

*Black Molly*

Jet-black in colour and seeming to have a velvety texture, this Livebearer is easy to identify. If given plenty of space, the male's dorsal fin will become large. It is an active fish, and requires plenty of vegetable matter in its diet. The Black Molly is sensitive to changes in the water temperature.

*Male Black Molly*

SPECIES DETAILS

**Size** 70 mm long
**Origin** Mexico
**Sexing** Males have gonopodium and may have large dorsal fin; females larger
**Ease of keeping** Moderately easy
**Food** All foods; plenty of green foods; algae
**Breeding method** Livebearer
**Breeding potential** Easy

AQUARIUM CONDITIONS

**Water** Medium-hard; 1 tsp salt per 5 litres
**Temperature** 24°C
**Tank type** Community
**Swimming level** All levels

# Poecilia reticulata

*Guppy, Millions Fish*

Within this species there are many colour strains, no two fishes ever being exactly alike. The female Guppy's body coloration is much more subdued than that of her mate, although her caudal fin may be vivid. The caudal fin may take many forms including fan, veil or pintail (see p. 18). Gravid (pregnant) females require a heavily planted tank; this also protects the fry from being eaten by the parents.

AQUARIUM CONDITIONS

**Water** Medium-hard
**Temperature** 24°C
**Tank type** Community
**Swimming level** All levels

SPECIES DETAILS

**Size** 30 mm long
**Origin** Trinidad
**Sexing** Males have gonopodium; females are duller and larger
**Ease of keeping** Easy
**Food** All foods; green foods
**Breeding method** Livebearer
**Breeding potential** Easy; prolific; needs heavily planted tank

*Male Guppy*

# Poecilia velifera

*Sailfin Molly*

This fish's body is dark green in colour but there is also an albino/gold variety as shown here. Each scale is dotted. The Sailfin Molly gets its name from the male's impressive dorsal fin, which is erected in order to court females or to deter rival males.

*Male Sailfin Molly*

### SPECIES DETAILS

**Size** 120 mm long
**Origin** Central America
**Sexing** Males have gonopodium and large dorsal fin
**Ease of keeping** Moderately easy
**Food** All foods; algae
**Breeding method** Livebearer
**Breeding potential** Easy

### AQUARIUM CONDITIONS

**Water** Hard; add 1 tsp sea-salt per 5 litres
**Temperature** 24°C
**Tank type** Community
**Swimming level** All levels

# Xiphophorus hybrid

*Swordtail*

Many colour varieties of Swordtail are now established, although the most well-known is red. There are also forms with exaggerated finnage and double swords. The lower edge of the male's caudal fin is extended into a long "sword". Before the sword develops, you can sex young fishes by observing the anal fin; the female's is fan-shaped, the male's rodlike. It isn't unknown for females to change into males, especially when old or affected by parasites.

*Male Swordtail hybrid*

### SPECIES DETAILS

**Size** 100 mm long
**Origin** Central America
**Sexing** Males have "sword" and gonopodium; females are deeper-bodied with fan-shaped anal fin
**Ease of keeping** Easy
**Food** All foods; green foods
**Breeding method** Livebearer
**Breeding potential** Easy; prolific

### AQUARIUM CONDITIONS

**Water** Medium-hard
**Temperature** 24°C
**Tank type** Community
**Swimming level** All levels

# Xiphophorus maculatus hybrid

*Platy*

By hybridization fish-keepers have developed a large number of colour strains of the Platy. In some cases the body and fins are a single colour; in others the body is unicolour and the fins are black; whilst others are multicoloured. The Platy is a short, stocky fish and doesn't carry a "sword". The dorsal fin rays are clearly visible. The Platy is a peaceful, hardy fish.

*Female Platy hybrid*

SPECIES DETAILS

**Size** 50 mm long
**Origin** Central America
**Sexing** Males have gonopodium
**Ease of keeping** Easy
**Food** All foods
**Breeding method** Live-bearer
**Breeding potential** Easy; prolific

AQUARIUM CONDITIONS

**Water** Medium-hard
**Temperature** 24°C
**Tank type** Community
**Swimming level** All levels

# Xiphophorus variatus hybrid

*Variatus Platy*

Most of these fishes have dark-lined vertical markings on the body and an orange-red caudal fin. However, varieties with bright body colours are found, especially where hybridization with *X.helleri* has occurred. In shape and size, this Livebearer is a cross between the Sword-tail and the Platy. Although fishes in this genus have upturned mouths, they will feed at all levels.

AQUARIUM CONDITIONS

**Water** Medium-hard
**Temperature** 24°C
**Tank type** Community
**Swimming level** All levels

SPECIES DETAILS

**Size** 62 mm long
**Origin** Mexico
**Sexing** Males have gonopodium; females have fan-shaped anal fin and are duller
**Ease of keeping** Easy
**Food** All types; green foods
**Breeding method** Livebearer
**Breeding potential** Easy; prolific

*Male Variatus Platy hybrid*

# Killifishes

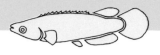

Killifish

Some remarkable kinds of Killifishes are "Annual Fishes"; they are necessarily short-lived because their natural habitats in Africa, Asia and South America disappear during the dry spell. To combat this annual event the fishes lay their fertilized eggs in mud, where they remain dormant until the following wet season when, stimulated by the new waters, the young fishes hatch out.

Despite the fact that they are short-lived, such Killifishes have much to recommend them as potential aquarium fishes. They are brightly coloured, have interesting reproduction methods and, being relatively small in size, can be kept in smaller aquariums. The reason for their bright colours is that they inhabit tannin-discoloured waters and so need gaudy hues for species recognition.

Because their fertilized eggs can survive almost complete dehydration and, depending on species, can be stored for lengthy periods in a dormant state, they have the advantage that you need only add water for an almost instant fish nursery! Another bonus is that it is very easy to exchange the fertilized eggs with other fishkeepers. Many of the Killifishes are egg-scatterers. In the wild their adhesive eggs often become trapped in bushy plants, but in the aquarium you can substitute artificial, nylon spawning mops (see p. 250).

Most of the fishes in this group swim in the upper levels of the aquarium. Many of them have a tendency towards aggressiveness.

## Aphyosemion gardneri

*Steel-blue Aphyosemion*

Despite the common name, this species has a number of colour variants. In the past, several newly discovered variants were mistaken for "new" fishes. This created understandable confusion when they were subsequently found to be the same species. However, the characteristic type has a blue body with bright red vertical markings. The edges of the dorsal, anal and caudal fins are yellow. The Steel-blue Aphyosemion is aggressive, and will attack smaller fishes.

*Male Steel-blue Aphyosemion*

### SPECIES DETAILS

**Size** 75 mm long
**Origin** W. Africa
**Sexing** Females are duller
**Ease of keeping** Easy
**Food** All foods
**Breeding method** Egg-scatterer
**Breeding potential** Easy; use nylon spawning mops

### AQUARIUM CONDITIONS

**Water** Soft; peat-filtered
**Temperature** 20°C
**Tank type** Species
**Swimming level** Upper levels
**Special needs** Shade

# Aplocheilus dayi

*Ceylon Killifish*

Looking like a miniature Pike, the Ceylon Killifish has a red body with bluish-green iridescences, and is best seen with some light coming through the front glass of the tank. Each iridescent scale seems to be individually picked out. The fish has red patterning in the fins, although it is less evident in the female. When young, both sexes have black dots on the body, but the male loses these with maturity. This species will spawn either in a peat layer on the floor of the aquarium or in nylon spawning mops.

*Male Ceylon Killifish*

SPECIES DETAILS

**Size** 70 mm long
**Origin** Sri Lanka
**Sexing** Females are duller with black dots
**Ease of keeping** Moderately easy
**Food** All foods
**Breeding method** Egg-burier or scatterer
**Breeding potential** Moderately easy; provide peat or spawning mops

AQUARIUM CONDITIONS

**Water** Soft; peat-filtered
**Temperature** 24°C
**Tank type** Community of large fishes or species
**Swimming level** Upper levels

# Jordanella floridae

*American Flagfish*

The fish gets its common name from the alternate rows of red and iridescent blue-green dots along its olive-green body, which are said to resemble a flag. Although its chunky looks aren't typical of the Killifish shape (see p. 70), this species is a member of the same taxonomic group, the Cyprinodonts. It is said to spawn by laying eggs in a depression in the gravel, but fishkeepers have found that it will also spawn in bushy plants. Male fishes tend to be pugnacious.

*Male American Flagfish*

SPECIES DETAILS

**Size** 70 mm long
**Origin** Florida, Mexico
**Sexing** Females have dark blotch on the rear of the dorsal fin
**Ease of keeping** Moderately easy; may be kept outdoors in summer
**Food** All foods; algae
**Breeding method** Egg-depositor or scatterer
**Breeding potential** Easy; use nylon spawning mops

AQUARIUM CONDITIONS

**Water** All types
**Temperature** 18–24°C
**Tank type** Community
**Swimming level** All levels

# Nothobranchius rachovi

*Rachov's Nothobranch*

The attractive orange body of this Killifish is striped with red and turquoise. The caudal fin has bright stripes of red, turquoise, orange and black. Rachov's Nothobranch is inclined to be aggressive in a community tank.The species is an egg-burier, and the eggs should be left in semi-moist peat for six to eight weeks before hatching is activated by immersing in water (see p. 251).

SPECIES DETAILS

**Water** Soft; peat-filtered
**Temperature** 22–6°C
**Tank type** Species
**Swimming level** Upper levels and midwater

AQUARIUM CONDITIONS

**Size** 50 mm long
**Origin** E. Africa
**Sexing** Females are duller
**Ease of keeping** Moderately easy
**Food** Live and dried foods
**Breeding method** Egg-burier
**Breeding potential** Moderately difficult

*Male Rachov's Nothobranch*

# Pachypanchax playfairi

*Playfair's Panchax*

This fish has a brown back, yellow underside, and green and red flanks. Although not difficult to breed, Playfair's Panchax may cause the beginner some concern as the male's scales stand up during spawning. In other species this can be a sign of disease (see p. 228), but in this case it is a natural characteristic and nothing to worry about. It will breed amongst any floating plants such as *Najas* or *Myriophyllum* (see p.169), or you can use artificial nylon spawning mops to make egg collection easier.

*Male Playfair's Panchax*

SPECIES DETAILS

**Size** 75 mm long
**Origin** E. Africa, Madagascar, Zanzibar
**Sexing** Females are duller
**Ease of keeping** Moderately easy
**Food** All foods
**Breeding method** Egg-scatterer
**Breeding potential** Easy

AQUARIUM CONDITIONS

**Water** All types
**Temperature** 24°C
**Tank type** Community or species
**Swimming level** Upper levels
**Special needs** Plants

# Catfishes

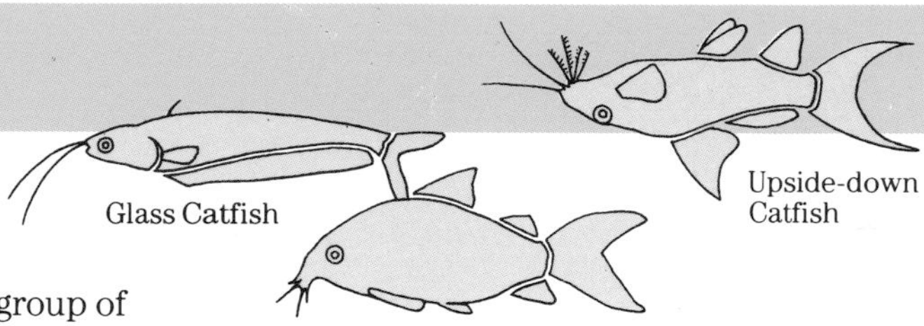


The majority of the Catfish group of fishes are native to South America and Africa. Most are medium-sized and aquarium-suitable.

Unlike most other fishes, the species in this group have no scales covering their skin. Instead, many have large, overlapping bony plates called *scutes*, whilst others, such as the Upside-Down Catfish, have no apparent protective covering over their skin and are often referred to as Naked Catfishes.

Many Catfishes can use atmospheric air, which they gulp in after a lightning dash to the surface. But most aren't natural upper-level swimmers – equipped with whisker-like growths called barbels around downturned mouths, Catfishes usually forage for food at the bottom of the aquarium. Despite the natural assumption that they are scavengers, don't expect them to thrive on other fishes' leftovers. Some species are nocturnal and need fast-sinking food to be given last thing at night. Such fishes have poorly developed eyes, and locate their food by means of their barbels. To protect these delicate organs, the gravel should be well rounded.

## Brochis splendens

*Short-bodied Catfish*

Depending on the light, this fish may appear metallic blue or emerald green. It is often confused with *Corydoras aeneus* (see p.74) but if the two are put side by side then the differences become apparent. *Brochis* has a longer-based dorsal fin, is deeper in the body and has a more pointed head.

*Group of Short-bodied Catfishes*

### SPECIES DETAILS

**Size** 75 mm long
**Origin** S. America
**Sexing** Females may be larger and plumper
**Ease of keeping** Easy
**Food** All foods; vegetable matter
**Breeding method** Egg-depositor
**Breeding potential** Moderately difficult

### AQUARIUM CONDITIONS

**Water** All types
**Temperature** 24°C
**Tank type** Community
**Swimming level** Lower levels
**Special needs** Rounded gravel

# Corydoras aeneus

*Bronze Catfish*

As its common name suggests, this Catfish has a plain brown body with a greenish-bronze iridescence. Its two rows of scutes are particularly noticeable. It has an amusing habit of rolling its eyes in a most unexpected manner. This fish is naturally gregarious, and should *not* therefore be kept singly. When breeding, the male fertilizes the eggs as they are held between the female's pelvic fins and she then transfers them to a chosen site.

*Pair of Bronze Catfishes*

SPECIES DETAILS

**Size** 75 mm long
**Origin** S. America
**Sexing** Females are plumper behind pectoral fins
**Ease of keeping** Easy
**Food** All foods
**Breeding method** Egg-depositor
**Breeding potential** Moderately difficult

AQUARIUM CONDITIONS

**Water** All types
**Temperature** 24°C
**Tank type** Community
**Swimming level** Lower levels
**Special needs** Rounded gravel

# Corydoras julii

*Leopard Catfish*

The body colour of this fish is grey with a mauve sheen. Several *Corydoras* have similar markings to this species and are sometimes sold as Leopard Catfish. It is doubtful if the majority of those imported are correctly named. The head markings in *C.julii* are distinct black spots, whereas in other species they are reticulated.

*Juvenile Leopard Catfish*

SPECIES DETAILS

**Size** 65 mm long
**Origin** S. America
**Sexing** Females are plumper behind pectoral fins (view from above for easier sex differentiation)
**Ease of keeping** Easy
**Food** All foods
**Breeding method** Egg-depositor
**Breeding potential** Moderately difficult; adding cooler water often stimulates breeding

AQUARIUM CONDITIONS

**Water** All types
**Temperature** 20–4°C
**Tank type** Community
**Swimming level** Lower levels
**Special needs** Rounded gravel

# Corydoras melanistius

*Black-spotted Corydoras*

It isn't easy to distinguish the Black-spotted Corydoras from others in the same genus, in particular the Leopard Catfish (see opposite below). However, the basic colouring of *C.melanistius* is a yellowy beige, and the pattern of markings isn't as distinct. The fish shown here is a juvenile, so it isn't yet showing its final colouring.

*Juvenile female Black-spotted Corydoras*

SPECIES DETAILS

**Size** 65 mm long
**Origin** S. America
**Sexing** Females are plumper *behind* pectoral fins; males are widest *at* the pectoral fin
**Ease of keeping** Easy
**Food** All foods
**Breeding method** Egg-depositor
**Breeding potential** Moderately difficult; adding cooler water often stimulates breeding

AQUARIUM CONDITIONS

**Water** All types
**Temperature** 24°C
**Tank type** Community
**Swimming level** Lower levels
**Special needs** Rounded gravel

# Corydoras reticulatus

*Reticulated Corydoras*

The pattern of dark markings from which the Reticulated Corydoras gets its name doesn't become distinct until the fishes have fully matured, although females never become quite as brightly coloured as their mates. The upper parts of the fish are an iridescent brown, but the belly is much paler. Like other Corydoras, this fish's two pairs of barbels are quite short. Feed it regularly to prevent it digging in the gravel.

*Female Reticulated Corydoras*

SPECIES DETAILS

**Size** 60 mm long
**Origin** S. America
**Sexing** Females are plumper behind pectoral fins
**Ease of keeping** Easy
**Food** All foods
**Breeding method** Egg-depositor
**Breeding potential** Moderately difficult

AQUARIUM CONDITIONS

**Water** All types
**Temperature** 24°C
**Tank type** Community
**Swimming level** Lower levels
**Special needs** Rounded gravel

# Dianema urostriata

*Stripe-tailed Catfish*

The two main features of this brown fish are long barbels and a black and white striped caudal fin. For a Catfish, it has a less flattened ventral area, as it doesn't spend all of its time on the tank floor.

*Stripe-tailed Catfish*

SPECIES DETAILS

**Size** 120 mm long
**Origin** S. America
**Sexing** No visible differences
**Ease of keeping** Easy
**Food** All foods
**Breeding method** Unknown
**Breeding potential** Not known to have bred in an aquarium

AQUARIUM CONDITIONS

**Water** All types
**Temperature** 24°C
**Tank type** Community
**Swimming level** Midwater and lower levels

# Hypostomus species

*Suckermouth Catfish*

This very large species has an elongated, spotted, brown body, a broad, flattened head and, of course, the distinctive ventral sucking mouth. Formerly known as *Plecostomus plecostomus*, it does a good job of keeping down algae growth in the aquarium. However, if it isn't given enough green foods it may retaliate by nibbling at your soft-leaved aquarium plants!

*Suckermouth Catfish*

SPECIES DETAILS

**Size** 250 mm long
**Origin** S. America
**Sexing** No visible differences
**Ease of keeping** Moderately easy
**Food** Vegetable matter
**Breeding method** Unknown
**Breeding potential** Not known to have bred in an aquarium

AQUARIUM CONDITIONS

**Water** All types
**Temperature** 24°C
**Tank type** Community
**Swimming level** Lower levels

# Kryptopterus bicirrhus

*Glass Catfish*

One glance at this fish will tell you how it got its common name: it is almost totally transparent. Its dorsal fin is a rudimentary single ray. The Glass Catfish should be kept in a small group. It must have a movement of water through the tank, usually achieved by extracting water from the filter end, and returning it clean at the other end (see p. 127).

*Glass Catfish*

SPECIES DETAILS

**Size** 90 mm long
**Origin** Far East
**Sexing** No visible differences
**Ease of keeping** Moderately difficult
**Food** Live foods preferred
**Breeding method** Unknown
**Breeding potential** Not known to have bred in an aquarium

AQUARIUM CONDITIONS

**Water** All types
**Temperature** 24°C
**Tank type** Community
**Swimming level** Midwater
**Special needs** Movement of water through tank

# Synodontis nigriventris

*Upside-down Catfish*

This brown fish is an aquarium curiosity as it swims upside down to protect itself from above-water predators. The normal fish coloration pattern of dark top and lighter bottom is reversed, thereby making it difficult to detect. It feeds by taking floating foods from the surface whilst upside down. Lockable, erectile spines in dorsal and pectoral fins make netting difficult.

*Upside-down Catfish*

SPECIES DETAILS

**Size** 60 mm long
**Origin** Central Africa
**Sexing** No visible differences
**Ease of keeping** Moderately easy
**Food** All foods
**Breeding method** Unknown
**Breeding potential** Not known to have bred in an aquarium

AQUARIUM CONDITIONS

**Water** All types
**Temperature** 24°C
**Tank type** Community
**Swimming level** Upper levels

# Loaches

Members of the Cobitidae family, Loaches originate in the Far East and Eurasia. They are mainly small to medium-sized, aquarium-suitable fishes.

Many species, such as the *Botia*, have a very flat ventral surface. This keeps the fish in close contact with the streambed. As the water can't easily get under the body and lift it off the gravel, it prevents the fish being swept away by fast-moving water. Other species have a worm-like body shape which allows them to investigate every nook and cranny amongst the rocks and plants.

Some fishes, such as the Coolie Loach, called *Acanthopthalmus* (literally "thorn-eye"), have an erectile spine beneath the eye. A number of marine fishes have similar spines, but those of the Loaches differ because they aren't poisonous. Like Catfishes, Loaches are also equipped with taste-sensitive barbels, which supplement their visual recognition. This is important as the water can be quite muddy and visibility low at riverbed level, where the Loaches usually dwell.

With the exception of the small *Botia sidthimunki*, Loaches are quite shy and only active at twilight or during the night. If you want to view Loaches it is therefore a good ploy to subdue the aquarium lighting, and thus trick the fishes into thinking it is time to come out. Another way is offer them a meal of worms, which attract them like a magnet. Because of their nocturnal habits, Loaches are difficult to study, and therefore very little is known about their breeding methods.

# Acanthopthalmus kuhli

*Coolie Loach*

Despite the unorthodox worm-like shape of these black-banded, golden-brown fishes, all the normal fins are present (dorsal and anal fins are set far back). They have four pairs of short barbels. They spend most of their time wriggling around the base of plant stems looking for food.

*Pair of Coolie Loaches*

SPECIES DETAILS

**Size** 110 mm long
**Origin** S.E. Asia
**Sexing** No visible differences
**Ease of keeping** Moderately easy
**Food** All foods, especially worms
**Breeding method** Egg-scatterer
**Breeding potential** Rarely known to have bred in an aquarium

AQUARIUM CONDITIONS

**Water** All types
**Temperature** 24°C
**Tank type** Community or species
**Swimming level** Lower levels
**Special needs** Retreats; plants

# Botia macracantha

*Clown Loach, Tiger Botia*

*Clown Loach*

A bright orange body marked with three broad, black bands (one passing across the eye) makes the Clown Loach the most colourful *Botia*. Its scales are so small that the skin appears to be naked. This fish likes the company of its own species, and a solitary specimen might well pine away if kept in isolation in a community aquarium.

## SPECIES DETAILS

**Size** 125 mm long
**Origin** S.E. Asia
**Sexing** No visible differences
**Ease of keeping** Moderately easy, but may be prone to White Spot disease (see p.231)
**Food** All foods, especially worms
**Breeding method** Unknown
**Breeding potential** Not known to have bred in the aquarium

## AQUARIUM CONDITIONS

**Water** All types (after acclimatization)
**Temperature** 24°C
**Tank type** Community or species
**Swimming level** Lower levels, but may make occasional excursions into midwater
**Special needs** Plants; retreats

# Botia sidthimunki

*Chain Loach, Dwarf Loach*

This fish has a dark brown, chain-like pattern laid over a bright golden body. Its ventral surface isn't quite flat, which suggests that it isn't a completely bottom-dwelling species. In fact, it is less shy than other Loaches and spends much time in midwater. When threatened, it will raise the sharp spines in front of its eyes. The Chain Loach is best kept in a group. Although it is an active fish, it likes to rest on a leaf or log by perching on its pelvic fins, as shown here.

*Chain Loach*

## SPECIES DETAILS

**Size** 55 mm long
**Origin** S.E. Asia
**Sexing** No visible differences
**Ease of keeping** Moderately easy
**Food** All foods, especially worms
**Breeding method** Unknown
**Breeding potential** Not known to have bred in an aquarium

## AQUARIUM CONDITIONS

**Water** Soft, medium-hard
**Temperature** 24°C
**Tank type** Community, but small fishes only or species
**Swimming level** Midwater and lower levels
**Special needs** Plants; resting places

# Other tropical egglaying species

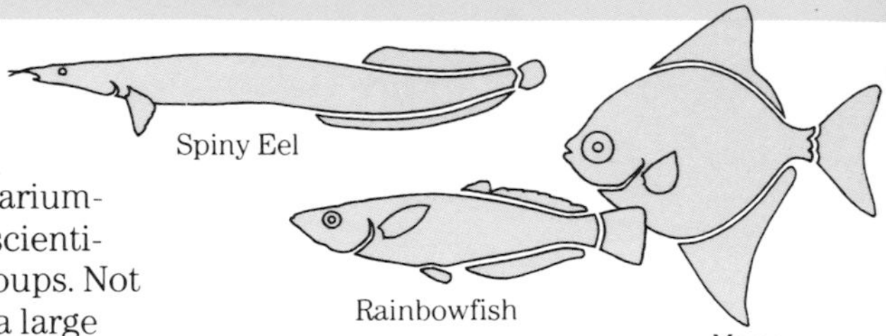


There are many species of aquarium-suitable fishes which don't fit scientifically into any of the larger groups. Not surprisingly, fishes from such a large number of genera come from a very wide range of locations and environments, and their differing body shapes, finnage and colours are very appealing, especially for the non-conformist fishkeeper who wants to keep something different and sees these "oddballs" as a challenge. The species covered in this section fall into one of three groups: Rainbowfishes, Spiny Eels and Monofishes.

Rainbowfishes can be identified by their two separate dorsal fins; unlike the adipose fins of fishes like Characins, both dorsal fins have supporting rays.

Spiny Eels bury themselves in the gravel just leaving their sharply pointed noses exposed.

The Mono or Fingerfish, found in estuarine and salt waters in the Far East, is very reminiscent of the freshwater Angelfish from South America (see p. 59), but lacks the filamentous pelvic fins and isn't so graceful in its movements.

## Bedotia geayi

*Madagascar Rainbowfish*

Varying in colour from yellow-brown to blue-green, the body contour of this Rainbowfish is quite symmetrical. The long anal fin and equally long second dorsal fin are red with black borders. There are two dorsal fins, the first of which is usually carried folded down. The upturned mouth shows that this fish is an upper-level swimmer and feeder. As young fishes will only accept moving food, turn the airstone up (see p. 123) to create circulating currents. The Madagascar Rainbowfish should be kept in a small group.

*Male Madagascar Rainbowfishes*

### SPECIES DETAILS

**Size** 100 mm long
**Origin** Madagascar
**Sexing** Females are duller
**Ease of keeping** Easy
**Food** Floating foods
**Breeding method** Egg-scatterer
**Breeding potential** Moderately easy, but food must be kept moving for fry

### AQUARIUM CONDITIONS

**Water** Hard
**Temperature** 24°C
**Tank type** Community or species
**Swimming level** Upper levels

# Macrognathus aculeatus

*Spiny Eel*

This species has the typical eel shape, a narrow head and a long nose. It is mainly brown in colour, with four or five equally spaced, brown eye-spot markings on the dorsal fin. The Spiny Eel is nocturnal.

*Spiny Eel*

SPECIES DETAILS

**Size** 350 mm long
**Origin** India, Far East
**Sexing** No visible differences
**Ease of keeping** Moderately easy
**Food** Worm foods
**Breeding method** Egg-scatterer
**Breeding potential** Not known to have bred in an aquarium

AQUARIUM CONDITIONS

**Water** All types; add 1 tsp salt per 5 litres
**Temperature** 24–6°C
**Tank type** Species
**Swimming level** Midwater and lower levels

# Mastacembelus argus

*Spiny Eel*

A nocturnal fish, this Spiny Eel has a speckled brown, elongated body. The anal, dorsal and caudal fins often merge into one continuous fin which provides the fish with a great degree of mobility – both forwards and backwards in the water.

*Spiny Eel*

SPECIES DETAILS

**Size** 250 mm long
**Origin** Far East
**Sexing** No visible differences
**Ease of keeping** Moderately easy
**Food** Worm foods
**Breeding method** Unknown
**Breeding potential** Not known to have bred in an aquarium

AQUARIUM CONDITIONS

**Water** Hard; add 1 tsp sea-salt per 5 litres
**Temperature** 26°C
**Tank type** Species
**Swimming level** Lower levels

# Melanotaenia nigrans

*Australian Rainbowfish*

Blue-mauve scales with dark edges create a net-like pattern on the Australian Rainbowfish's body. All the fins are a reddish colour, and the two dorsal fins and the anal fin are elongated, often trailing past the beginning of the caudal fin. This fish inhabits brackish waters.

### AQUARIUM CONDITIONS

**Water** Hard; add 1 tsp sea-salt per 5 litres
**Temperature** 24°C
**Tank type** Community
**Swimming level** Upper levels and midwater
**Special needs** Space

### SPECIES DETAILS

**Size** 100 mm long
**Origin** Australia
**Sexing** Males have darker-edged fins
**Ease of keeping** Easy
**Food** All foods
**Breeding method** Egg-scatterer
**Breeding potential** Moderately easy

*Male Australian Rainbowfish*

# Monodactylus argenteus

*Mono, Malayan Angelfish, Fingerfish*

The main body colour of this fish is silver with orange fins. There are two black bars, one through the eye, the other across the body in front of the dorsal fin. Mature Monos need sea-salt added to the water.

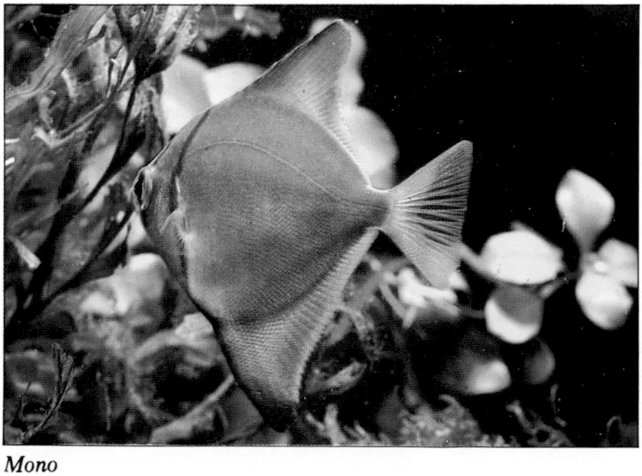

*Mono*

### SPECIES DETAILS

**Size** 150 mm long
**Origin** Coastal regions of Indo-Pacific
**Sexing** No visible differences
**Ease of keeping** Moderately easy
**Food** All foods
**Breeding method** Unknown
**Breeding potential** Not known to have bred in an aquarium

### AQUARIUM CONDITIONS

**Water** Hard; add 1–2 tsp sea-salt per 5 litres
**Temperature** 24°C
**Tank type** Species
**Swimming level** All levels

# COLDWATER FRESHWATER SPECIES

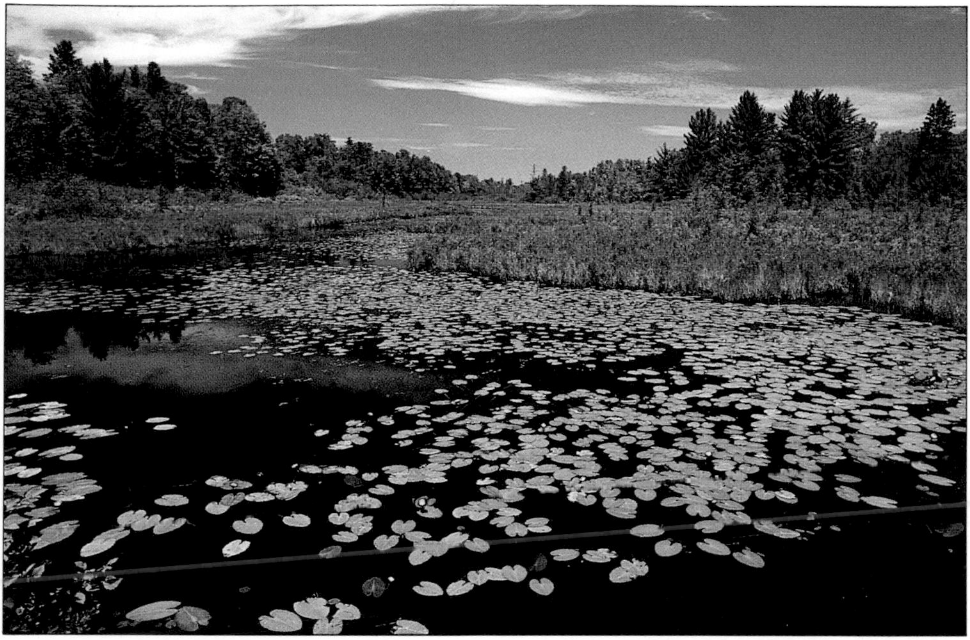

Fishes that can be kept in unheated tanks, and are native to rivers and streams in the Far East and Europe are referred to as coldwater freshwater species. Such fishes were the first type to be raised in captivity by man, originally for food purposes. This system is therefore the oldest form of fishkeeping. Although the Goldfish is still the most commonly kept coldwater freshwater fish, other species from temperate climates, such as Koi, are also popular. Many of them display colours that rival tropical species.

An added attraction of this system is that the interest can be extended outdoors beyond the aquarium to the garden pond (see p. 279), as some of the fishes will become too large to be confined in a tank. Another advantage is that the fishes are easily obtainable, will breed readily and eat commercially available foods.

For a coldwater freshwater aquarium you won't require any heating equipment or a thermostat. But, ironically, the temperature level will still require careful monitoring, especially in the summer, because the oxygen level falls as the water becomes warmer, affecting the fishes' health. When this happens you will have to increase aeration and siphon off some of the water, replacing it with fresh cold water. A drawback of this system is that coldwater freshwater fishes require more space than tropical species, so you must either have a larger tank or keep fewer specimens.

# Goldfishes

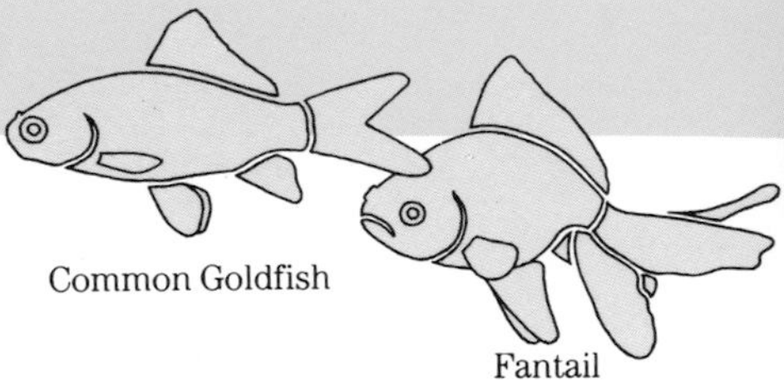


Starting with the dull-brown wild species, *Carassius auratus*, Chinese fish-keepers developed the metallic-looking orange variety known as the Goldfish centuries ago. The Chinese and Japanese developed many varieties with strangely shaped bodies and fins, and these eventually spread over the world. The Goldfish not only started off the whole fishkeeping hobby, but has also survived the competition from tropical freshwater and marine fishes. Much of the continued popularity of Goldfishes is due to the fact that the young fishes can be kept indoors in an aquarium and later, as they grow larger (up to 200 mm), they can be transferred outside to a garden pond. Unlike the tropical species, these fishes don't need the aquarium's water temperature to be maintained within narrow limits.

The order in which the fishes are arranged in this section shows the progressive development from the hardy singletail specimens such as the Common Goldfish, Shubunkins and Comets, to the more delicate, exotic-looking Fantails, Veiltails, Moors, Orandas, Lionheads and Celestials. The fishes' scales vary from the reflective "metallic" type through the "nacreous" (mother-of-pearl sheen) to "matt".

Goldfishes feed at all levels of the aquarium, and will accept most foods. However, their diet should contain a high carbohydrate level.

## Common Goldfish

The best body colour in the Common Goldfish is metallic orange-red without any silver at all. A sturdy fish, it has a long-based dorsal fin and a slightly forked, rounded caudal fin. There is also a colour variety, known as the London Shubunkin, which has less reflective scales, and a basically blue body.

*Pair of Common Goldfishes*

SPECIES DETAILS

**Size** Up to 200 mm long
**Origin** China
**Sexing** Males have breeding tubercles on gill-covers
**Ease of keeping** Easy
**Food** All foods
**Breeding method** Egg-scatterer
**Breeding potential** Easy

AQUARIUM CONDITIONS

**Water** All types
**Temperature** 0–20°C
**Tank type** Community
**Swimming level** All levels

# Bristol Shubunkin

This variety has nacreous scales, which may be a mixture of blue, purple, brown, yellow, red, orange and black. Although smaller than the London Shubunkin (see p. 84), it has much more developed fins. The large-lobed caudal fin is carried erect, and the sizable dorsal fin is as high as the long body is deep.

*Group of Bristol Shubunkins*

SPECIES DETAILS

**Size** Up to 125 mm long
**Origin** China
**Sexing** Males have breeding tubercles on head and gill-covers
**Ease of keeping** Easy
**Food** All foods
**Breeding method** Egg-scatterer
**Breeding potential** Easy

AQUARIUM CONDITIONS

**Water** All types
**Temperature** 0–20°C
**Tank type** Community
**Swimming level** All levels

# Comet

The orange-red Comet is usually of metallic form, and there is also a metallic silvery variety that has a deep red cap on the head. Smaller than the Common Goldfish, it has a particularly attractive caudal fin which is deeply forked and may be almost as long again as the body. The long pelvic and pectoral fins are pointed. It is a fast swimmer, but can't sustain the speed.

*Comet*

SPECIES DETAILS

**Size** 110 mm long
**Origin** China
**Sexing** Males have breeding tubercles on head and gill-covers
**Ease of keeping** Easy
**Food** All foods
**Breeding method** Egg-scatterer
**Breeding potential** Easy

AQUARIUM CONDITIONS

**Water** All types
**Temperature** 8–20°C
**Tank type** Community
**Swimming level** All levels

# Fantail

A nacreous or metallic orange "twintail" (see p. 18) the Fantail has a rounded, egg-shaped body. In good specimens, the top edge of the caudal fin stands up well and doesn't droop below the horizontal.

*Fantail*

SPECIES DETAILS

**Size** Up to 90 mm long
**Origin** China
**Sexing** Males have breeding tubercles on head and gill-covers
**Ease of keeping** Easy; keep indoors in winter
**Food** All foods
**Breeding method** Egg-scatterer
**Breeding potential** Easy

AQUARIUM CONDITIONS

**Water** All types, very clean
**Temperature** 8–20°C
**Tank type** Community
**Swimming level** All levels

# Veiltail

There are both nacreous and metallic forms and red-capped variants of the usually orange Veiltail. It has an elaborate, square-cut caudal fin which hangs gracefully in folds, and flowing pectoral and pelvic fins. The fish shown here is a commercial variety.

*Red-capped Veiltail*

SPECIES DETAILS

**Size** Up to 90 mm long
**Origin** China
**Sexing** Males have breeding tubercles on head and gill-covers
**Ease of keeping** Easy; keep indoors in winter
**Food** All foods
**Breeding method** Egg-scatterer
**Breeding potential** Easy

AQUARIUM CONDITIONS

**Water** All types, very clean
**Temperature** 8–20°C
**Tank type** Community
**Swimming level** All levels

# Telescope Moor

With the normal Veiltail body shape and finnage (see p. 86), this variety differs from the Veiltail in two respects: its "telescope" eyes and its velvety, jet-black body. Ideally, there should be no gold colour showing through the black, although this may occur in juvenile fishes and in poor-quality adult specimens.

*Telescope Moor*

SPECIES DETAILS

**Size** Up to 120 mm long
**Origin** China
**Sexing** Males have breeding tubercles on head and gill-covers
**Ease of keeping** Easy; keep indoors in winter
**Food** All foods
**Breeding method** Egg-scatterer
**Breeding potential** Easy

AQUARIUM CONDITIONS

**Water** All types, very clean
**Temperature** 8–20°C
**Tank type** Community
**Swimming level** All levels

# Oranda

This metallic variety has a raspberry-like "hood" over the entire head. Orandas are usually red and may have some white patches, although there are white fishes with small red hoods called "Tancho Orandas".

*Oranda*

SPECIES DETAILS

**Size** Up to 120 mm long
**Origin** China
**Sexing** Males have breeding tubercles on head and gill-covers
**Ease of keeping** Easy; keep indoors in winter
**Food** All foods
**Breeding method** Egg-scatterer
**Breeding potential** Easy

AQUARIUM CONDITIONS

**Water** All types, very clean
**Temperature** 8–20°C
**Tank type** Community
**Swimming level** All levels

# Lionhead

The orange-red head and body closely follow the physical characteristics of the Oranda (see p. 87), but in this variety there is no dorsal fin. The anal and caudal fins are short and stiffly held.

*Group of young Lionheads*

SPECIES DETAILS

**Size** Up to 120 mm long
**Origin** China
**Sexing** Males have breeding tubercles on head and gill-covers
**Ease of keeping** Easy; keep indoors in winter
**Food** All foods
**Breeding method** Egg-scatterer
**Breeding potential** Easy

AQUARIUM CONDITIONS

**Water** All types, very clean
**Temperature** 8–22°C
**Tank type** Community
**Swimming level** All levels

# Celestial

Ideally, this metallic orange fish has rudimentary anal and caudal fins, no dorsal fin and upturned eyes. As it may find locating and competing for food difficult, it should only be kept with other Celestials.

*Celestial*

SPECIES DETAILS

**Size** Up to 110 mm long
**Origin** China
**Sexing** Males have breeding tubercles on head and gill-covers
**Ease of keeping** Easy; keep indoors in winter
**Food** All foods
**Breeding method** Egg-scatterer
**Breeding potential** Easy

AQUARIUM CONDITIONS

**Water** All types, very clean
**Temperature** 8–22°C
**Tank type** Single variety
**Swimming level** All levels

# Koi

Koi

Scientifically speaking, this species is a form of Carp, *Cyprinus carpio*. However, it is often referred to as "Nishiki Koi", which is Japanese for Brocaded Carp.

Although outwardly very similar in body form to the Common Goldfish, Koi grow very much larger and have a single barbel at each corner of the mouth. Koi have traditionally been cultured to be seen from above — that is, as they swim in a pool. This is due to the way they are kept in their country of origin, Japan, where garden pools are very popular.

The fishes are divided into single-colour, two-colour and multi-coloured varieties, and are further distinguished by their scale development. Some have a few large scales (known as *Doitsu*), others a pine-cone effect (known as *Matsuba*), and a third group have extra metallic speckling (gold types are known as *Kin-rin*, and silver as *Gin-rin*). Prize-winning specimens may be one metre long and cost several thousands of pounds. Such fishes can't be produced in the aquarium; only very juvenile forms should be kept indoors. Once Koi grow longer than about 120 mm, they should be moved outdoors to larger, natural living quarters. If the pond is likely to freeze over in the winter months, a minimum water depth of 1.5–2 metres must be provided for safe overwintering.

Koi will breed quite readily in a pond. Some fishkeepers find it better to collect their eggs after spawning and hatch them in an indoor aquarium.

# Two-colour Koi

*Kohaku and Hariwaki Koi*

The white variety with red markings (above) is a *Kohaku*, whilst the fish with gold and silver patterning (below) is a *Hariwaki*. Both have the typical carp-shaped body, with the greatest depth occurring just forward of the long-based dorsal fin. Their downturned mouths show that they are bottom-feeders, but they will accept floating pellet food.

*Kohaku Koi (above) and Hariwaki Koi (below)*

SPECIES DETAILS

**Size** Up to 250 mm as juveniles

**Origin** Japan

**Sexing** Males have breeding tubercles on pectoral fins; females are fatter

**Ease of keeping** Easy

**Food** All foods

**Breeding method** Egg-scatterer

**Breeding potential** Easy

AQUARIUM CONDITIONS

**Water** All types, well-filtered

**Temperature** 0–20°C

**Tank type** Species

**Swimming level** All levels

# Three-colour Koi

*Asagi and Taisho Sanke*

The upper fish is an *Asagi* variety, with a light-blue body marked in orange and black, and *Doitsu* scales (see p. 89). The lower fish has the traditional three colours of the *Sanke* variety; this one is a *Taisho Sanke* – red and black on a white skin. (The *Showa Sanke* variety is red and white on a black skin.)

*Asagi Koi (above) and Taisho Sanke Koi (below)*

SPECIES DETAILS

**Size** Up to 250 mm long as juveniles
**Origin** Japan
**Sexing** Males have breeding tubercles on pectoral fins, females are fatter
**Ease of keeping** Easy
**Food** All foods
**Breeding method** Egg-scatterer
**Breeding potential** Easy

AQUARIUM CONDITIONS

**Water** All types, well-filtered
**Temperature** 0–20°C
**Tank type** Species
**Swimming level** All levels
**Special needs** Space

# Mongrel Koi

This yellow and orange fish doesn't fit into any of the colour standards recognized by the specialist Koi societies. It can't therefore be entered for showing, but this doesn't mean it won't make an attractive addition to your aquarium or pond. Another point in its favour is that it will be relatively cheap to purchase compared to a show-quality specimen.

*Mongrel Koi*

SPECIES DETAILS

**Size** Up to 250 mm long as juveniles
**Origin** Japan
**Sexing** Males have breeding tubercles on pectoral fins, females are fatter
**Ease of keeping** Easy
**Food** All foods
**Breeding method** Egg-scatterer
**Breeding potential** Easy

AQUARIUM CONDITIONS

**Water** All types, well-filtered
**Temperature** 0–20°C
**Tank type** Species
**Swimming level** All levels
**Special needs** Space

# Other coldwater species

By far the greatest number of species kept in contemporary coldwater tanks come from North America, but there are also many European fishes worth keeping. It used to be a commonly held belief that coldwater species (other than Goldfishes) were dull-coloured and, since many were mainly known as sport for anglers, required extra large aquariums. Fortunately, nowadays nothing could be further from the truth: Sunfishes, Shiners and Basses are all suitably sized for the indoor aquarium. Some of them rival tropical species for colour, and have equally interesting spawning behaviour. They feed at all levels.

Many coldwater fishes, such as the Sunfishes, are easy to breed, but others are more difficult. The Bitterling, for example, depends on the presence of freshwater Mussels (as the eggs hatch inside them), but it is well worth furnishing the aquarium to suit its needs.

Despite the fact that you need less technical equipment, keeping coldwater species may be slightly more difficult than freshwater tropicals. This is because these fishes require proportionally larger tanks and extremely well-oxygenated water. Another problem may be that of keeping the fishes cool enough in high summer. In thundery weather, turn the aeration up or siphon some water from the tank and replace with colder water.

## Elassoma evergladei

*Pigmy Sunfish*

As its common name suggests, this is a tiny fish. The body is normally yellow-green in colour with black and metallic-looking, irregular markings. However, when the male is in breeding condition, it becomes a deep blue-black colour with iridescent markings, as shown here. The Pigmy Sunfish is territorial and will thrive in a species tank. It can be kept in an unheated indoor tank, and may even be transferred outdoors to a pond (see p. 279) in the summer.

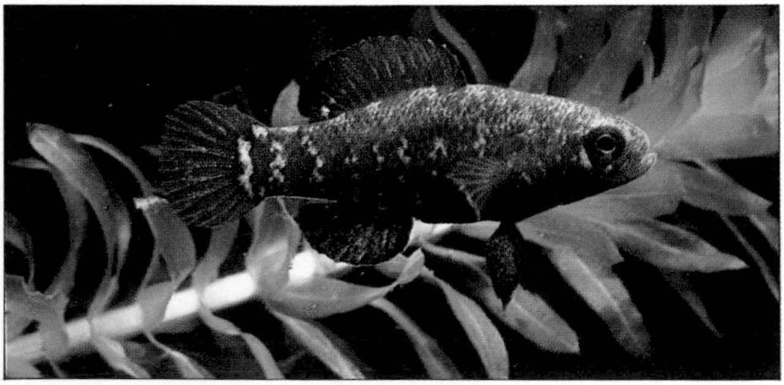

*Male Pigmy Sunfish in breeding condition*

SPECIES DETAILS

**Size** 37 mm long
**Origin** Florida
**Sexing** Males turn black at breeding time; females deeper-bodied
**Ease of keeping** Moderately easy
**Food** All foods; worms
**Breeding method** Egg-scatterer
**Breeding potential** Moderately easy

AQUARIUM CONDITIONS

**Water** Hard
**Temperature** 10–22°C
**Tank type** Species
**Swimming level** All levels
**Special needs** Plants; retreats

# Enneacanthus chaetodon

*Black-banded Sunfish*

Formerly known as *Mesogonistius chaetodon*, this small Sunfish has bright orange rays on the front edge of its dorsal and pelvic fins, and four vertical black bars across its grey-yellow body. It lays its eggs in a shallow depression in the gravel.

*Female Black-banded Sunfish in breeding colours*

SPECIES DETAILS

**Size** 100 mm long
**Origin** N. America
**Sexing** Females are more brightly coloured at breeding time
**Ease of keeping** Moderately easy
**Food** All foods
**Breeding method** Egg-depositor
**Breeding potential** Difficult

AQUARIUM CONDITIONS

**Water** All types
**Temperature** 8–22°C
**Tank type** Species
**Swimming level** All levels
**Special needs** Plants; retreats

# Lepomis gibbosus

*Pumpkinseed*

With its yellow-green body covered with blue-green iridescences and red speckles, it is easy to see how the Pumpkinseed got its common name. Good specimens also have a bright red throat and belly. It has an "earflap" extension behind the gill-cover, which is characteristic of its family. Normally a peaceful fish, the male guards the eggs and fry fiercely.

AQUARIUM CONDITIONS

**Water** All types; well-filtered
**Temperature** 8–22°C
**Tank type** Species
**Swimming level** All levels
**Special needs** Space

SPECIES DETAILS

**Size** 200 mm long
**Origin** N. America
**Sexing** Females are duller and plumper
**Ease of keeping** Moderately easy
**Food** All foods
**Breeding method** Egg-depositor
**Breeding potential** Moderately difficult

*Male Pumpkinseed*

# Notropis lutrensis

*Red Shiner*

*Male Red Shiner*

This slim fish shows blue and purple body coloration. The head and fins are red (hence the common name), and there is a dark blue triangular mark behind the gill-cover. All the colours are intensified at breeding time, and the male develops white tubercles on his head, sides and fins. The Red Shiner needs plenty of swimming space in well-oxygenated water (see p. 123).

SPECIES DETAILS

**Size** 80 mm long
**Origin** N. America
**Sexing** Males have white tubercles on head
**Ease of keeping** Moderately easy
**Food** All foods
**Breeding method** Egg-scatterer
**Breeding potential** Moderately difficult

AQUARIUM CONDITIONS

**Water** All types; well-filtered
**Temperature** 10–22°C
**Tank type** Community
**Swimming level** All levels

# Rhodeus sericeus amarus

*Bitterling*

Normally both sexes of this fish are silver-grey in colour, but in the breeding season the male Bitterling becomes brilliant violet and green. Unusually, the eggs hatch inside a freshwater Mussel. The female lays them there by means of a long tube (*ovipositor*), and the male's fertilizing milt is then drawn in as the mollusc breathes.

*Male Bitterling*

SPECIES DETAILS

**Size** 90 mm long
**Origin** Europe
**Sexing** Males change colour at breeding time; females have long ovipositor
**Ease of keeping** Moderately easy
**Food** All foods
**Breeding method** Egg-depositor
**Breeding potential** Difficult; must have freshwater Mussels (hard to keep alive)

AQUARIUM CONDITIONS

**Water** All types
**Temperature** 10–22°C
**Tank type** Species
**Swimming level** All levels

# Umbra pygmaea

*Eastern Mudminnow*

The body colour of the Eastern Mudminnow is rather a drab brown with dark markings, the male being darker than the female. This species is able to utilize atmospheric air by means of its swim-bladder (see p. 21) should the oxygen level in the water become depleted. It lays its eggs in the gravel, in caves or in other plant-built retreats. The Mudminnow gained its common name because of its habit of burrowing tail-first into the river mud.

*Male Eastern Mudminnow*

SPECIES DETAILS

**Size** 75 mm long
**Origin** N. America
**Sexing** Females are lighter in colour
**Ease of keeping** Moderately easy
**Food** All foods; live worm foods
**Breeding method** Egg-depositor
**Breeding potential** Moderately easy

AQUARIUM CONDITIONS

**Water** All types
**Temperature** 8–20°C
**Tank type** Species
**Swimming level** Lower levels
**Special needs** Retreats

# Zacco platypus

*Pale Chub*

The sides of this fish are marked with broad bright blue-green stripes which taper and fade before the belly. Males are larger than females and have an anal fin that is very well-developed with a long lobe, which may be used during spawning. An active species, the Pale Chub prefers fast-flowing waters and so a well-aerated, fast-filtered aquarium with a lid (to prevent it jumping out) is essential.

*Female Pale Chub*

SPECIES DETAILS

**Size** 180 mm long
**Origin** Far East
**Sexing** Males have longer anal fin, breeding tubercles on jaw and gill-cover
**Ease of keeping** Moderately easy
**Food** All foods; algae
**Breeding method** Unknown
**Breeding potential** Not known to have bred in an aquarium

AQUARIUM CONDITIONS

**Water** All types; well-filtered
**Temperature** 8–20°C
**Tank type** Community
**Swimming level** All levels

# TROPICAL MARINE SPECIES

Saltwater fishes from the coral reefs of the Pacific, Caribbean and Mediterranean are known as tropical marines. The attraction of this most recent aquarium system lies in the fishes' vast variety of colours and shapes. They vary in size, but larger specimens may be prohibitively expensive. Although there are fewer suppliers of marine fishes than fresh-water types, most species are reasonably easy to obtain.

Tropical marine fishkeeping is the most demanding branch of the hobby because of the nature of the fishes, and the difficulty in maintaining the right environment to support them. The fishes are more difficult to breed than fresh-water species, they are more expensive and tend to be aggressive. Also, as most are sensitive to chemical changes in the water, the conditions must be kept stable. Strong aeration (see p. 123) and filtration in the form of an undergravel filter (see p. 128–9) are essential for this, along with a heater (see p. 131–2). The tank itself must be impervious to saltwater, and contain no metal as this will corrode. Although plants can't be kept in the tank, corals are a colourful alternative.

NOTE: In this section, the standard mix of synthetic sea-water referred to in "Aquarium conditions" is a compromise between that used by your dealer and that given on the manufacturer's instructions. For interest, where known an indication of the specific gravity (see p. 147) of the water in the fishes' natural habitat is also given.

# Anemonefishes and Damselfishes

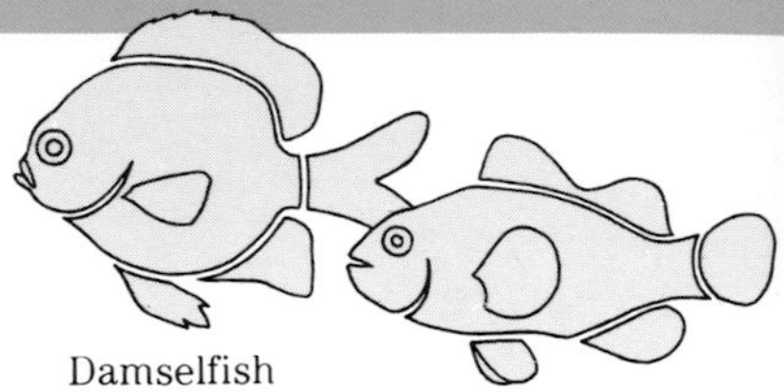


The fishes in this group originate from the coral reefs of the Indo-Pacific, and are hardy and extremely colourful. As they swim at all levels in the water, they maximize the use of space in your aquarium. Because of their small size, they incur low airfreight charges, and so are amongst the least expensive marine species.

Anemonefishes, or Clownfishes, are so-called because they make their home in the tentacles of the Sea-anemone (see p. 110). Fortunately, unlike other fishes Anemonefishes aren't stung by their host as the Sea-anemone's stinging cells aren't activated by the fishes' skin mucus.

Damselfishes don't live in Sea-anemones, but frequent the coral heads. They swim with a perky, bobbing action, whereas Anemonefishes have a more distinctive "waddle" in their swimming movement. Damselfishes are very colourful when young, but may lose the clarity of their colour patterns with advancing age.

Unfortunately, both types of fishes can be very aggressive, especially towards their own kind; this means that perhaps only one or two of each species can be kept in the same aquarium.

Most of these fishes spawn in a similar fashion to the freshwater Cichlid fishes, laying eggs on a pre-selected site.

## Abudefduf oxyodon

*Blue-velvet Damselfish, Blue-banded Sergeant Major*

The velvety blue-black body from which this Damselfish gets its name is crossed by a yellow-white line between the dorsal fin and the head. Electric-blue lines on the head and rear part of the body fade with age.

*Blue-velvet Damselfish*

### SPECIES DETAILS

**Size** 110 mm long
**Origin** Pacific Ocean
**Sexing** No visible differences
**Ease of keeping** Easy; aggressive
**Food** All foods
**Breeding method** Egg-depositor
**Breeding potential** Rarely aquarium-bred

### AQUARIUM CONDITIONS

**Water** Standard mix
**Temperature** 25°C
**Tank type** Community, but only one of this species
**Swimming level** All levels
**Special needs** Retreats

# Amphiprion ocellaris

*Common Clownfish*

Easily identifiable with its bright orange body and three white bands, the Common Clownfish may have acquired its name because of its rather ungainly waddling swimming action. Ideally, you should provide the fish with a Sea-anemone (see p.110) as a safe sanctuary – a large one will house several small fishes.

## AQUARIUM CONDITIONS

**Water** Standard mix
**Temperature** 25°C
**Tank type** Community
**Swimming level** All levels
**Special needs** Sea-anemone for refuge

## SPECIES DETAILS

**Size** 80 mm long
**Origin** Indo-Pacific
**Sexing** No visible differences
**Ease of keeping** Easy
**Food** All foods
**Breeding method** Egg-depositor
**Breeding potential** Rarely aquarium-bred

*Common Clownfish*

# Dascyllus albisella

*Hawaiian Humbug*

This species has a very dark blackish body, marked with three white spots, one on its forehead and one on each side of its black body. In young fishes the spots may be larger. The Hawaiian Humbug can be aggressive.

*Hawaiian Humbug*

## SPECIES DETAILS

**Size** 120 mm long
**Origin** Hawaii
**Sexing** No visible differences
**Ease of keeping** Easy; aggressive if breeding
**Food** All finely chopped meaty foods
**Breeding method** Egg-depositor
**Breeding potential** Rarely aquarium-bred

## AQUARIUM CONDITIONS

**Water** Standard mix
**Temperature** 25°C
**Tank type** Community
**Swimming level** All levels
**Special needs** Coral retreats

# Angelfishes and Butterflyfishes

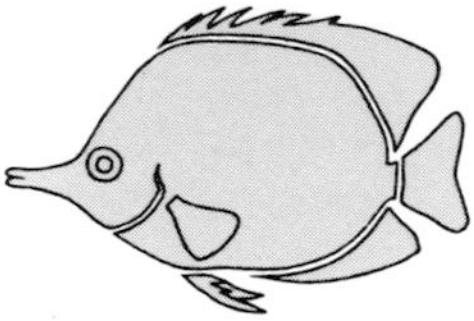
Butterflyfish

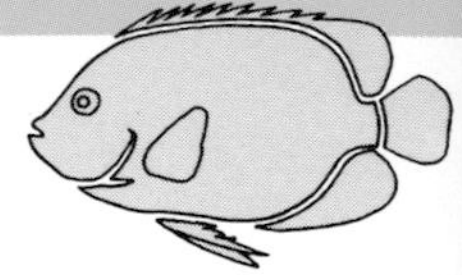
Angelfish

The colouring of this group of fishes is necessarily bright to stand out against the coral heads in the brilliantly lit surroundings of their native Indo-Pacific reefs. The colours and patterns, many of which are for camouflage or species recognition purposes, are quite staggering. Both Angelfishes and Butterflyfishes have deep, laterally compressed bodies. They feed at all levels.

Angelfishes range in size from the small Pigmy Angelfishes of the *Centropyge* genus to the larger *Pomacanthus* genus containing the majestic Emperor Angelfish. All have small mouths relative to their body size. Those with specialized feeding habits (e.g. sponge-feeders) are particularly hard to maintain in captivity. Young Angelfishes are quite different in colour to the adult fishes, most being a dark blue with white or yellow markings (see p. 16). The family is territorial and will be aggressive towards other fishes.

Butterflyfishes lack the sharp-spined gill-cover of the Angelfishes, but are just as beautiful. Several have elongated snouts, ideally suited to the extraction of food from crevices in the coral reef. Fishes in this family are somewhat delicate, and therefore aren't recommended for the absolute beginner.

## Centropyge loriculus

*Flame Angelfish*

This species has a bright orange-red body marked with black vertical stripes, and dark-tipped anal and dorsal fins. Unlike other Angelfishes, the fish's markings don't change with age. Similar in shape to many Damselfishes, it can be identified by the spine on its gill cover. Although rare, the Flame Angelfish is hardy.

*Flame Angelfish*

### SPECIES DETAILS

**Size** 100 mm long

**Origin** Central and Western Pacific Ocean

**Sexing** No visible differences

**Ease of keeping** Moderately easy

**Food** All finely chopped foods, including plenty of green foods

**Breeding method** Unknown

**Breeding potential** Rarely, if ever, aquarium-bred

### AQUARIUM CONDITIONS

**Water** Standard mix

**Temperature** 25°C

**Tank type** Community of similar-sized fishes

**Swimming level** All levels

# Euxiphipops navarchus

*Blue-girdled Angelfish*

The light-edged blue bars running across this Angelfish's yellow body give it its common name. These bars break up the "fishy" silhouette and camouflage the eye, a primary target for aggressive members of its own group. It is quite a shy species and so requires hiding places (rocky "caves" or plants) in its tank.

*Blue-girdled Angelfish*

SPECIES DETAILS

**Size** 200 mm long
**Origin** Pacific Ocean
**Sexing** No visible differences
**Ease of keeping** Easy, but aggressive
**Food** All finely chopped foods; green foods
**Breeding method** Unknown
**Breeding potential** Rarely, if ever, aquarium-bred

AQUARIUM CONDITIONS

**Water** Standard mix
**Temperature** 25°C
**Tank type** Community
**Swimming level** All levels
**Special needs** Retreats

# Holacanthus tricolor

*Rock Beauty*

This extremely large fish has a yellow body with a blue-edged black patch, which extends from the gill-cover to the caudal fin. It has red edges to its anal and dorsal fins. Unfortunately, the Rock Beauty can be susceptible to skin diseases (see p. 227).

*Rock Beauty*

SPECIES DETAILS

**Size** 600 mm long
**Origin** Atlantic Ocean, Caribbean
**Sexing** No visible differences
**Ease of keeping** Moderately difficult; aggressive
**Food** All finely chopped foods; green foods
**Breeding method** Unknown
**Breeding potential** Rarely, if ever, aquarium-bred

AQUARIUM CONDITIONS

**Water** Standard mix
**Temperature** 25°C
**Tank type** Community
**Swimming level** All levels
**Special needs** Retreats

# Pomacanthus imperator

*Emperor Angelfish*

Young specimens of this species have blue bodies with white markings. As the fish matures, the stripes change into wavy yellow horizontal lines. The eye is disguised beneath a light-edged blue patch and there is a disruptive pattern of blue behind the gills. The caudal fin is yellow and the anal fin blue. The Emperor Angelfish is a territorial species.

*Emperor Angelfish*

SPECIES DETAILS

**Size** 400 mm long
**Origin** Indo-Pacific, Red Sea
**Sexing** No visible differences
**Ease of keeping** Easy, but aggressive
**Food** All finely chopped foods; green foods
**Breeding method** Unknown
**Breeding potential** Rarely, if ever, aquarium-bred

AQUARIUM CONDITIONS

**Water** Standard mix
**Temperature** 25°C
**Tank type** Community
**Swimming level** All levels
**Special needs** Retreats

# Chaetodon lunula

*Red-striped Butterflyfish*

The diagonal red-brown markings on the yellow body give this fish its name. It has a black patch over its eye, and another above the gill cover. The two blotches are separated by a broad white band. Its fins have red-brown stripes and black margins.

*Red-striped Butterflyfish*

SPECIES DETAILS

**Size** 200 mm long
**Origin** Indo-Pacific, Red Sea
**Sexing** No visible differences
**Ease of keeping** Moderately easy; aggressive
**Food** All finely chopped foods; green foods
**Breeding method** Unknown
**Breeding potential** Rarely, if ever, aquarium-bred

AQUARIUM CONDITIONS

**Water** Standard mix
**Temperature** 25°C
**Tank type** Community
**Swimming level** All levels
**Special needs** Retreats

# Chaetodon octofasciatus

*Eight-banded Butterflyfish*

This white fish has eight vertical dark bands across its light, silver-coloured body and caudal fin (hence its name). Its yellow and white anal and spiky-rayed dorsal fins have black margins, and its pelvic fins are bright yellow. It will eat polyps and Sea-anemones (see p. 110), so don't keep it in a tank with live corals or invertebrates.

SPECIES DETAILS

**Water** Standard mix
**Temperature** 25°C
**Tank type** Community
**Swimming level** All levels

AQUARIUM CONDITIONS

**Size** 200 mm long
**Origin** Indo-Pacific, Red Sea
**Sexing** No visible differences
**Ease of keeping** Moderately easy
**Food** All finely chopped
**Breeding method** Unknown
**Breeding potential** Rarely, if ever, aquarium-bred

*Eight-banded Butterflyfish*

# Chelmon rostratus

*Copper-banded Butterflyfish, Long-nosed Butterflyfish*

The five black-edged, orange-yellow bands on this thin, whitish fish give it its name. It has an eye-spot at the top of one band, and the broad yellow band on its caudal peduncle is split by a dark stripe. Unfortunately, this Butterflyfish isn't hardy.

*Copper-banded Butterflyfish*

SPECIES DETAILS

**Size** 150 mm long
**Origin** Indo-Pacific
**Sexing** Males may have steeper head profile
**Ease of keeping** Difficult; aggressive
**Food** Small live foods
**Breeding method** Unknown
**Breeding potential** Rarely, if ever, aquarium-bred

AQUARIUM CONDITIONS

**Water** Standard mix; stable conditions
**Temperature** 25°C
**Tank type** Species or community if large tank
**Swimming level** Midwater and lower levels

# Forcipiger longirostris

*Long-snouted Coralfish, Forcepsfish*

Similar in shape to *Chelmon rostratus* (see p.101), this fish has a plain yellow body, a black forehead and an eye-spot on the anal fin – thus it is well camouflaged. Its long, thin snout enables it to extract marine crustaceans from crevices in the coral reef it inhabits.

*Long-snouted Coralfish*

SPECIES DETAILS

**Size** 180 mm long
**Origin** Indo-Pacific
**Sexing** No visible differences
**Ease of keeping** Difficult; aggressive
**Food** Live worm foods
**Breeding method** Unknown
**Breeding potential** Rarely, if ever, aquarium-bred

AQUARIUM CONDITIONS

**Water** Standard mix; stable conditions
**Temperature** 24°C
**Tank type** Species or community if large tank
**Swimming level** Midwater and lower levels

# Heniochus acuminatus

*Wimplefish, Pennantfish, Poor Man's Moorish Idol*

This fish has a white body crossed by three black bands, one covering its eye, and a bright yellow caudal fin. Its long dorsal fin rays resemble a pennant.

*Wimplefish*

SPECIES DETAILS

**Size** 250 mm long
**Origin** Indo-Pacific, Red Sea
**Sexing** No visible differences
**Ease of keeping** Moderately difficult
**Food** Live foods; finely chopped green foods
**Breeding method** Unknown
**Breeding potential** Rarely, if ever, aquarium-bred

AQUARIUM CONDITIONS

**Water** Standard mix
**Temperature** 25°C
**Tank type** Species or large community
**Swimming level** All levels

# Other tropical marine species

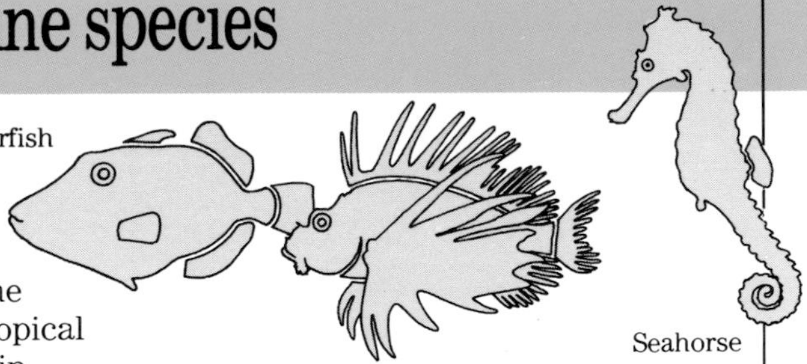


This section covers a selection of the most popular aquarium-suitable tropical marine species. These vary greatly in body form, colouring and behaviour.

The Surgeonfishes have vividly coloured, laterally compressed bodies which are equipped with sharp, erectile spines ("scalpels") on their caudal peduncle. Likewise, Triggerfishes have an unusual feature – the first spine of their dorsal fin is modified into a sharp spike. The fishes can erect and lock these at will, either to attack other fishes or to lock themselves into a crevice. Lionfishes are also able to inflict damage and poison by means of the beautiful, sharp spines that make up their dorsal, ventral and anal fins. Juvenile Wrasses often perform a cleaning service on other fishes, whilst Seahorses have curious swimming and resting attitudes and an unusual breeding method.

Not all the fishes in this section are compatible: some large fishes will eat smaller companions or invertebrates; and slow-moving fishes won't tolerate more active specimens in the same tank.

## Acanthurus leucosternon

*Powder Blue Surgeonfish*

*Powder Blue Surgeonfish*

The oval, laterally compressed body of this Surgeonfish is a delicate powder-blue colour, from which it gets its name. It has a dark blue-black head, and a white throat. Its blue-edged dorsal fin and its pectoral fin are bright yellow. This species has retractable "scalpels" with which it will defend itself if necessary.

### SPECIES DETAILS

**Size** 300 mm long
**Origin** Indo-Pacific
**Sexing** No visible differences
**Ease of keeping** Moderately difficult
**Food** All foods, green foods
**Breeding method** Unknown
**Breeding potential** Rarely aquarium-bred

### AQUARIUM CONDITIONS

**Water** Standard mix; strongly aerated (S.G. 1.020–1.024)
**Temperature** 25°C
**Tank type** Community
**Swimming level** All levels
**Special needs** Space

# Apogon nematopterus

*Pyjama Cardinalfish, Polka Dot Cardinalfish*

Also known as *Sphaeramia nematopterus*, this Cardinalfish has a white body with brown and yellow markings, and two dorsal fins (the front one is spotted). It is a nocturnal, shoaling fish.

*Pyjama Cardinalfish*

### SPECIES DETAILS

**Size** 75 mm long
**Origin** Indo-Pacific
**Sexing** No visible differences
**Ease of keeping** Easy
**Food** All finely chopped foods; green foods
**Breeding method** Mouth-brooder
**Breeding potential** Rarely aquarium-bred

### AQUARIUM CONDITIONS

**Water** Standard mix (S.G. 1.025)
**Temperature** 25°C
**Tank type** Community of peaceful fishes
**Swimming level** All levels

# Balistapus undulatus

*Undulate Triggerfish, Orange-green Triggerfish*

The green body of this fish is covered with wavy orange lines. Its pelvic fins are absent, being reduced to primitive stumps. Like all Triggerfishes, this species can lock its dorsal fin erect to prevent capture. It has strong jaws and will eat invertebrates. At rest it adopts a head-down attitude or lies on the sand.

*Undulate Triggerfish*

### SPECIES DETAILS

**Size** 350 mm long
**Origin** Indo-Pacific
**Sexing** No visible differences
**Ease of keeping** Easy; aggressive
**Food** All finely chopped foods; green foods
**Breeding method** Egg-depositor
**Breeding potential** Rarely aquarium-bred

### AQUARIUM CONDITIONS

**Water** Standard mix (S.G. 1.023)
**Temperature** 25°C
**Tank type** Community
**Swimming level** All levels

# Bodianus rufus

*Spanish Hogfish*

The top half of this fish is coloured red-violet, while the rest of the body and the lower fins are bright yellow. Like many of the Wrasses, the juvenile form acts as a cleanerfish to other fishes (see p. 106). The Spanish Hogfish feeds on small crustaceans, and live foods. At night it lies on or in the sand. It swims partly by means of its pectoral fins.

### AQUARIUM CONDITIONS

**Water** Standard mix (S.G. 1.020–1.023)
**Temperature** 25°C
**Tank type** Community
**Swimming level** All levels
**Special needs** Soft sand base to the aquarium

### SPECIES DETAILS

**Size** 100 mm long (bigger in nature)
**Origin** Caribbean
**Sexing** No visible differences
**Ease of keeping** Easy
**Food** Preferably live foods; all finely chopped foods; green foods
**Breeding method** Egg-scatterer
**Breeding potential** Rarely aquarium-bred

*Spanish Hogfish*

# Gramma loreto

*Royal Gramma, Fairy Basslet*

Vivid colours characterize the Royal Gramma: its head and the front half of its body is bright purple, whilst the rear is an equally bright yellow. It has a dark spot in its dorsal fin and a diagonal dark stripe running through its eye. The Royal Gramma is a territorial cave-dweller, and so only a few should be kept in the same aquarium.

### AQUARIUM CONDITIONS

**Water** Standard mix (S.G. 1.025)
**Temperature** 25°C
**Tank type** Community
**Swimming level** All levels
**Special needs** Retreats

### SPECIES DETAILS

**Size** 75 mm long
**Origin** Caribbean
**Sexing** Males have more pointed fins and may be larger
**Ease of keeping** Moderately easy, but sensitive to changing water conditions
**Food** Live foods; finely chopped foods
**Breeding method** Nest-builder; eggs laid on the roof of caves
**Breeding potential** Rarely aquarium-bred

*Female Royal Gramma*

# Hippocampus kuda

*Yellow Seahorse, Oceanic Seahorse*

Often black when imported, this Seahorse will become yellow once it is settled in the aquarium. It has a hard body covering and no caudal or anal fins. Like all Seahorses, it swims in an upright position, and may anchor itself when at rest with its prehensile "tail".

*Group of Yellow Seahorses*

SPECIES DETAILS

**Size** 50 mm long
**Origin** Indo-Pacific
**Sexing** Males have brood pouch
**Ease of keeping** Difficult
**Food** Live foods
**Breeding method** Female deposits eggs in male's breeding pouch
**Breeding potential** Rarely aquarium-bred

AQUARIUM CONDITIONS

**Water** Standard mix (S.G. 1.023)
**Temperature** 25°C
**Tank type** Species
**Swimming level** All levels

# Labroides dimidiatus

*Cleanerfish, Blue Streak, Bridled Beauty*

The slender body of this Wrasse is marked with horizontal black and light-blue bands. Its mouth is right at the tip of its head. It is called a Cleanerfish because of its sociable action of removing parasites and dead scar tissue from the sides, and very often the gills, of other larger fishes. It does this in order to feed off the parasites, which form its staple diet. In nature, its "customers" often visit the coral reef to be cleaned. When they see the Cleaner-fish they tilt their heads and remain inactive, signalling to the Cleanerfish to begin.

*Cleanerfish*

SPECIES DETAILS

**Size** 100 mm long
**Origin** Indo-Pacific
**Sexing** No visible differences
**Ease of keeping** Moderately easy
**Food** Parasites; meats
**Breeding method** Unknown
**Breeding potential** Rarely aquarium-bred

AQUARIUM CONDITIONS

**Water** Standard mix (S.G. 1.023–1.025)
**Temperature** 25°C
**Tank type** Community
**Swimming level** All levels
**Special needs** Other fishes

# Opistognathus aurifrons

*Yellowheaded Jawfish, Jack-in-the-Box*

As its common name suggests, the head of this species is yellow, and its body is a delicate blue colour. The Yellow-headed Jawfish spends the day standing vertically in the entrance to its burrow. It will make the burrow in a layer of sand or crushed coral, and will retreat into it at night, closing the entrance behind it with a pebble or shell.

*Yellowheaded Jawfish*

SPECIES DETAILS

**Size** 100 mm long
**Origin** Caribbean
**Sexing** No visible diferences
**Ease of keeping** Moderately easy
**Food** Finely chopped meaty foods
**Breeding method** Mouthbrooder
**Breeding potential** Rarely aquarium-bred

AQUARIUM CONDITIONS

**Water** Standard mix (S.G. 1.020–1.023)
**Temperature** 25°C
**Tank type** Species
**Swimming level** Lower levels

# Pterois volitans

*Lionfish, Scorpionfish, Turkeyfish*

The body colour of this species is red-brown with white bands. It has long, grooved fin rays with venom glands at their base, so avoid handling. The Lionfish feeds on live fishes, usually six a day!

*Lionfish*

SPECIES DETAILS

**Size** 350 mm long
**Origin** Indo-Pacific, Red Sea
**Sexing** No visible differences
**Ease of keeping** Moderately easy
**Food** Live fishes; meats
**Breeding method** Unknown
**Breeding potential** Rarely aquarium-bred

AQUARIUM CONDITIONS

**Water** Standard mix (S.G. 1.025)
**Temperature** 25°C
**Tank type** Community, but large fishes only
**Swimming level** All levels

# Synchiropus splendidus

*Mandarinfish*

This species is very garishly marked, with blue-green scribbly lines and dots covering its gold body. The fins, including its two dorsals, are similarly coloured and have blue edges. It has gold-spotted gill-covers, and slightly protruberant eyes. Mandarinfishes are intolerant of other fishes and should therefore be kept in a species tank.

*Male Mandarinfish*

## SPECIES DETAILS

**Size** 75 mm long
**Origin** Indo-Pacific
**Sexing** Males have longer rays on first dorsal fin
**Ease of keeping** Difficult
**Food** Reluctant feeder; try live worm foods
**Breeding method** Unknown
**Breeding potential** Rarely, if ever, aquarium-bred

## AQUARIUM CONDITIONS

**Water** Standard mix (S.G. 1.020–1.023)
**Temperature** 25°C
**Tank type** Species
**Swimming level** Lower levels

# Zanclus cornutus

*Moorish Idol, Toby*

Often confused with the Wimplefish (see p. 102), the Moorish Idol has a more "painted" face and "horns" over the eyes. Two black bars cross its white and yellow body and its caudal fin is black. It has a very long, trailing dorsal fin. It isn't a hardy species.

*Moorish Idol*

## SPECIES DETAILS

**Size** 230 mm long
**Origin** Indo-Pacific
**Sexing** No visible differences
**Ease of keeping** Difficult
**Food** Reluctant feeder; try finely chopped foods; algae
**Breeding method** Unknown
**Breeding potential** Rarely, if ever, aquarium-bred

## AQUARIUM CONDITIONS

**Water** Standard mix (S.G. 1.020–1.023)
**Temperature** 25°C
**Tank type** Species
**Swimming level** All levels

# COLDWATER MARINE SPECIES

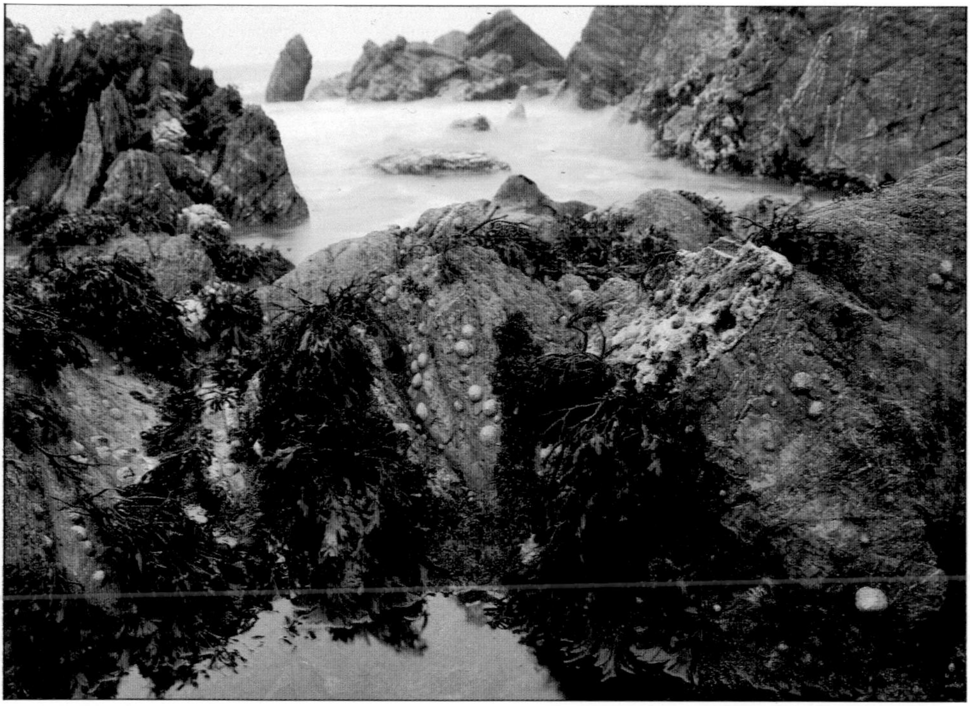

Regrettably, the keeping of marine life-forms such as Blennies, Sea-anemones and Tubeworms from local, non-tropical locations has come to be regarded as the poor relation of fishkeeping. Whilst it must be admitted that the variety and colours of these species can't compare with those from warmer tropical waters, they can form an interesting "budget" collection. Coldwater marines require temperatures between 8–20°C, and therefore a heater isn't necessary.

Requiring little technical equipment, the coldwater marine aquarium can be set up and maintained on a shoestring budget (for example, using a suitable secondhand tank, see pp. 192–7). Livestock can be gathered for free at the seashore, but make sure you disturb the collecting sites as little as possible and don't over-collect particular species. It is wise to choose specimens that are 25–75 mm in length, and from one location only, especially if you are collecting both fishes and invertebrates, as if they thrive together in nature they aren't likely to trouble each other in an aquarium. Although you can use locally collected water in your tank, it may be polluted so it is advisable to make up the standard synthetic saltwater mix (see p. 144). Most coldwater marines will eat meaty foods or algae.

Although these species breed easily in their natural habitat, they are less likely to do so in an indoor tank and it may be difficult to select suitable pairs.

# Blennius gattorugine

*Tompot Blenny*

The cylindrical-shaped brown body of this fish is crossed by six darker brown bars. It has a "crest" above its eyes, and its long dorsal fin has both hard and soft rays. Like other Blennies, the Tompot doesn't have scales, but is covered by a thick skin.

*Tompot Blenny*

SPECIES DETAILS

**Size** 200 mm long

**Origin** Mediterranean, N.E. Atlantic

**Sexing** Females are smaller

**Ease of keeping** Moderately easy

**Food** All foods; algae

**Breeding method** Secretive egg-depositor

**Breeding potential** Variable (see p. 109)

AQUARIUM CONDITIONS

**Water** Standard mix

**Temperature** 8–20°C

**Tank type** Species or same-size community

**Swimming level** Lower levels

# Tealia felina

*Dahlia Sea-anemone*

These mainly red and white invertebrates have a tubular body, multiple, flowing tentacles, and a disc-like sucker foot. Their breeding method may vary depending on the conditions in which they are living.

*Dahlia Anemone*

SPECIES DETAILS

**Size** 20–150 mm diameter

**Origin** N. Sea, Atlantic coasts

**Sexing** No visible differences

**Ease of keeping** Easy

**Food** Meaty foods placed within the tentacles

**Breeding method** Livebearer, egg-scatterer, or division

**Breeding potential** Variable (see p. 109)

AQUARIUM CONDITIONS

**Water** Standard mix

**Temperature** 8–20°C

**Tank type** Invertebrates

**Swimming level** Lower levels

# Asterias rubens

*Common Starfish*

*Common Starfish*

The five limbs of the Starfish are brown-yellow in colour, and are covered in suckers. They taper and may turn up towards the tip. If one of its limbs is damaged, the Starfish is able to generate a new one. It is a bottom-dweller, feeding on molluscs, sponges, crustaceans and the sediment found on the seabed.

### SPECIES DETAILS

**Size** Up to 200 mm diameter
**Origin** N. Atlantic, Mediterranean
**Sexing** No visible differences
**Ease of keeping** Easy
**Food** Meaty foods, chopped shellfishes
**Breeding method** Egg-scatterer
**Breeding potential** Variable (see p. 109)

### AQUARIUM CONDITIONS

**Water** Standard mix
**Temperature** 8–20°C
**Tank type** Species
**Swimming level** Lower levels

# Serpula vermicularis

*Tubeworm*

This species has a sludge-green, calcareous (chalky) outer tube up to 5 mm thick. When threatened, or at night, it will retract its pinky white tentacles and close the hinged "lid" at the top of the tube. Tubeworms are hardy, requiring little attention or even feeding, although the aquarium shouldn't be too well-filtered as this may remove their microscopic foods.

### AQUARIUM CONDITIONS

**Water** Standard mix
**Temperature** 8–20°C
**Tank type** Community, but not with Starfishes
**Swimming level** Lower levels

### SPECIES DETAILS

**Size** 20 mm long
**Origin** N. Sea, Atlantic
**Sexing** No visible differences
**Ease of keeping** Easy
**Food** Planktonic algae
**Breeding method** Egg-scatterer
**Breeding potential** Variable (see p. 109)

*Tubeworm*

# 4

# THE INDOOR AQUARIUM

There are certain basic pieces of equipment without which no indoor aquarium is complete whatever the fishkeeping system. The most obvious of these is, of course, the tank, but you will also need to investigate the various types of aeration and filtration systems, heating and lighting equipment and the chemical composition of water. All the elements covered in this chapter affect the aquarium conditions, and since providing the right environment is essential to your fishes' well-being, it is worth spending time deciding exactly what suits you and your fishes best.

# TANKS

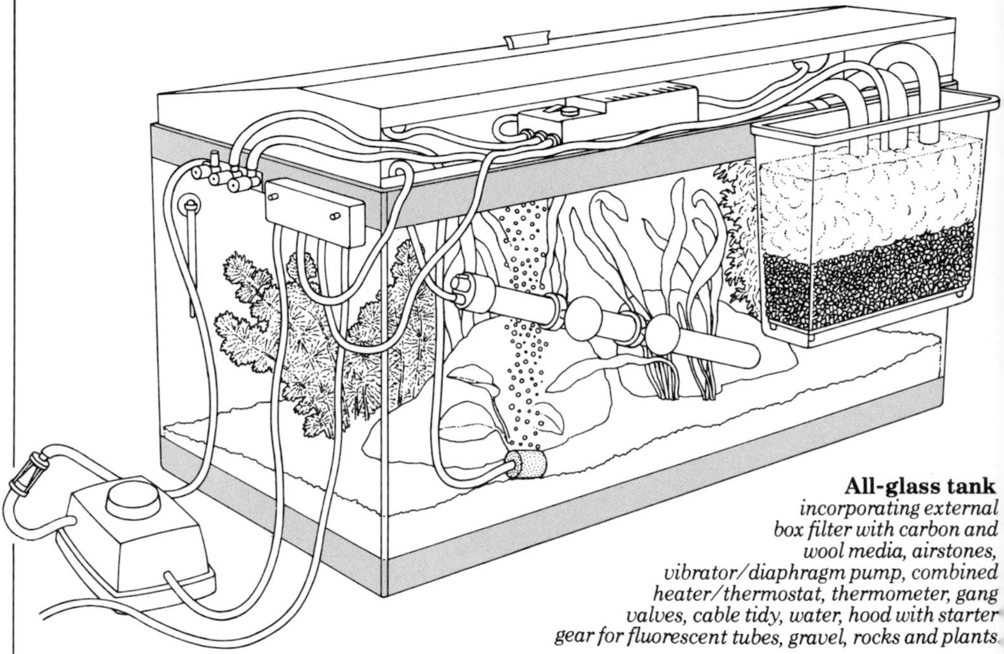

**All-glass tank** *incorporating external box filter with carbon and wool media, airstones, vibrator/diaphragm pump, combined heater/thermostat, thermometer, gang valves, cable tidy, water, hood with starter gear for fluorescent tubes, gravel, rocks and plants.*

One of the most important pieces of aquarium equipment, fish tanks are currently available in a wide variety of sizes and materials. Your ideal tank will depend primarily on the type and number of fishes you intend keeping.

Coldwater fishes require more oxygen than tropical species and since oxygen enters the tank at the water surface, the surface area of the tank must be large. It is therefore important to calculate the number of fishes a tank can support before purchasing it. Overstocking a tank can cause your fishes discomfort and may result in death, so if in doubt buy a bigger tank or keep fewer fishes.

The shape of the tank doesn't only affect the oxygen levels in the water, it also governs the amount of swimming room the fishes can enjoy: the deeper the tank the more space there is. However, whereas a large surface area for oxygen renewal is essential to the fishes' survival, extra swimming space isn't, so it is therefore more important to have a long, wide tank than a deep one, hence the popularity of the traditional, rectangular aquarium.

You must also consider the material from which the tank is made — metal-framed glass, all-glass or plastic. If you intend to keep marine fishes don't choose a tank containing any metal parts since the salt water will corrode them.

Once you have decided on the right size, shape and construction of tank, you have one more choice to make — that of a suitable, safe location for it in your home.

# Choosing the right tank

Choosing a tank isn't just a matter of picking the model you think will look most attractive in your living room. Because the tank contains the fishes' atmosphere, it must be big enough to hold enough water to sustain the intended inmates. In fact, the physical size of the tank you choose is critical to your success as a fishkeeper – if you buy a tank that is too small for your chosen fishes, they will, literally, suffocate.

**Tank type and oxygen**

Fishes depend on oxygen dissolved in the aquarium water in order to survive, and therefore any particular volume of water can only support a given number of fishes. In a crowded room we can obtain relief by going outside, or opening a window. But unlike us, fishes can't move if their environment is unhealthy, and therefore they depend very heavily on their owner to look after their interests by keeping the number of aquarium inhabitants to a safe level.

**Oxygen and tank size**

Although it may seem that the bigger the tank, the more fishes you can keep in it, unfortunately this isn't the case. The main factor in assessing the number of fishes you can keep in a particular tank is the surface area of the water, as this is the only way that fresh supplies of oxygen can reach the fishes (p. 122). As soon as a deficiency of dissolved oxygen occurs, fresh oxygen enters the water almost immediately, but to make the rate of replacement as rapid as possible, the aquarium must have a large water surface area. Another reason for the importance of the size of the water surface to the fishes' well-being is that this is also the place where any unwanted gases can get out of the water.

**What shape of tank should I choose?**

You can have any shape of aquarium you like, bearing in mind that the number of fishes in the aquarium depends on the surface area, *not* the depth, *nor* the overall volume of water. Tall, narrow aquariums may look fashionable and dramatic, but they won't hold any more fishes than a tank half their height.

The "standard" shape of most aquariums is based on the double cube, with the longest side being horizontal. As the length of that side increases, the water-depth dimension is usually increased proportionately too. This gives a well-balanced viewing "window", rather than a narrow slit effect. The front-to-back dimension of a standard tank is usually between 30 and 38 cms. The only exception to this is in the sort used for raising young fry, where a broader, shallower design is preferred for optimum oxygen supplies.

**What capacity tank will my fishes need?**

Two factors affect oxygen supply and demand, and consequently the capacity of the tank required. First, the temperature – since cooler water holds more oxygen than warm water. And second, different fishes need different amounts of oxygen – coldwater freshwater and coldwater marine fishes require more than tropical freshwater fishes, and tropical marine fishes need even more.

**How do different needs affect tank choice?**

There is a simple way of calculating what size of aquarium the various types of fishes need – you allow a certain number of square cms of water surface area to each cm of body length in the fishes. The only problem with this method is that you have to remember three sets of figures: one for freshwater tropicals, one for freshwater coldwater fishes, and one for tropical marine types, as shown in the diagram on the right.

On a straight "most fishes per given aquarium size" basis, freshwater tropicals win hands down, followed by freshwater coldwater fishes and finally tropical marines. In practical terms, the aquariums for the three different types should be of lengths not smaller than 60 cm, 90 cm and 120 cm respectively, each with a water depth of about 40 cm. The reason for this rule is that, in addition to providing an adequate water surface area for optimum oxygen replacement, you must also take into account the condition of the water. The larger the volume of water, the more stable the water conditions will remain, and this is most critical in tropical marine aquariums.

**How should the tank be constructed?**
Not only is tank size important – the construction must suit the purpose too. Water exerts considerable pressure on a tank's glass walls, and therefore as tank size (length and depth) increases, so should the thickness of the glass used. In all-glass tanks the glass must be thick enough to act as its own support because there are no iron frames to provide extra strength. For small aquariums up to 45 cms in length, 4 mm-thick glass will suffice. For a 90 cm tank choose 6.5 mm-thick glass. And a tank over 120 cm long should have 10 mm-thick glass and a crossbar at the top to prevent the front panel bowing out under the pressure of the water.

**What is the tank capacity?**
The amount of water your tank will hold depends on its dimensions. Figures for typical tanks are given on p. 280.

## TANK/FISH RATIO

To calculate the size of tank your fishes will need take the figure below and multiply it by their total body length in cms. As an example, a tank 90 cm long by 30 cm front-to-back, with a water surface area of 2700 $cm^2$, will hold 90 cm of freshwater tropical fishes, 36 cm of freshwater coldwater fishes, or 22.5 cm of tropical marines.

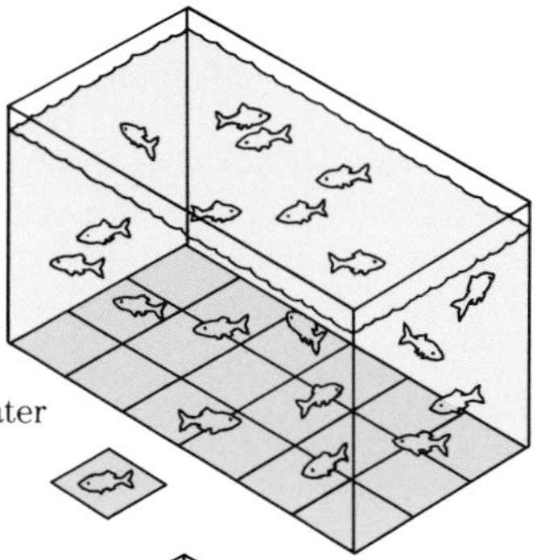

Freshwater tropical 30 $cm^2$

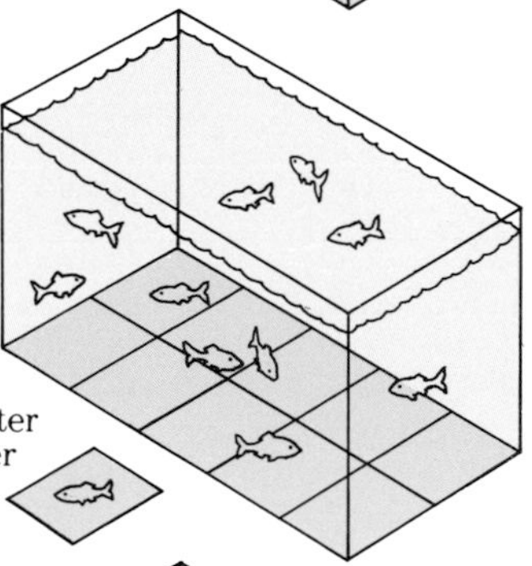

Freshwater coldwater 75 $cm^2$

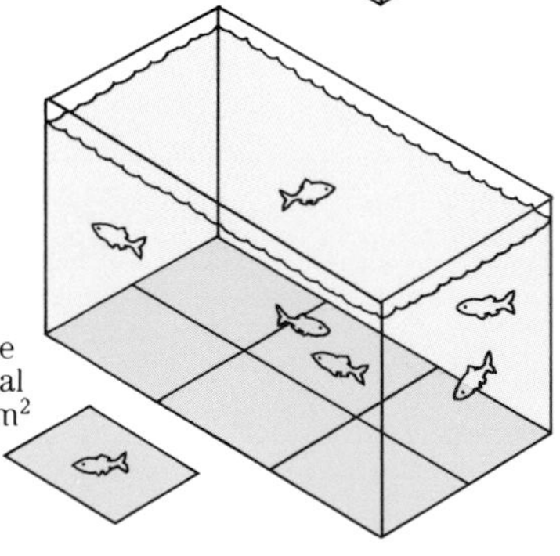

Marine tropical 120 $cm^2$

# Tank construction

Aquariums can be made from almost any material: concrete, wood, fibreglass, metal-framed glass panels, extruded acrylic, and all-glass.

**Metal-framed glass tanks**

Until the 1960s, most tanks had a pressed-steel or angle-iron frame in which glass panels were held in place with putty. With the introduction of viable all-glass tanks, iron-framed tanks declined in popularity.

**All-glass tanks**

The invention of silicone rubber adhesives in the 1960s meant that glass panels could be stuck together edge-to-edge, without any risks of a leak. Without their thick frames, tanks became slim, attractive pieces of furniture, rather than ugly bits of hardware. The arrival of the all-glass tank had another revolutionary effect on fishkeeping. The old iron-framed tanks couldn't be used for keeping marine fishes, as the saltwater corroded the frames too quickly, but silicone-bonded glass proved impervious to saltwater, and so marine fishkeeping became much more feasible.

**Plastic tanks**

One-piece, extruded plastic tanks used to have several disadvantages: they were small, they suffered from discoloration and scratching, and if over-powerful lamps were used in their reflector hoods they would buckle. Modern acrylics appear to have overcome these problems, but to preserve their clarity you must handle and clean them carefully.

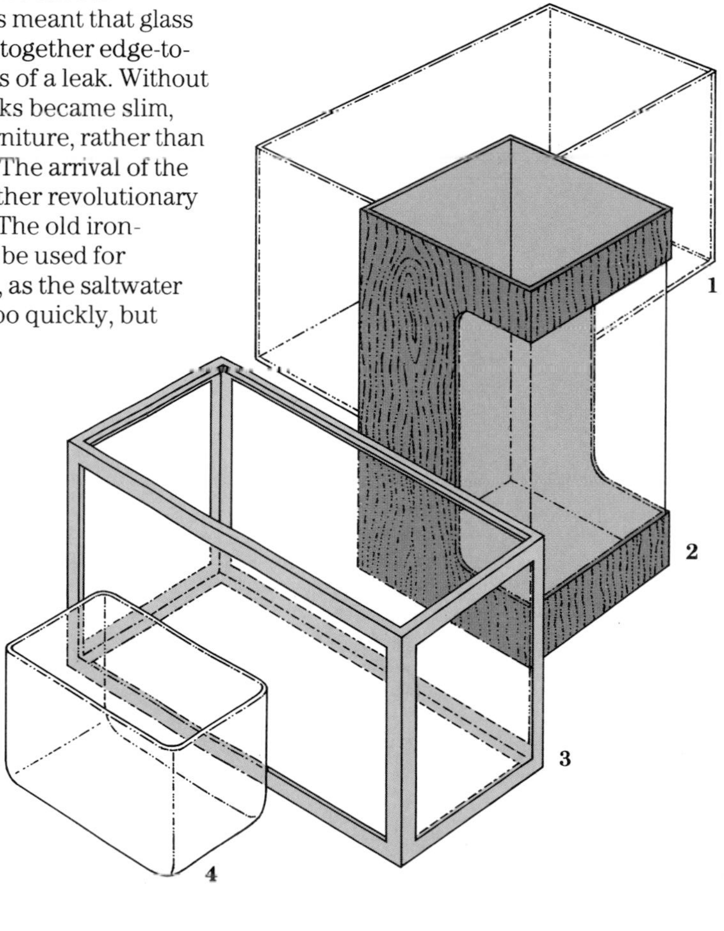


**1 *All-glass tank***
*The most common type, the all-glass tank is long-lasting, and has no rust problems.*

**2 *Custom tank***
*This tank is wooden (sealed with polyurethane paint), with a glass viewing window.*

**3 *Edged tank***
*Some all-glass tanks have the bare edges protected with a plastic trim.*

**4 *Moulded acrylic tank***
*One-piece tanks give a distorted view of the fishes, and are therefore only suitable as breeding aquariums.*

# Buying and siting a tank

Pet shops may sell a few standard tanks, but you will get the widest choice if you visit a specialist aquatic dealer. Other sources include the mail-order suppliers which advertise in the hobby magazines.

**Standard types**
Providing that the space you have selected for your aquarium is of a certain "standard" size, then you should have no problem buying a tank from a dealer. "Off-the-peg" tanks are sold either in linear length sizes (usually in multiples of 30 cm), or capacities (for example, 100 litres). One advantage of buying a standard tank is that this type is usually designed with a matching reflector-hood. In contrast, if you choose a custom-built, "odd-sized" aquarium, you may have a problem finding a hood that fits.

**Custom-built types**
If you have a "non-standard" vacant site, then you may consider having a tank tailor-made to fit, or even making one yourself. Although designing and building your own aquarium is unlikely to cut your costs – ready-made all-glass tanks are very good value for money – you will get a great sense of achievement. Seal wood-and-glass and concrete-and-glass tanks with polyurethane paint before use. This will waterproof the wood, and prevent chemicals in the concrete from leaching into the water.

*Choosing the glass*
The thickness of glass required is affected by the size and design of tank. In all-glass tanks it should be at least 2 mm

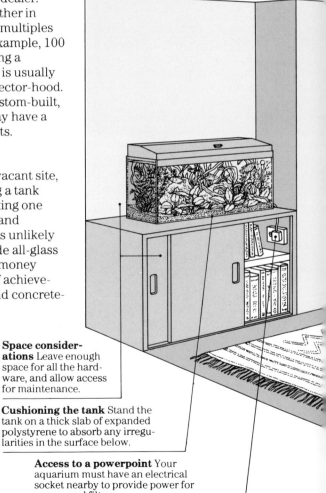

**CHOOSING A LOCATION**
Make sure that you have chosen a suitable position: the site should be draught-free, out of direct sunlight, and in an attractive and convenient place. The aquarium should stand on a very firm, level base, with the weight distributed across the floor joists. Place stands with legs on a thick wood or metal plate to protect the flooring.

**Space considerations** Leave enough space for all the hardware, and allow access for maintenance.

**Cushioning the tank** Stand the tank on a thick slab of expanded polystyrene to absorb any irregularities in the surface below.

**Access to a powerpoint** Your aquarium must have an electrical socket nearby to provide power for pumps and filters.

thicker than the glass used in framed tanks. As the size of the tank increases, the thickness of the glass must increase: 6 mm-thick glass is suitable for 60 cm-long tanks, but this should be increased by 2 mm for every extra 30 cm in length or 10 cm in water depth. You can save a little money if you make the back, sides and bottom panels from glass of an inferior optical quality, but don't penny-pinch on the cost of optically correct glass for the front panel.

**Secondhand tanks**

It may be worth your while looking for secondhand tanks, especially if you are expanding into a fish-house and appearances aren't of prime importance. Check that such tanks are of sound construction. A defective, leaking tank can be sealed with silicone rubber sealant. As long as all the tanks in the fish-house are fitted with cover-glasses, you can do away with reflector-hoods and use long fluorescent tubes to light several tanks.

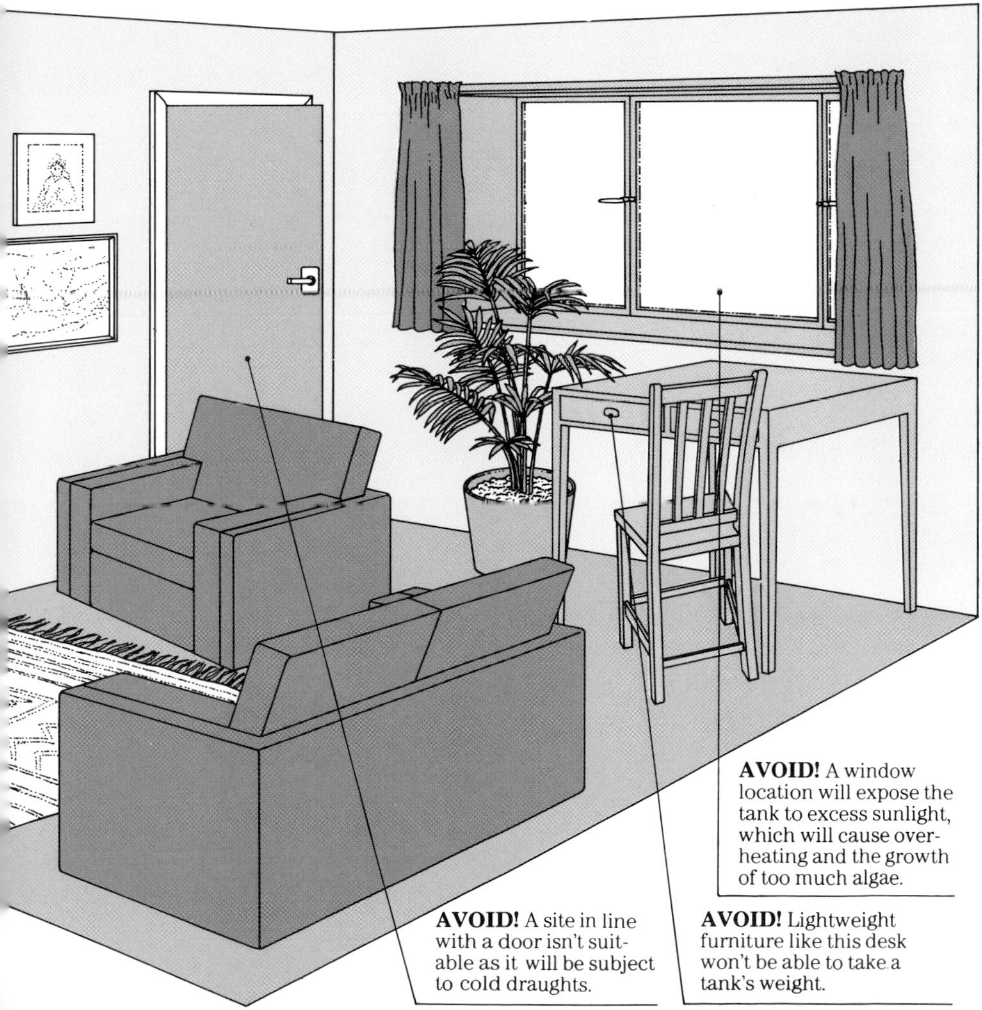

**AVOID!** A window location will expose the tank to excess sunlight, which will cause overheating and the growth of too much algae.

**AVOID!** A site in line with a door isn't suitable as it will be subject to cold draughts.

**AVOID!** Lightweight furniture like this desk won't be able to take a tank's weight.

**Choosing a site that avoids light and temperature problems**

You must avoid siting your aquarium on a window sill, or any place where it will receive a lot of direct sunlight. An ideal site will receive a little morning sun (this can be beneficial when breeding), but not too much hot summer sun. However, it doesn't matter if the location is a dark recess; in fact, this can be to your advantage, as you can adjust the lighting to match the aquarium's requirements without outside influences affecting it.

Avoid hallways and passages that are liable to temperature fluctuations because of the regular opening and closing of doors. Also a violently slammed door might frighten the fishes or, worse still, crack the glass.

**Providing access to services**

You will always have to make some compromise on the site of your tank – for example, aquariums are nearly always some distance from the water supply. However, there is one factor that you must never ignore. You *must* arrange for it to be near an electrical outlet socket in order to run the lights, filtration and heating equipment.

You shouldn't fit your aquarium into a tight space: there must be room above it to lift the hood for feeding, and to get your arm in to re-arrange plants or wield a net, and you should leave enough space at the sides and underneath for the airpump, filters and tubs of fish food. If the sight of the "hardware" offends you, cover the space with moveable, decorative panelling.

**A WINDOW SITE**

Avoid siting a tank on a windowsill as this will have several detrimental effects:

- An excessive amount of green algae will grow, which will soon cover the whole aquarium.
- The tank may overheat in the summer, leading to oxygen deficiency in the water. Coldwater fishes will suffer particularly badly.
- The aquarium may lose heat too quickly in winter or if the window is open – this can cause the water temperature to drop, thus harming tropical species.

## MOVING A TANK

Never try to move a full tank, not even to reposition it slightly. Set-up tanks are very heavy, and moving them can be dangerous: the glass may crack under stress and cause a flood.

**Moving procedure**

If you are moving some distance, especially in winter, you should transport tropical species in a heat-conserving box (see p. 29). Otherwise, follow this procedure:

**1** First, disconnect the electricity supply.
**2** Next, drain off about one-third of the water into a snap-top bucket.
**3** Remove the plants and put them into the bucket.
**4** Catch the fishes and put them in the bucket with the plants.
**5** Put the lid on the bucket.
**6** Set aside one-third of the water in a large jerrycan.
**7** Now drain off the remaining water.
**8** With smaller tanks, make sure that the rocks can't topple over during transit. Tanks over 60 cm in length must be emptied completely.
**9** When reassembling the set-up straightaway, try to use as much of the original water as possible to avoid stressing the fishes.

# AERATION AND FILTRATION

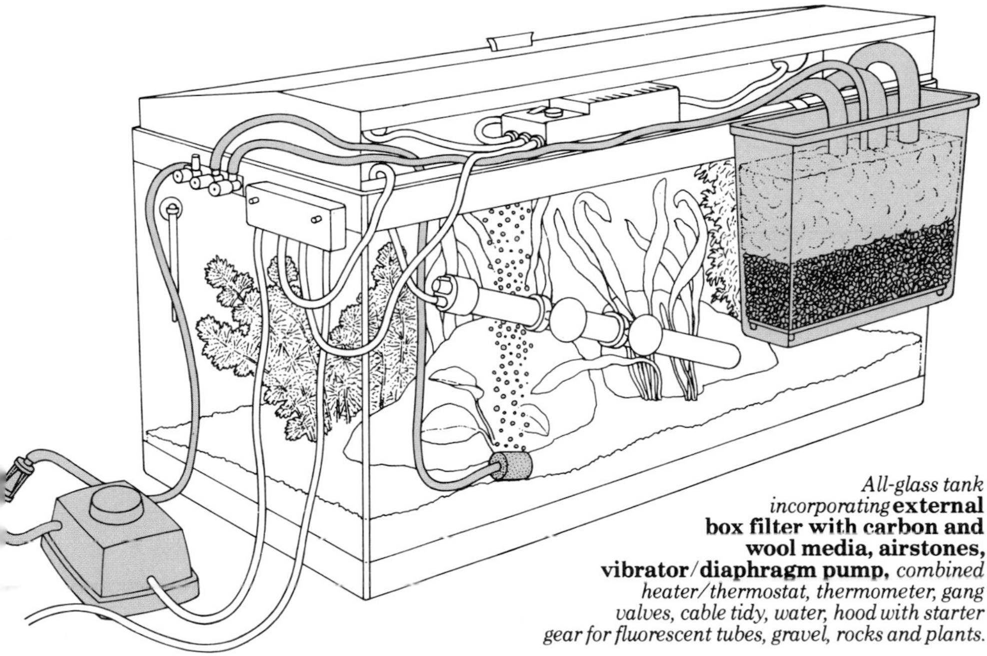

*All-glass tank incorporating* **external box filter with carbon and wool media, airstones, vibrator/diaphragm pump,** *combined heater/thermostat, thermometer, gang valves, cable tidy, water, hood with starter gear for fluorescent tubes, gravel, rocks and plants.*

It is absolutely essential that the water in the aquarium is kept clean and fresh at all times if your fishes are to be healthy. Fortunately, this doesn't mean that it must be completely changed every few days. Instead, it can be cleaned by means of artificial aids – filters and airpumps.

No matter how attractive your aquarium looks, the water will soon become contaminated by dissolved waste products from the fishes. This dirt, visible or not, must be removed, and this is the purpose of filtration.

Not only does the water need to be clean; it must also contain a healthy amount of oxygen. This is particularly important in coldwater aquariums during the summer months, when the dissolved oxygen level falls as the water warms up. An artificial system is therefore needed to introduce more oxygen into the water and, at the same time, to disperse some of the carbon dioxide. (This process is known as aeration.)

Aeration isn't obligatory, and some aquariums function well without it. However, such tanks are generally understocked, have plenty of healthy plants, and are in tip-top condition. I would therefore advise all beginners to invest in an airpump. Even if you don't want it for aeration purposes (for example, if you are using an air-operated filter which produces aeration anyway), you can use the pump to operate other useful equipment, such as a brine-shrimp hatcher or a protein skimmer.

# How do aeration and filtration work?

Although you won't need any technical skill to operate your aeration and filtration systems, it is helpful to understand how the processes work.

## Aeration

This is achieved by means of an airpump. By creating a constant stream of airbubbles in the aquarium, water is circulated around the tank. In this way the bottom layers of the tank water are brought into contact with the atmosphere at the water surface, where more oxygen is absorbed into the water and carbon dioxide driven off.

As well as increasing the oxygen content of the water, aeration has an additional advantage – it equalizes the temperature throughout the aquarium.

Because aeration facilitates the absorption of oxygen, it effectively increases the surface area of the water. However, don't assume you can therefore keep more fishes than you calculated originally. Aeration is only an artificial aid and, should the system fail, the tank conditions will revert to the original unaerated state, and the fishes may become overcrowded and gasp for air.

## Filtration

The purpose of the filter is to remove dirty materials from the aquarium. For this reason regular maintenance is very important as a neglected filter becomes a box of concentrated dirt through which all the aquarium water is continuously passed. Filters work in one of three ways:

- Mechanical filtration removes suspended materials from the water.
- Chemical filtration removes dissolved materials from the water.
- Biological filtration (see p.128) uses beneficial bacteria to convert toxic substances into relatively harmless ones.

## Mechanical and chemical filter methods

The material through which the water is passed is known as the "filter medium", and is situated inside a box or cylinder. This container may be sited inside or outside the aquarium. In the majority of systems, more than one medium is used so that mechanical and chemical filtration occur simultaneously.

*Mechanical filtration*

To strain suspended dirt from the water, a reasonably tightly packed material is required. This is usually a man-made substance like nylon floss or synthetic foam. Only buy foam designed for aquarium use as others may be toxic.

*Chemical filtration*

The best material to remove dissolved solids is activated carbon. It has a large surface area, which readily soaks up dissolved minerals and chemicals, such as fish urine, on its surface (adsorption). Unfortunately, carbon will also adsorb useful substances, so it mustn't be put in the tank when medications are in use.

After a period of time the carbon will no longer adsorb any more dissolved material. When this happens it must be replaced. To find out if it is still sufficiently active, add a few drops of dye (such as methylene blue) near to the inlet of the siphon tube. If the dye reappears at the output of the filter, the carbon needs replacing.

## Water composition and filters

Filters can be used to change the chemical composition of the aquarium water (see p. 146). For example, peat can be used to maintain acidic water conditions, and ion-exchange resins can reduce the hardness of the water.

# The airpump

Aeration is provided by compressed air, using electrically driven airpumps. These conform to two basic designs: the vibrator/diaphragm type or a piston/rotary-vane type.

When choosing an airpump, its size needn't be related to the dimensions of your tank – even the smallest model will provide enough air for a 60-cm-long aquarium. Instead, you should consider how much air you need to produce overall. The larger the pump, the more air it produces. So if you intend to use it to operate filters, airstones, and other equipment like brine-shrimp hatchers, then a relatively large pump is the best choice. Larger pumps have built-in adjustable air controls to regulate the output; with a small pump you have to use airline clamps.

## AIRSTONES

Simple aeration can be achieved by passing the air from the airpump through a submerged block of porous material (special sintered glass or hardwood for example) to break up the airflow into minute bubbles. Airstones can become blocked after lengthy periods of use, although these blockages can sometimes be cleared by boiling.

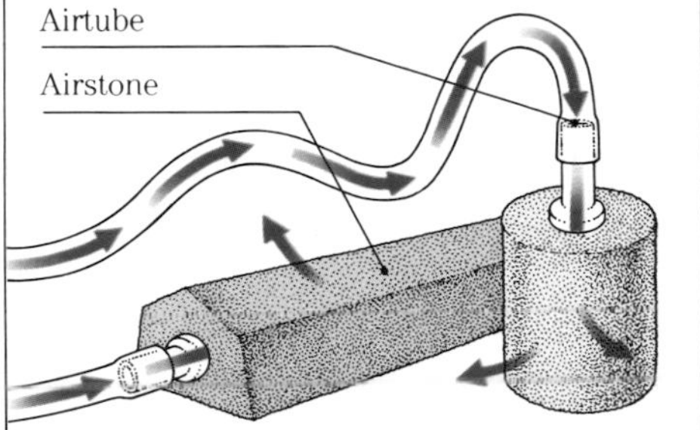


**KEY**
**Air**

***Piston/rotary vane pump***
*These models are among the more expensive pumps, particularly the larger, higher-capacity ones. They are most suitable for the marine aquarium (where stronger aeration and filtration are needed), and for freshwater fish-keepers with fish-houses.*

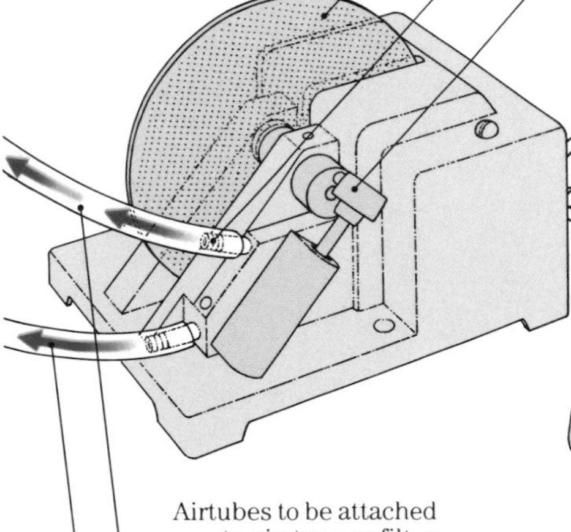


***Vibrator/diaphragm pump***
*This pump type can be noisy and requires periodic cleaning. Top-of-the-range models have a built-in control which allows the amount of air to be varied.*

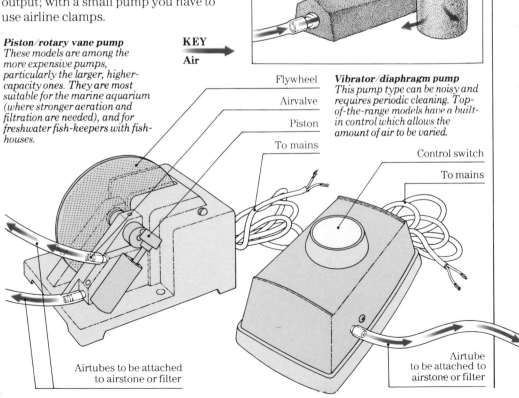

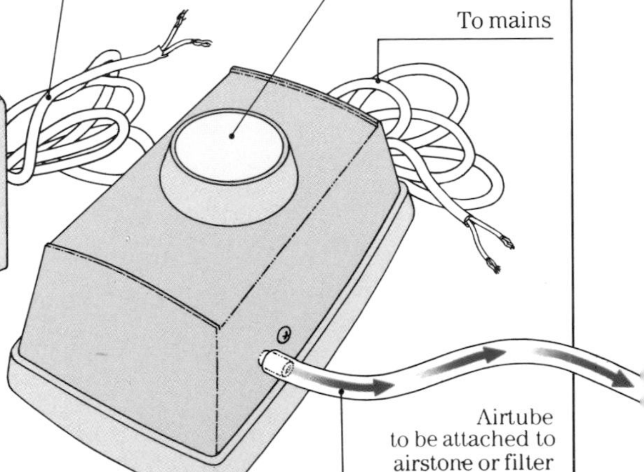

# Air-operated filters

The most usual way of returning water to the tank after it has passed through the filter medium is by means of an "airlift" built into the filter; this takes the form of a pipe which is fitted inside the filter and rises vertically out of it to the water surface. Air is introduced into the bottom of this semi-submerged pipe, mixing air-bubbles with the clean water in the filter, and because this mixture is lighter than water, it rises in the pipe and the clean water comes out of the top. This process is ideal for small aquariums.

### External filters

These filters can be used on all fully furnished, planted aquariums. Because they are accessible, they can be cleaned easily, without disturbing the plants and fishes. They hang on the outside of the tank. You may need to cut the hood to accommodate the inlet and outlet pipes. In order to maintain a flow of water across the aquarium, you should return the water to the opposite side of the tank to the filter by fitting an extension tube on the filter outlet pipe.

### Priming the filter

An external filter won't start working until it is filled with water – primed. To do this submerge the inlet siphon tube in the tank. When it is full of water, cap the end of the tube with your finger or a starting stick to prevent air from entering, and place it in the filter box. As

## AIR-OPERATED FILTER DESIGN

### How an external filter works

When the filter is situated outside the tank, dirty water is fed into it by means of a simple siphon tube. This should be an inverted "U" shape, arranged over the edge of the aquarium with one end under the water and the other end in the filter box. The water is passed up the tube, down through the filter medium (usually layers of carbon and filter wool, see p. 122), and then returned up the second part of the tube, cleaned, to the tank. These filter boxes are normally hung at the back or side of the tank.

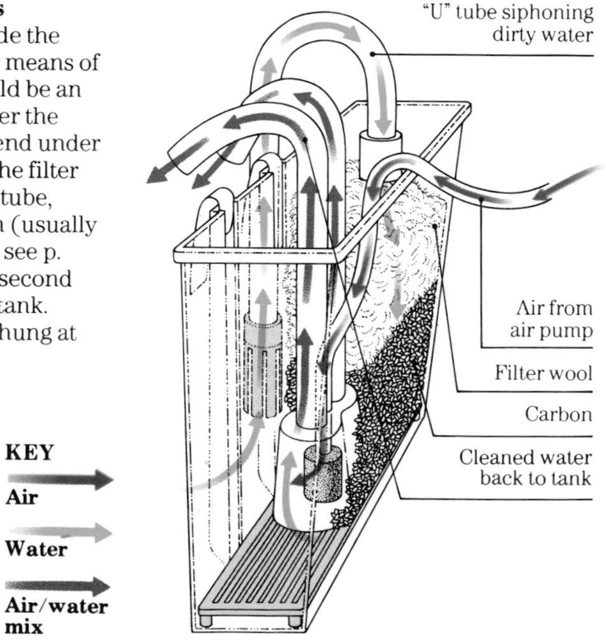


***Box filter***
*The external, air-operated filter has several advantages: it is inexpensive, efficient, and its design means that it is very easy to clean.*

soon as you uncap the tube, the water will begin to siphon into the filter.

**Cut-off mechanism**

You don't need to worry about external filters overflowing, even if the filter medium becomes so tightly clogged that no water can pass through it. The reason for this is that as soon as the water in the filter box reaches the same level as the water in the aquarium, the siphon action stops. Then, as the airlift draws water through the box, the water level in the filter box falls, and siphoning restarts.

**Internal filters**

These filters have a major drawback – cleaning and maintainance is difficult, and as the filter is removed, dirty water can spill back into the tank. To prevent this, try placing a plastic bag under the entire filter. However, they are excellent for fry tanks, as the tiny fishes can't be drawn into the filter medium.

Internal filters are no problem to install – they just sit on the bottom of the aquarium in any convenient spot. Make sure that space is left around them, so that the water can reach the filter. To prevent the filter from floating upwards, place a few pebbles in the box before adding the filter medium. Alternatively, fit suction caps to anchor it to the side of the aquarium.

**Other uses for filters**

An external filter box can be pressed into use for other purposes if desired. Hung on the inside of the aquarium so that the box is under the water, it can segregate a single bullying fish from other fishes, or be a handy Brine Shrimp hatchery.

**How a simple internal filter works**

A simple form of filter consists of a piece of foam sponge mounted on an air lift. Dirty water enters through the holes in the sponge, and the suspended particles become trapped. The air then forces the clean water out through the tube adjacent to the airlift.

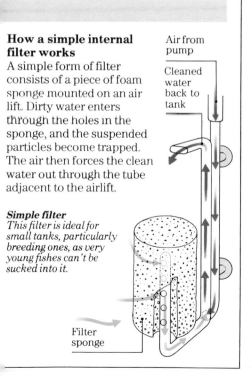

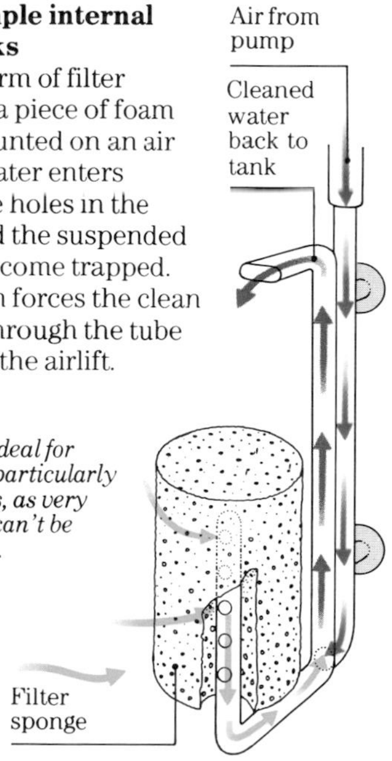


***Simple filter***
*This filter is ideal for small tanks, particularly breeding ones, as very young fishes can't be sucked into it.*

## PROTEIN SKIMMING

This air-operated system is a variation on chemical filtration, extracting dissolved proteins and other waste products that discolour and pollute the water. It should only be used in a marine aquarium as an occasional supplement to an undergravel filter (see pp. 128–9). The protein skimmer consists of an internal airlift with a collecting chamber at the top. As the rising column of bubbles reaches the surface, the foaming residue (containing proteins and other organic substances) is deposited in the collecting chamber, where it settles out into a yellow liquid. When a sufficiently large volume of this liquid has been collected, it can be poured away.

# Power filters

In place of the simple airlift used in air-operated filters (see p. 124), most power filters move the water by means of an electric motor which drives a small centrifugal pump. This is the basic principle of cannister-type filters. There is also another type, the open-box model, which uses a motorized, magnetically coupled impeller to move the water. Of the two types, the cannister filter is the more powerful, so which you choose will depend on the size of your tank.

The main advantage these filters have over air-operated types is that the volume of water flowing through them is much greater, and so the water in the tank is filtered faster. Many fishkeepers believe that their fishes grow stronger as a result of living in the stronger water currents set up by this form of filtration. Power filters are therefore particularly suitable for large aquariums, and for clearing a dirty tank quickly. They are especially recommended for use in a coldwater aquarium where the fishes require well-oxygenated water.

Innovative designs include a type that aerates the water as it passes through the filter, and a model that uses the water to cool the motor (or conversely, uses the motor's heat to keep the water warm). There is also a type that uses diatomaceous earth (powdered fossils) as a filter medium. It should only be used periodically to supplement your usual filter.

## HOW BOX FILTERS WORK

Motorized box filters are hung on the outside of the tank. A motorized magnet is fitted beneath the filter box, and a second magnet is attached to an impeller inside the box. As the exterior magnet revolves, it affects the internal magnet and so makes the impeller turn and the water circulate. To ensure a flow of water across the tank, either the inlet siphon or return tube can be extended to draw dirty water from, or return clean water to, the opposite side of the tank.

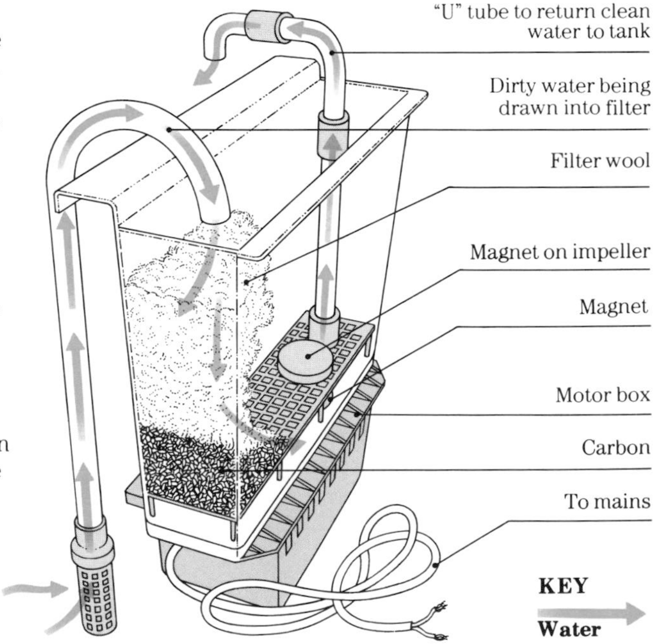


***Motorized box filter***
*The advantage of this system is that, as it can filter more water than air-operated types, it can be used in larger tanks.*

## HOW CANNISTER FILTERS WORK

In this type of pump the water is circulated by a small centrifugal pump driven by an electric motor. Because the waterflow is so great, many cannister models (both internal and external types) are provided with spray-bar attachments to help dissipate the returning water over the whole length of the aquarium.

**External models**
These filters may be attached to the tank side by means of a special hanger, or fitted remotely, via long water hoses. You must make sure that all hose connections are secure, as a leak from a disconnected hose can empty a tank very quickly.

**Internal models**
Submersible cannister models are available for "in-tank" use; they have the motor in the top section, above the filter medium. Many have a built in facility for introducing air into the returning water stream to improve aeration in the tank.

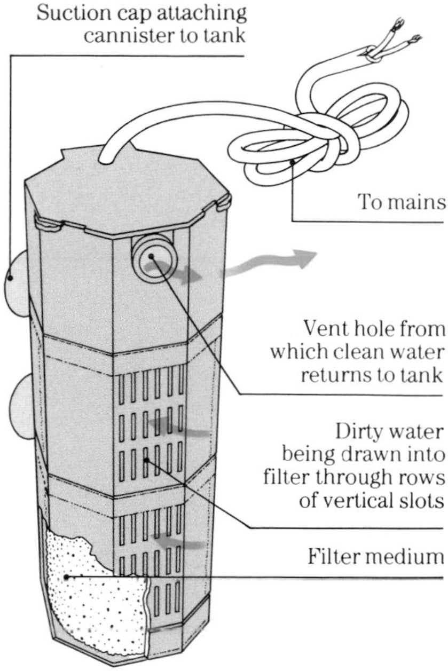


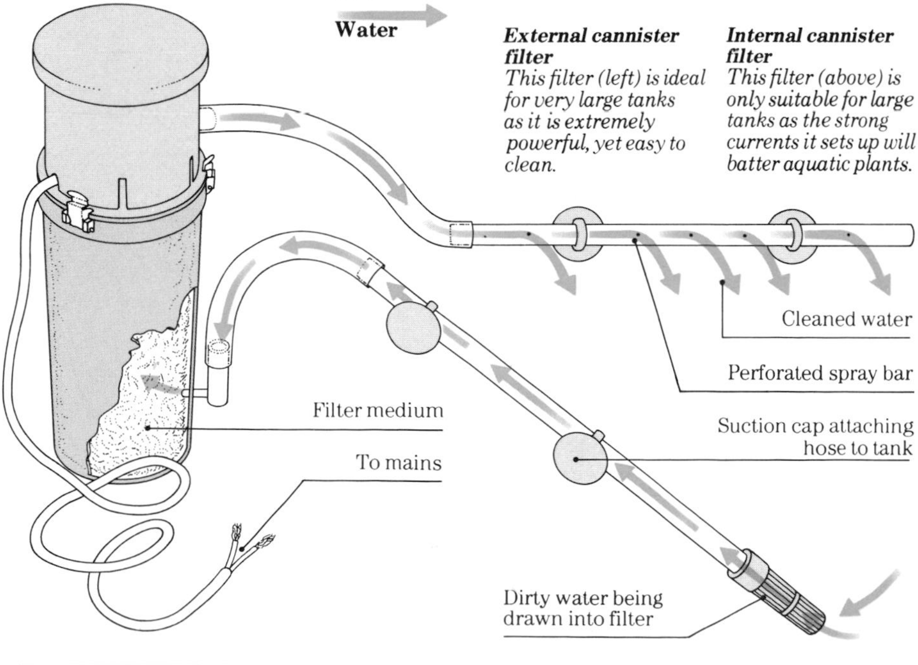


***External cannister filter***
*This filter (left) is ideal for very large tanks as it is extremely powerful, yet easy to clean.*

***Internal cannister filter***
*This filter (above) is only suitable for large tanks as the strong currents it sets up will batter aquatic plants.*

# The process of biological filtration

The main difference between biological filtration and mechanical and chemical filtration methods (see p. 122) is that it neutralizes toxic substances in the water, whereas they actually remove them.

Biological filtration is particularly effective in dealing with ammonia, which is produced by the bacteria that feed on all decaying substances (such as fish wastes and uneaten foods), and is also excreted by the fish during respiration. In this system, a colony of beneficial bacteria is introduced into the aquarium to break down the ammonia into a relatively harmless nitrate compound. The bacteria establish themselves on the

## HOW AN UNDERGRAVEL FILTER WORKS

In this system, there isn't a filter medium. Instead, a slotted plate is installed underneath the gravel, where it creates a large aerobic bed for the bacteria to develop on by passing well-oxygenated water through the gravel. An airlift (see p. 124) either forces water out of the gravel (which is replaced with more water flowing down into it), or pushes the water under the gravel so that it rises up through it. The latter process is known as "reverse-flow" filtration.

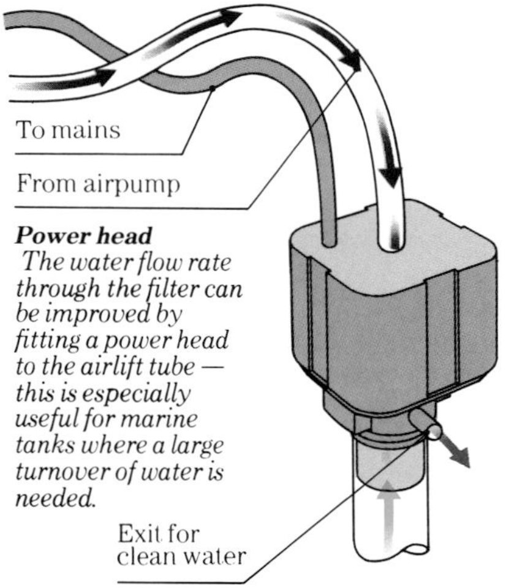


***Power head***
*The water flow rate through the filter can be improved by fitting a power head to the airlift tube — this is especially useful for marine tanks where a large turnover of water is needed.*

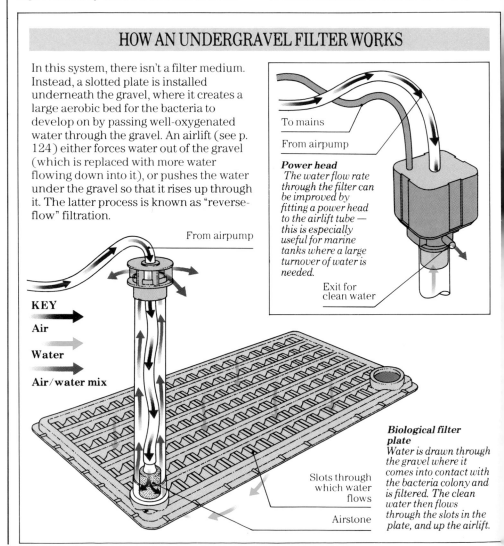

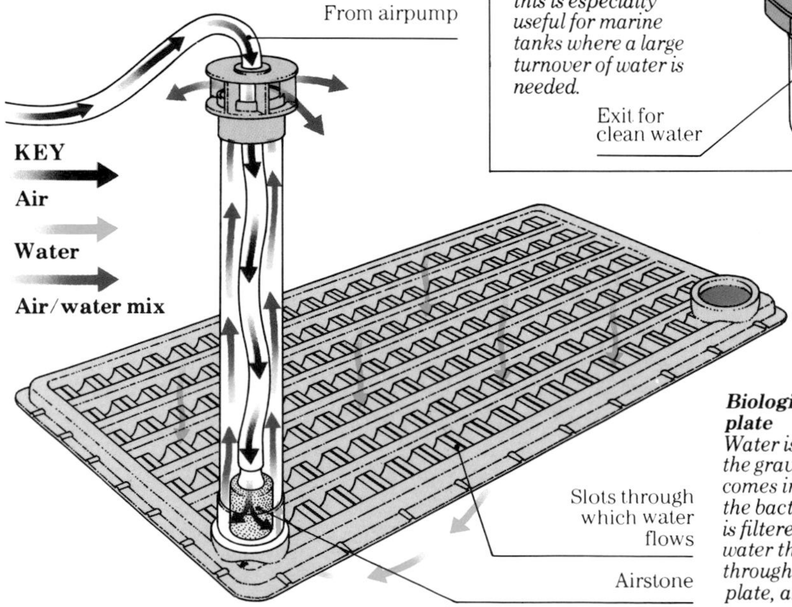


***Biological filter plate***
*Water is drawn through the gravel where it comes into contact with the bacteria colony and is filtered. The clean water then flows through the slots in the plate, and up the airlift.*

surface of the aquarium gravel, but they can be encouraged to multiply and colonize the rest of the gravel if the area is kept oxygen-rich (aerobic). A slotted plate is therefore fitted at the tank bottom so that well-oxygenated water can be circulated through the gravel. As the water containing the ammonia meets the bacteria, it is converted in a two-stage process. First, *Nitrosomonas* bacteria convert the ammonia into a waste product called nitrite which, in turn, is changed into nitrate by *Nitrobacter* bacteria. Although nitrate may inhibit the fishes' growth, it is far less toxic than ammonia. Both nitrite and nitrate can be utilized by plants as food.

This filter type has several advantages over other systems. It is unobtrusive, requires no long hoses, and is quiet in operation. It can be used in all aquariums and is obligatory for marine collections.

**Installing an undergravel filter**

The filter plate should fit the base of the tank as closely as possible, so that the greatest possible filtration area can be covered. If you use a commercial filter, you may need to cut the plate or join several together to get an exact fit. Alternatively, you can make your own plate by cutting slots in the ridges of a piece of corrugated roofing plastic (see p. 193).

**1** Place the filter plate flat on the floor of the bare tank and, if you like, fix it down with sealant.

**2** Fit the airlift in a corner (two can be used in large tanks), and seal to the top of the plate with sealer. To move the water more quickly, put an airstone (see p. 123) in the airlift tube. In freshwater aquariums an undergravel filter will operate quite satisfactorily from a modest air-pump (see p. 123). However, marine tanks need a larger turnover of water, so use a power filter (see p. 126) or a motorized impeller fitted in the waterflow circuit.

## PLANT FILTRATION

Another form of filtration, which is both biological and chemical in action, uses plants or algae to remove ammonia, carbon dioxide and nitrates from the water. In marine aquariums, algae tray filters can be used, but they require plenty of space above the tank and may be inconvenient to fit.

In freshwater systems, a separate, planted tank is set up alongside the tank containing the fishes. Water is siphoned from the aquarium, through the planted tank and pumped back again. The planted tank should be brightly lit so that photosynthesis can occur.

Filtration takes place as the plants take in carbon dioxide and inorganic molecules from the water, and release oxygen and organic compounds in their place. This system is particularly useful in tanks containing herbivorous fishes which eat any plants in their environment. Also, it enables a supply of live aquatic foods to be safely reared away from hungry fishes. If disease strikes the main aquarium you should isolate the planted tank.

**3** Cover the filter with 2–3 cms of gravel, and top with nylon netting followed by another 4–5 cm gravel layer. The gravel needs to be deep to accommodate the plants' roots, and the netting will prevent the fishes from uncovering the plate.

**4** Although bacteria will develop spontaneously in a new aquarium, it will take time for the colony to multiply sufficiently to act as a filter. To speed the process up, add gravel from an established aquarium to seed the gravel bed with bacteria. To tell when the colony is established, test the water with a nitrite test kit. When the reading is at its lowest, the filter has matured. *Always* carry out these tests in marine aquariums before adding the fishes (see p. 146).

# HEATING

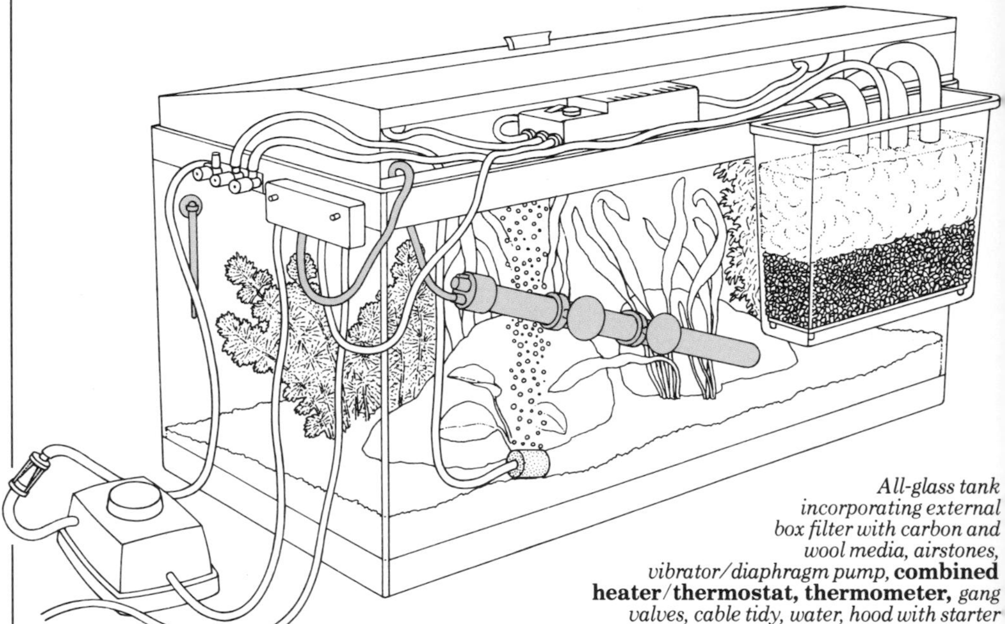

*All-glass tank incorporating external box filter with carbon and wool media, airstones, vibrator/diaphragm pump,* **combined heater/thermostat, thermometer,** *gang valves, cable tidy, water, hood with starter gear for fluorescent tubes, gravel, rocks and plants.*

Unlike other aquarium "hardware" such as tanks, aeration devices or filters, heaters are only necessary if you want to keep tropical species. Heating an aquarium isn't particularly costly as the water temperature in the tropics isn't especially hot – around 24°C, the temperature of a lukewarm bath. And the cost of maintaining the water at the correct temperature isn't high because once it has reached the correct heat, very little power is required to maintain it at that level.

The majority of single aquariums are heated by small, individual appliances run from the main electricity supply. However, if you have a multiple-tank collection kept in a separate "fish-house", you can use an alternative energy source such as paraffin to heat the entire room (see p. 135).

Today, aquarium temperature is thermostatically controlled; this is automatic and completely reliable. It doesn't mean that you can't control the temperature setting—thermostats can be adjusted by a few degrees either side of the "standard" 24° C mark if you want to alter the water temperature (for example during the treatment of diseases or to stimulate the fishes into breeding).

# Aquarium heaters

An aquarium heater consists of an electric element (just like those found in an electric fire) wound on a ceramic former enclosed in a heat-resistant, watertight, glass tube. A cable connects the tube via a watertight cap to the electricity supply. Heaters are fully submersible, and are made in "standard" sizes – 50, 75, 100, 125, 150, 200 and 300 watts.

An alternative design of "in-tank" heating takes the form of a heater cable buried beneath the aquarium gravel. This type of heater has the advantage that the fishes can't burn themselves on it.

An "outside-tank" heater, consisting of an electrically heated mat on which the aquarium stands, has been developed recently. However, it is still too early to gauge its success.

Low-voltage heating is another option. It is much safer than the standard sort, but it has several disadvantages: a transformer is necessary, any wiring is likely to be thicker than usual, and the heater is larger than the standard type.

**Selecting the right size of heater**

The heater you choose for your aquarium must be powerful enough to heat it properly. A rough guide is to allow 10 watts of power per five litres of water for a tank sited in a normally heated room. As an example, a tank that is 90 cms long, 30 cms wide and 38 cms deep has a capacity of 100 litres, so a heater rated at 200 watts would be suitable. In tanks that are 90 cms or more in length, heat distribution is better and more even if you spread the heater requirement between two separate units (in the case of a 90-cm-long tank, each heater should be rated at 100 watts), placed one at each end of the tank.

A large heater in a small tank will work quite normally, but there is a danger that if the thermostat malfunctions and

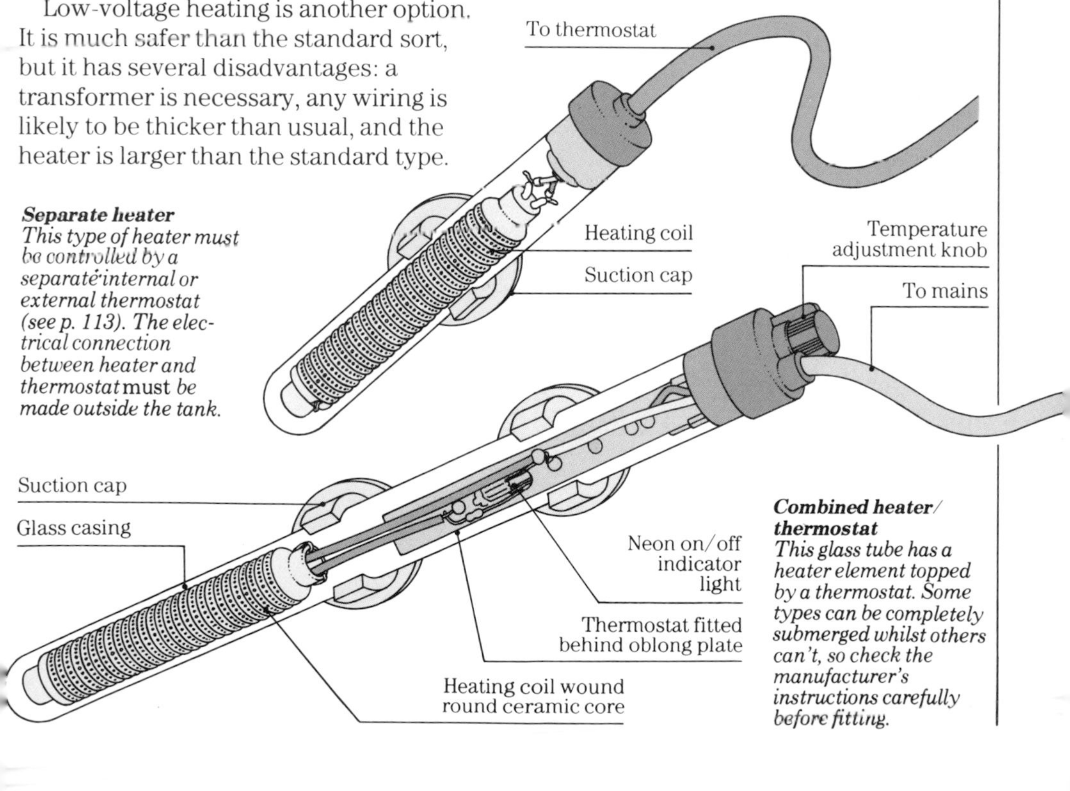


***Separate heater***
*This type of heater must be controlled by a separate internal or external thermostat (see p. 113). The electrical connection between heater and thermostat* must *be made outside the tank.*

***Combined heater/ thermostat***
*This glass tube has a heater element topped by a thermostat. Some types can be completely submerged whilst others can't, so check the manufacturer's instructions carefully before fitting.*

sticks, particularly in the "On" position, then the relatively small volume of water will quickly overheat, perhaps even boil, killing the fishes and cracking the aquarium glass. Also, because the temperature in a small tank fluctuates rapidly, a heater that is too large for the tank may harm the thermostat contacts by switching on and off too frequently, leading to thermostat malfunction.

Conversely, a small heater in a large tank may only just be able to cope with the heating load, and will be quite useless if the temperature in the room in which the aquarium is sited falls too low.

**SAFETY ADVICE**

- Don't switch the power on unless the heater is completely immersed. Heaters warm up very quickly out of water, so if you are holding one it will burn you. If, as a result, you drop it, it may explode!
- Outside thermostat knobs can be very tempting to inquisitive children. If your model doesn't have a cover over the control knob, set the temperature, then take the knob off completely.

## INSTALLING THE HEATING APPARATUS

There are two installation methods: one for combined heater/thermostats, the other for separate heaters that are controlled by an external thermostat.

**Combined units**
Generally, these units are placed upright in a corner of the tank, but if the water depth isn't sufficient, they can be placed at an angle across the back or side. The thermostatic section must always be above the heater element.

**Separate units**
Make sure that the thermostat's sensor-plate is flat against the glass. The *whole* of the plate must be against the glass for it to function properly, so if the tank's back or side glass is patterned or ribbed, you can't expect accurate control.

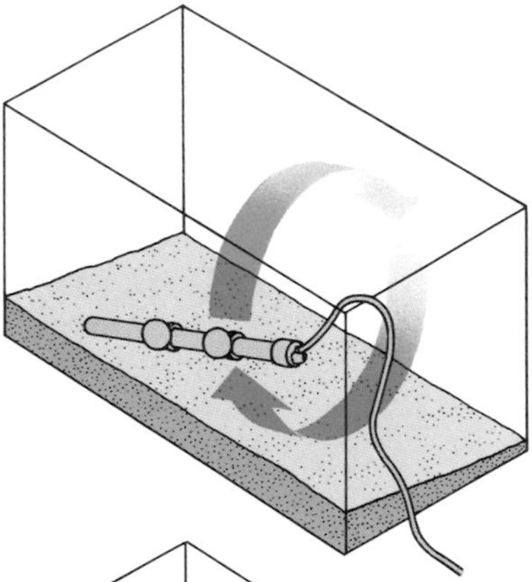

***Installing a combined unit***
*This unit is placed at an angle across the back glass of the tank and fixed in place with special plastic clips that are held onto the aquarium wall with suction pads. Make sure that the heater part is clear of the gravel to allow for water circulation. If it isn't, it will overheat, resulting in a cracked glass tube and live, electrified water.*

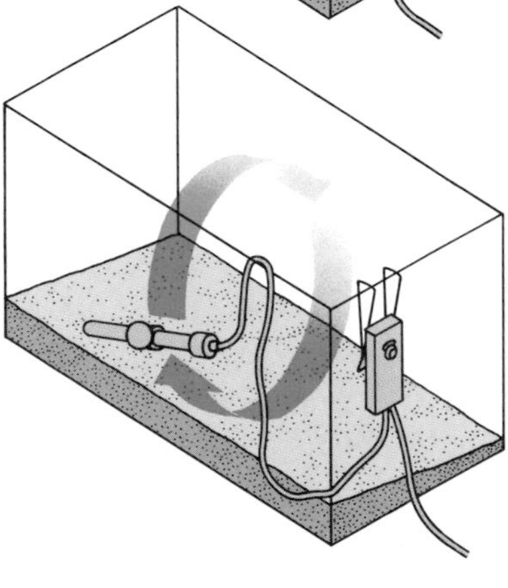

***Installing separate units***
*The heater should be mounted as low as possible in the tank, but still clear of the gravel. A separate heater must be controlled by a separate thermostat (see p. 133). The thermostat has enough wire on it to enable it to be mounted away from the heater. Cover the thermostat wire clip with air-line tubing, then mount it on the side of the tank.*

# Aquarium thermostats

A simple electro-mechanical device, the thermostat switches off the heat when the required temperature is reached, and switches it back on again when the temperature falls to a pre-specified level. Thermostats are usually factory-set to operate around the 24°C mark, but have provision for you to make adjustments.

## Internal thermostats

An internal thermostat consists of a bi-metallic strip fixed inside a watertight glass tube. When the water temperature alters, the strip bends and makes a contact with the heating element, switching the power on or off. There is usually a magnet incorporated in the design to ensure that the contacts connect quickly.

Heaters and thermostats used to be separate entities, but it is now common practice to house both the heating element and the thermostat mechanism in the same glass tube. This combined internal heater/thermostat is much simpler to connect to the electricity supply than a separate unit. Designs of combined units may vary: some are completely submersible, others aren't.

## Thermostat care

In general, modern thermostats don't require any maintenance. This is particularly true for internal, submersible

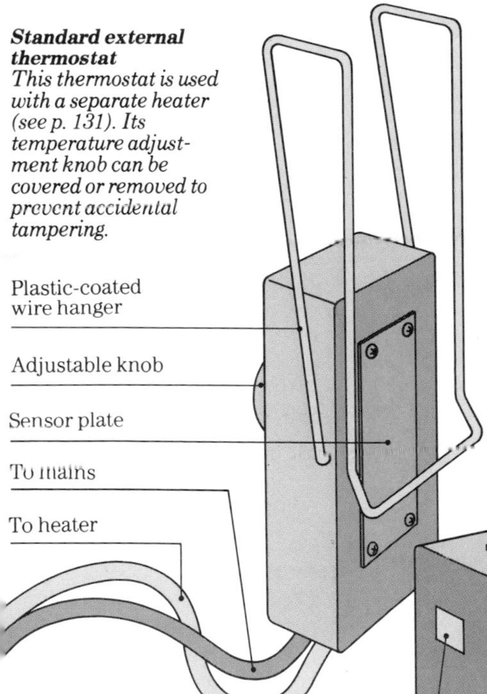


***Standard external thermostat***
*This thermostat is used with a separate heater (see p. 131). Its temperature adjustment knob can be covered or removed to prevent accidental tampering.*

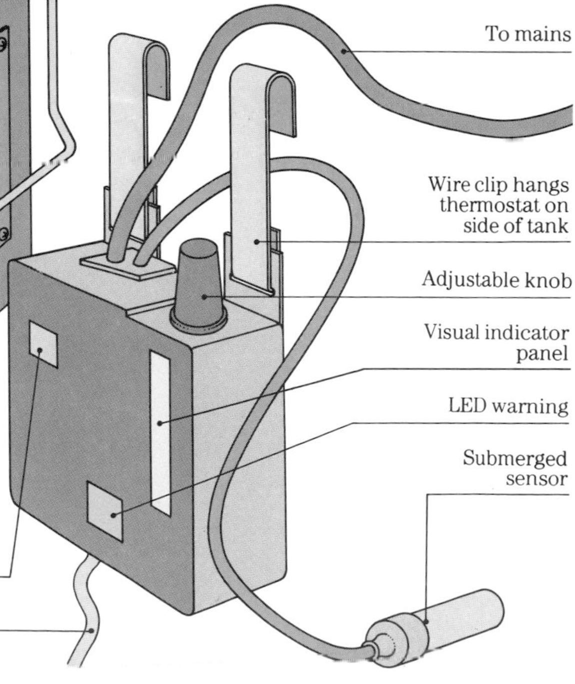


***Microchip external thermostat***
*Also used with a separate heater (see p. 131), this electronic thermostat senses temperature changes via a submerged probe.*

types as they are sealed units, and they either work properly or not at all. However, you should check their water-tight caps periodically for damage because they are attacked by snails.

With external clip-on thermostats, make sure that the sole-plate is firmly held against the glass. And with the electronic types, check that the temperature-sensing probe is fully immersed in the water.

Because they work on an electro-magnetic principle, thermostats are liable to malfunction through interference from magnetic algae scrapers, so keep any scrapers "parked" well away from them.

### External thermostats

Used with a separate internal heating element, external thermostats usually have a flat stainless steel sole-plate by which they sense the temperature changes of the water through the aquarium glass. A much more recent design uses micro-chip technology, sensing the temperature through a miniature probe hanging in the water. These thermostats are held on the outside of the aquarium by a spring wire clip. If you have a marine aquarium you *must* shield this clip from the saltwater to prevent corroding metal polluting the tank. To do this, cover the clip with plastic airline tubing.

### Adjusting a thermostat

All types of thermostat are pre-set, usually to a 24°C norm, but this temperature setting can be adjusted. You may want to increase the temperature for breeding purposes, or decrease it in order to keep fishes that prefer lower temperatures. You *must* switch off the power supply before making any adjustments to a thermostat.

## TEMPERATURE MEASUREMENT

There are many inexpensive, simple-to-use aquarium thermometers on the market. Internal types float upright, and are "anchored" to the glass with suction pads. They have a column of coloured liquid that shows the temperature, in a similar way to the familiar medical thermometer. Externally operated types adhere to the outside of the glass, and may be of either a mechanical dial or liquid crystal design. Direct sunshine may affect the readings of liquid crystal types, so install them on the shady side of your tank.

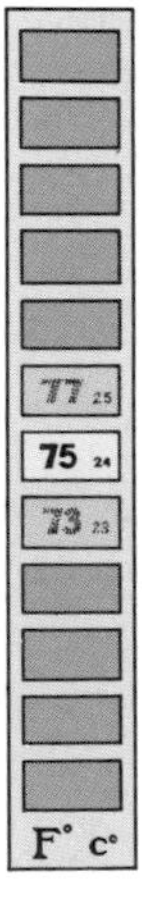


***Liquid crystal thermometer***
*This comes in the form of an adhesive strip (right) which you stick to the outside of the tank. The lighter coloured panel shows the temperature of the water.*

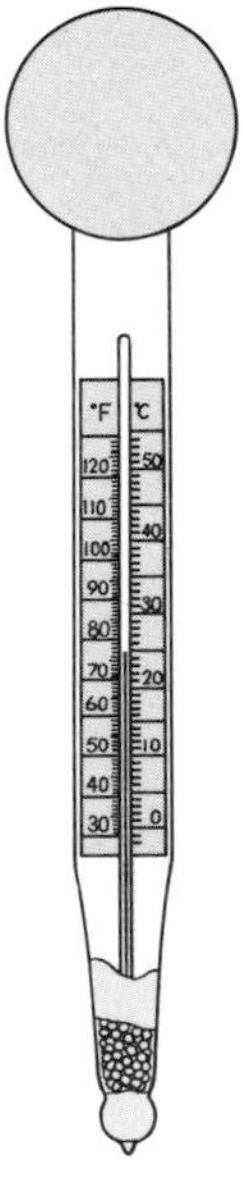


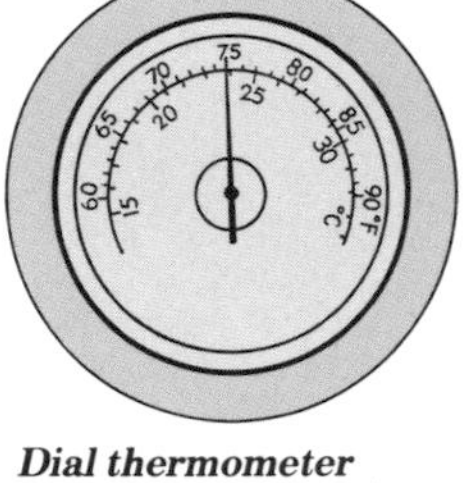

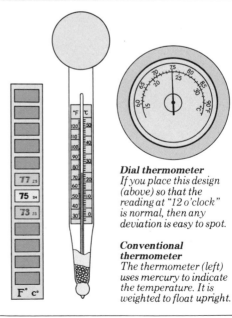

***Dial thermometer***
*If you place this design (above) so that the reading at "12 o'clock" is normal, then any deviation is easy to spot.*

***Conventional thermometer***
*The thermometer (left) uses mercury to indicate the temperature. It is weighted to float upright.*

# Conserving heat

Saving heat isn't just a project for the "shoestring" fishkeeper. If you have several tanks, you may well save money by heating the room they are in, rather than the individual tanks. And emergency heat conservation measures are important to protect your fishes in the event of a power failure.

**Space heating**

The simplest way to heat a fish-room is to use a paraffin (kerosene) heater. You will get a light film of oil on the water surface, but you can remove this by drawing across a sheet of absorbent paper. Alternatively, use a gas or electric heater equipped with a thermostatic control or extend your central heating system.

*Advantages*

One of the benefits of space heating is that you can keep fishes in any watertight container within the room. And individual tank temperature control is possible, for the higher parts of the room will be warmer than the lower. So, to increase the water temperature for particular fishes, you just move their tank up a shelf or move the fishes up to a higher tank. This is useful when trying to breed fishes, as an extra degree or so of heat often makes all the difference.

With space heating, there is no risk of fishes becoming chilled at water-change time. If you use a large fishless tank to hold the water for the next change, it will automatically be at the right temperature. Similarly, cold draughts won't get into the tanks when you lift the hoods.

*Disadvantages*

Space heating has two drawbacks: expense and comfort. The heat isn't very comfortable, and so you won't want to use the room for other purposes.

## SAVING HEAT

Whatever form of heating your aquarium uses, you can take steps to conserve the heat. The best way to do this is to lag the back and sides of the tank with polystyrene ceiling tiles, but this method is only suitable for tanks built into wall units or chimney breasts, when the cladding won't be seen. It isn't advisable for a free-standing aquarium as it will spoil its appearance.

If you have a space-heated fish-room where the room's decorative appearance isn't important, you can save heat by lining the walls with thicker polystyrene insulating material. This comes in sheets or rolls, and is glued in place like wallpaper.

**Emergency protection**

Don't panic if either the power or a heater fails as in any tank over 60 cms long the temperature will fall quite slowly – no more than a degree or so – during an average power failure. Also, the room itself will be warm, and if you can heat it by alternative means, the fall in the water temperature will be even less.

If the power is off for an appreciable period, wrap the tank in layers of blanket or newspapers to conserve what heat there is, and keep an eye on the water temperature. Many fishes are quite safe in temperatures as low as 18°C, as long as the temperature drop is gradual. If it falls drastically, then use a gas cooker, paraffin heater or coal fire to heat some water, put this into bottles and stand them in the tank to raise the temperature.

*Don't* pour enormous amounts of boiling water into the aquarium; this will upset the fishes, and may even crack the tank's glass panels.

# WATER

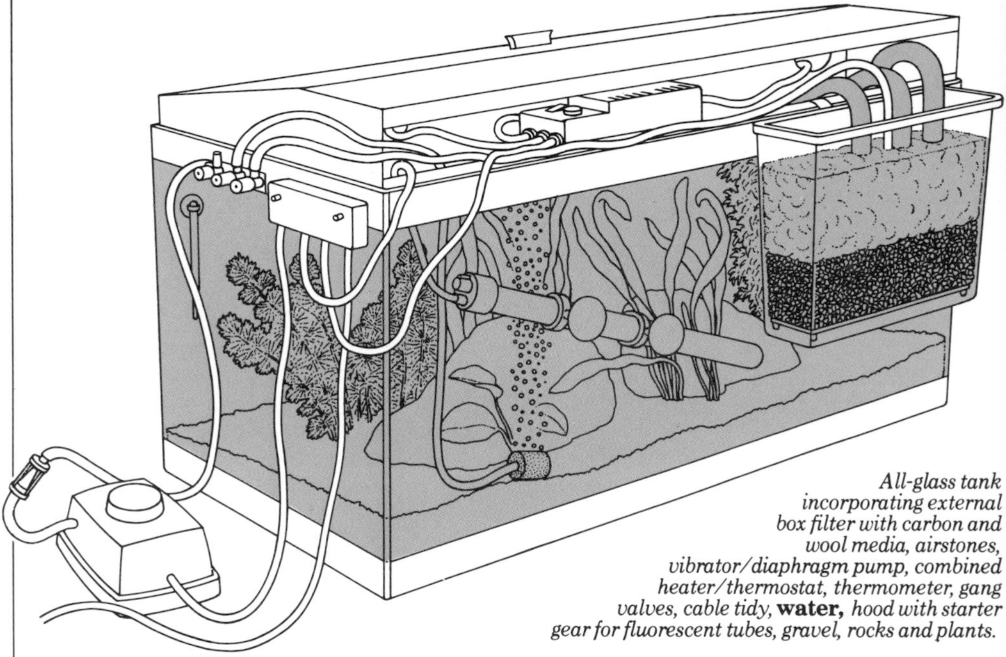

*All-glass tank incorporating external box filter with carbon and wool media, airstones, vibrator/diaphragm pump, combined heater/thermostat, thermometer, gang valves, cable tidy,* **water,** *hood with starter gear for fluorescent tubes, gravel, rocks and plants.*

The ingredient common to (and essential in) all aquariums, no matter what sort of aquatic creature is being kept, is water. And providing the correct water supply is often the key to successful fishkeeping. Water is the fishes' equivalent of the atmosphere that we breathe – it carries the dissolved oxygen to them and supports the weight of their bodies and the stems of plants. To survive, an aquatic organism must either drink water and excrete salt, or excrete water and absorb salt (see pp. 19–20), depending on whether it is a freshwater or a marine inhabitant.

Water is a common commodity – it covers approximately 70 percent of the earth's surface. The majority of this water is saltwater (that is, seawater); freshwater rivers and lakes form a much smaller proportion, perhaps three percent, of the overall water surface in the world, and of this, 75 percent is the frozen polar ice-caps, not liquid at all. However, when it comes to an aquarium, whether freshwater or marine, just tipping in some of this water won't do – it may well be polluted or the wrong quality for the fishes in your tank. This section explains what water is composed of, how its composition affects fishes and how to assess its quality and its suitability for your fishes.

# What is water?

To most people, water is simply plain $H_2O$ for freshwater, with the addition of some Sodium Chloride (Na Cl) in saltwater. However, water isn't just $H_2O$ – various minerals and trace elements are also present, and you need to understand how they affect the water composition in order to cater for aquarium-kept fishes that come from very differing locations, with consequently differing water conditions and properties.

Most of the earth's water is salt, and is found in the sea. This is a very stable environment, with the saltiness (salinity) varying only slightly from area to area. In contrast, the "minority" water – the fresh type – varies greatly in condition, depending on location. This means that while marine fishes can be kept, to all intents and purposes, in one type of water, freshwater fishes have different requirements which you have to take into consideration when preparing the water for their aquarium. To understand how the composition can vary, you must study the process that produces fresh water (see below).

A further complication is that bodies of water are often affected by industrial pollutants, and chemicals such as fluoride are frequently added to domestic water supplies.

**HOW IS FRESH WATER PRODUCED?**

Water vapour evaporates from the sea under the action of the sun, and passes into the atmosphere. During this process only pure water is moved – none of the dissolved salt in the sea is affected. As the water vapour moves over land, it condenses and falls as rain. Rain starts out as pure water, but the atmosphere it passes through and the surface it falls on affect it. During its journey back to the sea, water is "contaminated" by minerals in the ground, waste from industrial plants, and decaying vegetation.

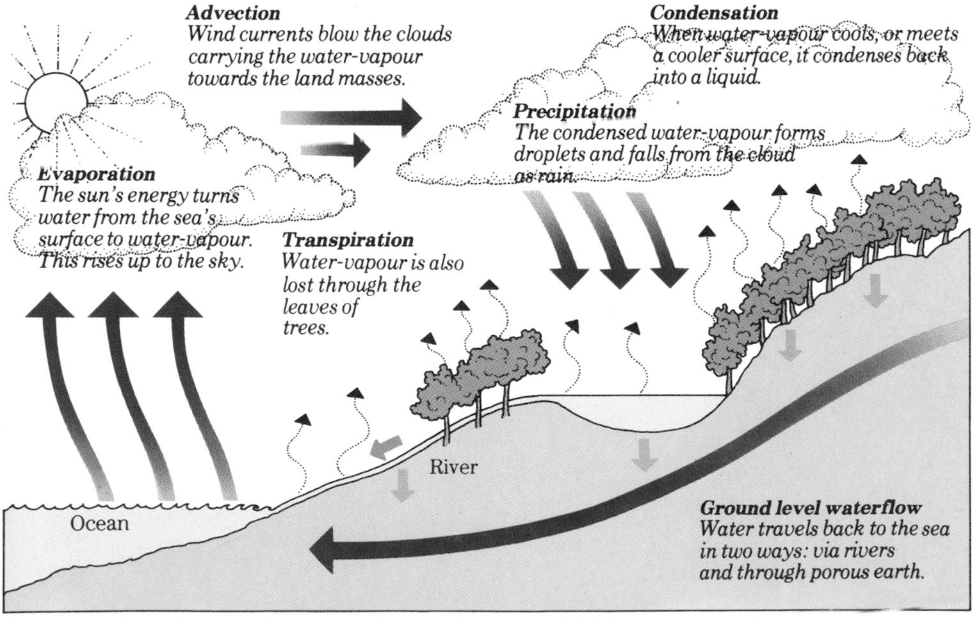

***Advection***
*Wind currents blow the clouds carrying the water-vapour towards the land masses.*
***Condensation***
*When water-vapour cools, or meets a cooler surface, it condenses back into a liquid.*
***Precipitation***
*The condensed water-vapour forms droplets and falls from the cloud as rain.*
***Evaporation***
*The sun's energy turns water from the sea's surface to water-vapour. This rises up to the sky.*
***Transpiration***
*Water-vapour is also lost through the leaves of trees.*
River
Ocean
***Ground level waterflow***
*Water travels back to the sea in two ways: via rivers and through porous earth.*

# Water and habitat

It is always the best policy to give aquarium fishes water which is as close as possible in quality to that which they are found in naturally.

## Freshwater habitats

Unlike marine fishes, the conditions that freshwater species are found in vary immensely.

*Mountain streams*

Cool and fast-moving, these streams are well-oxygenated. Few minerals are picked up from the insoluble rocks, and the water remains soft, but once it reaches the moors and heathland, the peaty soil may produce an acid reaction. Aquarium fishes found in this type of habitat include Danios (see pp. 40–1).

*Tropical streams, rivers and swamps*

Rivers that run through dense tropical rain-forest regions are generally soft and acidic; they are slower-flowing than the mountain streams and therefore less well-oxygenated. The majority of aquarium-kept tropical fishes come from such areas, including Barbs (see pp. 36–9), Characins (see pp. 45–52), Cichlids (see pp. 53–9), and Rasboras (see pp. 42–3). For further information on these habitats see pp. 139–40.

Rivers in non-rainforest locations such as the Central Americas are also slow-moving, but have harder water with alkaline reactions. Here, livebearers like the Swordtail (see p. 68), Molly (see p. 68) and Platy (see p. 69) live.

For information on swamps see p. 141.

*Seasonal tropical rivers*

Many pools and small rivers are seasonal – they dry out each year, and are only refilled by the monsoon rains. Conditions may vary quite dramatically: depending on the pool or streambed's composition, the water may remain soft and acid, or, during evaporation, become progressively harder. These waters are generally very slow-moving or stationary, and therefore low in oxygen. Seasonal pools are the habitat of the Killifishes (see pp. 70–2), who bury their fertilized eggs to survive the temporary drought.

*Lakes*

In large lakes such as those in Africa's Rift Valley very little water inflow occurs, and loss is by evaporation only. Consequently, the water is very hard and alkaline. The oxygen content is maintained to a good level by wind and wave action. Among the aquarium-kept fishes found here are many Cichlid species (see pp. 53–9).

## Brackish-water habitats

As rivers near the sea, they may become tidal, and the fishes have to adapt to increasing amounts of salt in the water. The fast-moving, well-oxygenated water also becomes harder and more alkaline. Species that live in these conditions are comfortable in saltwater or in freshwater with a little salt added. Aquarium-kept brackish-water tropical species include Monos (see p. 82) and some Mollies (see p. 67).

## Marine habitats

The sea is a very stable environment, with only a small deviation in salt content throughout the world – so the Red Sea, for example, is only slightly "saltier" than the Indian Ocean. The majority of aquarium-kept marine fishes (see pp. 96–108) come from one type of location – the coral reef (see p. 142). These species are suited to aquarium life, unlike fishes from deeper waters.

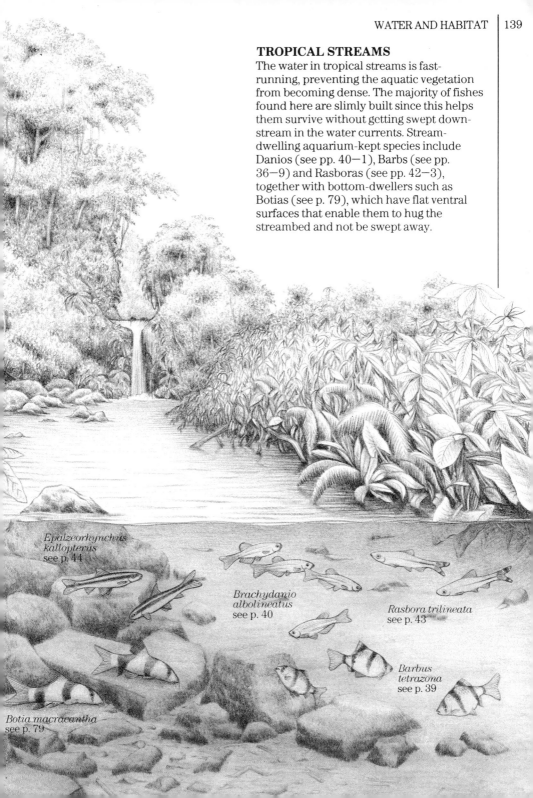

## TROPICAL STREAMS

The water in tropical streams is fast-running, preventing the aquatic vegetation from becoming dense. The majority of fishes found here are slimly built since this helps them survive without getting swept down-stream in the water currents. Stream-dwelling aquarium-kept species include Danios (see pp. 40–1), Barbs (see pp. 36–9) and Rasboras (see pp. 42–3), together with bottom-dwellers such as Botias (see p. 79), which have flat ventral surfaces that enable them to hug the streambed and not be swept away.

## TROPICAL RIVERS

Unlike streams, in large, slow-moving tropical rivers sediment isn't swept away, and so aquatic plants can gain a hold. Aquarium-kept fishes found in this type of habitat are often large and deep-bodied, like Hatchetfishes (see p. 46), which hang at the surface, waiting for the insects that the abundance of plants encourages. The reeds at the side of the river are inhabited by Angelfishes (see p. 59) and Pencilfishes (see p. 51). Moreover, because the riverbed is a "storehouse" of food, bottom-dwelling species such as *Corydoras* catfishes (see pp. 74–5) live here.

## SWAMPS

Swamp regions are formed by seasonal flooding in tropical areas. The suddenly submerged terrestrial life (plants, insects and small animals) provides food for fishes, but it also upsets the water conditions. Rotting vegetation can become a problem once the flood waters recede, and this factor, combined with the lack of movement, means that the water may become depleted of oxygen. The fishes have adapted to this: Bettas, Gouramis and certain Catfishes (see pp. 60—5) can breathe atmospheric oxygen, whilst other fishes bury themselves in the mud to escape the drought.

## CORAL REEFS

Seawater is alkaline, and so the chemical composition of the coral reef, formed by the accumulation of limestone from calcium-depositing animals, has little effect on the saltwater around it. Because the water is shallow, it is warm. And since sediment is cleared away by filter-feeding invertebrates such as sea-anemones and tubeworms and by constant wave action, the sea around a reef is unusually clear and well-lit by the sun. Thus, the bright colours of species such as *Amphiprion* and *Chaetodon* can be easily seen by other fishes.

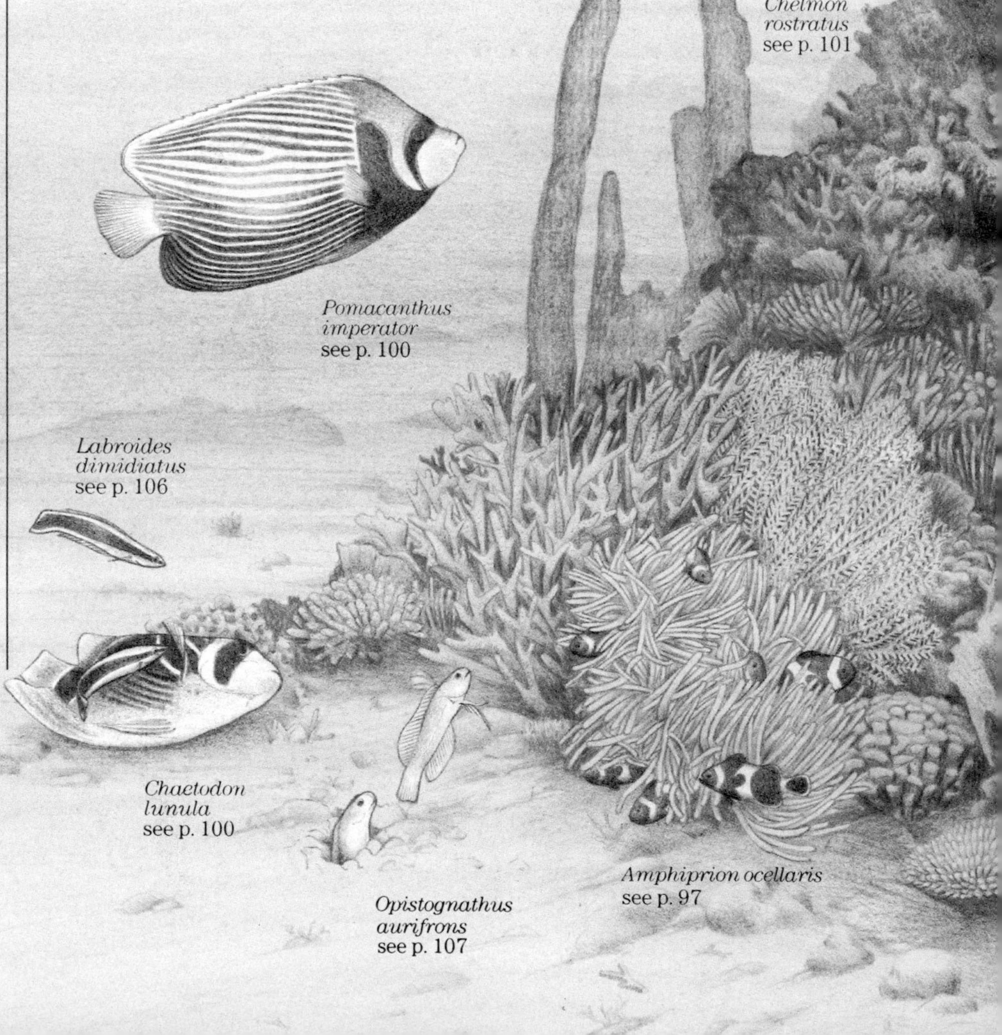

*Chelmon rostratus* see p. 101

*Pomacanthus imperator* see p. 100

*Labroides dimidiatus* see p. 106

*Chaetodon lunula* see p. 100

*Amphiprion ocellaris* see p. 97

*Opistognathus aurifrons* see p. 107

# LOCATION AND WATER QUALITY

The chart below summarizes the difference in quality between different bodies of water. When providing water for your fishes, your aim should be to match the quality of that found in their natural environment as closely as possible.

## TYPES OF FRESHWATER LOCATIONS

| Location | Water conditions | Temperature |
|---|---|---|
| **Mountain streams** | Reasonably pure, especially if the stream-bed is of a non-soluble nature such as granite. | Temperature may vary, even in tropical countries. For example, the headwaters of the Amazon regularly contain melted snow from the Andes. |
| **Tropical jungle streams and small rivers** | The water may be affected by decaying vegetation. | Although daytime water temperature may be steady, a large drop in temperature (as much as 10°C) may occur at night or when the sun sets. |
| **Larger rivers** | May be polluted by industry or affected by mud and silt. | The temperature is more stable than in smaller rivers. |
| **Large lakes** | The water conditions depend upon the lake-bed material. | The temperature is usually quite stable. |

## TYPES OF SALTWATER LOCATIONS

| Location | Water conditions | Temperature |
|---|---|---|
| **Coral reefs** | Depending on the geographical location, the salinity of the water will vary. For example, the Red Sea and the Caribbean are slightly saltier than the open ocean. | The temperature is quite stable because of the immense volume. |
| **Estuaries** | The salinity level varies according to the amount of dilution by freshwater from rivers. Also, if it rains, the water will become less saline, but if the tide is in, the proportion of saltwater to freshwater will rise. | The temperature of tropical estuaries is usually stable. However, in coldwater locations, it will vary according to the season. |
| **Isolated rockpools (at low tide)** | In these locations, the water conditions aren't normal. The salinity of the water may increase as some evaporates. | As water evaporates, the temperature may rise rapidly. |

# Sources of water

Despite agreeing that a natural supply is best, the majority of fishkeepers rely on water from the domestic supply for aquarium use. This is because city-dwellers have no access to large, unpolluted bodies of natural water, and transportation costs are prohibitive.

## Using fresh rainwater

Collecting enough rainwater for a large aquarium may prove difficult. Also, there is the added hazard that the rainwater is probably already polluted by industrial wastes. You can usually collect enough rainwater to fill a small aquarium or to use for softening harder water in a large tank, but you must take precautions to collect clean, unpolluted water only:

- Allow the first few minutes-worth of a rain shower to run to waste in order to wash away dirt and dust from the collecting area.
- Make sure that the rain is collected via non-toxic, non-metallic gutters, piping and containers. Plastic guttering and containers are ideal.

## Using tap water

Domestic water is supplied fit for human consumption, and will have been treated with chemicals to render it safe for that purpose. Some of these additives are dangerous to fishes, but their harmful effects can be eliminated:

- Before you start to fill the tank, run the tap for several minutes to discard water that has stood in copper piping and may have absorbed traces of the metal.
- Chlorine is usually driven out of the water when filling the aquarium, but to make sure that it has all been removed, you should use vigorous aeration in the tank for 12 hours.
- To remove heavy metallic deposits (such as copper and zinc) use water-treatment tablets or liquids available from your aquatic dealer.

## Sources of seawater

If you live near the sea, you will have access to a ready supply of natural seawater, but using natural seawater has hazards. You should only use such water if you can collect it well away from any likely source of pollution such as industrial discharges or sewage outfalls. And, whilst the natural water may contain vital trace elements of benefit to the fishes, there is always a risk of introducing disease. It is better therefore to use a synthetic substitute known as a "sea-mix".

### *Using a "sea-mix"*

Synthetic "sea-mixes" (available from your aquatic dealer) have been scientifically formulated to resemble natural seawater as closely as possible. You must use all the "sea-mix" in the packet so that the correct mixture is obtained. Mixes are available in different-sized bags to suit different tank capacities. Once you have mixed the water, check the density (see p. 147). Before introducing fishes to the synthetic seawater, you should aerate it vigorously for several hours. Then test the density again.

# Water quality

It is vital to keep water conditions as pure, clean and well-balanced as possible. To do this, you need an efficient filtration system coupled with regular partial water changes (see p. 218). Although you can keep the majority of freshwater fishes in tap water, you should know how to test the water quality and how to adjust its chemistry if necessary.

## Is the water too acid or alkaline?

The amount of hydrogen ions present in the water determine whether it is acidic or alkaline. This can be measured by the pH (*pondus hydrogenii*) scale. The most acidic reaction is 0, and the most alkaline reaction is 14. Water that gives a pH reading of 7 is "neutral" – neither acidic nor alkaline. Because the calibration is on a logarithmic scale, an increase or decrease of one unit of pH actually represents a change in the acidity or alkalinity of 10 times.

Subjecting fishes to a sudden change in pH value has a detrimental effect on their health, putting them under stress (see p. 214). As a result, their colours may fade, they may fail to assimilate oxygen properly, and slime may form on their skin.

## Is the water too hard or too soft?

The hardness of the water is only relevant to freshwater aquariums, as seawater has so much mineral content that hardness analysis would be immaterial. Freshwater collected from as near to its natural source (see p. 137) as possible is likely to be the softest. Once water begins its journey back to the sea, contamination occurs, and it becomes

### TESTING THE pH OF WATER

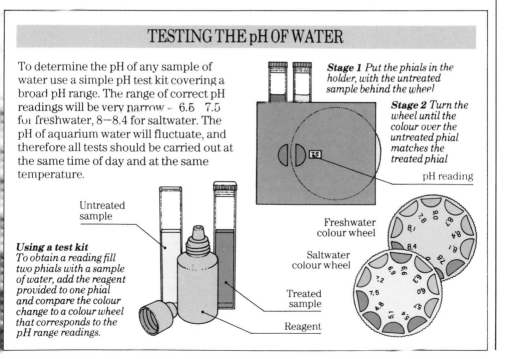

To determine the pH of any sample of water use a simple pH test kit covering a broad pH range. The range of correct pH readings will be very narrow – 6.5–7.5 for freshwater, 8–8.4 for saltwater. The pH of aquarium water will fluctuate, and therefore all tests should be carried out at the same time of day and at the same temperature.

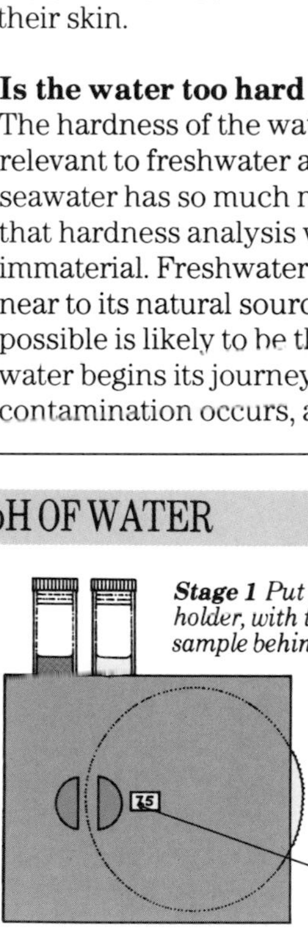


***Stage 1** Put the phials in the holder, with the untreated sample behind the wheel*

***Stage 2** Turn the wheel until the colour over the untreated phial matches the treated phial*

pH reading

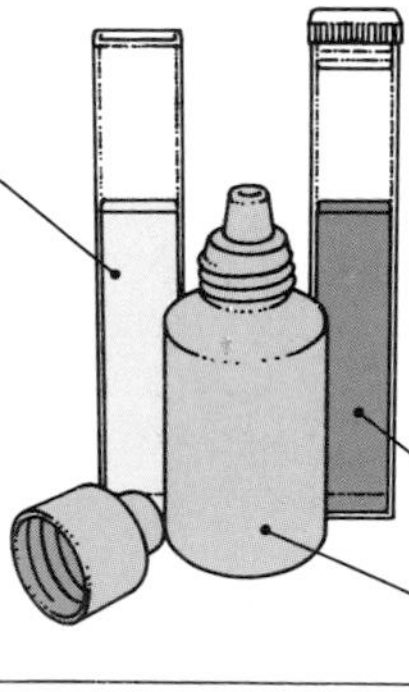

Untreated sample

Treated sample

Reagent

***Using a test kit***
*To obtain a reading fill two phials with a sample of water, add the reagent provided to one phial and compare the colour change to a colour wheel that corresponds to the pH range readings.*

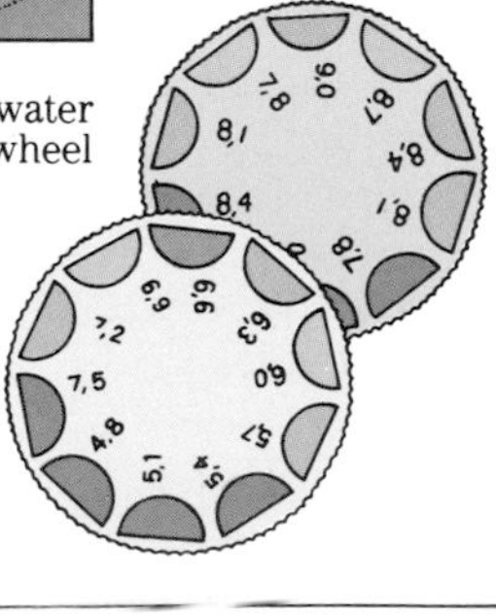

Freshwater colour wheel

Saltwater colour wheel

progressively harder through the action of dissolved minerals.

Fishes' requirements vary – some species come from naturally hard waters, others from soft (see p. 138). As far as possible, you should try to cater for these needs. As with pH, fishes can't tolerate major changes in hardness, and adverse effects may result.

*Altering hardness*

Hardness can be of two types – temporary or permanent. The temporary type is easily removed – just boil the water before you add it to the tank (allow it to cool too!). However, to remove the permanent type you will have to buy ion-exchange resins from your aquatic dealer and add them to your mechanical filter or water softener. Both types of hardness can be reduced by dilution; adding softer water will lower the hardness figure.

As with pH, you can measure water hardness using a test kit which employs a simple colour-change principle. These kits will test for either temporary or permanent hardness.

## Checking nitrite levels

In freshwater aquariums where biological filtration (see pp. 128–9) isn't used, fishes may be affected by ammonia or other nitrogenous poisoning. The reason for this is that although mechanical and chemical filters will remove suspended sediment and absorb some dissolved waste matters, they can't cope with all the pollutants present. You can check your aquarium water for excess toxins with a nitrite test kit which measures the amount of nitrogen (N) in mg per litre. One mg of nitrogen is equal to 3.3 mg of nitrite ($NO_2$), the critical level above which fishes' health may suffer. A level of 0.1 mg of nitrogen is harmless. A reading of up to 0.5 mg of nitrogen indicates that the breakdown of organic materials isn't complete.

## WATER DON'TS

- Don't make sudden changes to water conditions
- When making a regular change, always use water of the correct composition and temperature
- Don't change fishes from tank to tank unless the water conditions in each tank are exactly the same
- If you use rainwater as a source of soft water don't collect it from a dirty roof or in metallic containers; let it rain for several minutes before collecting any water; this allows for the dust and dirt to be washed away from the roof and gutters
- Don't let water anywhere near electrical equipment
- Don't use saltwater to top up evaporation losses in marine tanks, use freshwater instead as salts aren't lost during evaporation

*Factors producing a rise in nitrite levels*

Several different pollutants can increase the level of nitrites in freshwater aquariums:

- Large deposits of detritus on the aquarium floor.
- Overdirty, uncleaned filters.
- Failure to carry out regular partial water changes (see p. 218).
- Overfeeding, leading to uneaten food decaying in the tank.
- Overcrowded fishes – make sure that you judge the amount of space that your fishes need correctly (see p. 116).

*Correcting excess nitrite levels*

If you find that your aquarium has an excess level of nitrite you must rectify the problem immediately. Change about 80 percent of the water (see p. 218), making sure that all the detritus on the tank floor is removed. Match the temperature of the new water to that of the old.

## Checking density

The "saltiness" or strength of a sea-mix is measured in terms of density. To check the density of marine aquarium water you use a hydrometer (see right). This instrument floats higher or lower in the water, depending on the water's density. Always make readings at the same water temperature. Make sure that you read the value at the true waterline: the surface level of the water makes an upturned ring around the hydrometer stem that is called the miniscus; take a correct reading where the lowest point of the miniscus crosses the scale readings on the hydrometer stem.

## Adjusting water conditions for breeding

You don't have to alter the water conditions to breed freshwater fishes, but they will thrive better and breed more readily in water conditions which nearly approximate to that of their natural habitat (see pp. 138–42). The majority of freshwater fishes prefer softer, more acidic water than domestic tap water in hardwater areas. With most egg-laying species (see p. 238) all you have to do is raise the temperature a few degrees (see p. 249). Another simple trick that is particularly successful with *Corydoras* species (see pp. 74–5) is the addition of a fresh supply of water; this simulates seasonal changes in nature. More sensitive fishes such as the Discus (see p. 59) and Pencilfishes (see p. 51) will require water conditions nearer to that of their natural habitat before breeding, and it is also quite likely that their eggs won't develop properly if the water conditions aren't exactly right. You must acclimatize fishes to new water conditions slowly. Regardless of whether or not the breeding attempt is successful, the fishes must then be re-acclimatized back to the water conditions (if different) of the aquarium that they came from.

### MATERIALS THAT UPSET WATER QUALITY

- Certain metals (see p. 214) such as magnesium, copper, zinc
- Some rocks (see p. 157)
- Overfeeding – uneaten food will decay in the tank, raising the level of nitrites
- Some base materials (see p. 155)
- Poisons – pollutants drawn in from the atmosphere
- Neglected filters
- Pre-cast concrete decorations
- Coloured gravel

***Hydrometer***
*Used to check density, hydrometers are essential in marine tanks. This model has a built-in thermometer, enabling you to check the two most important water conditions at once.*

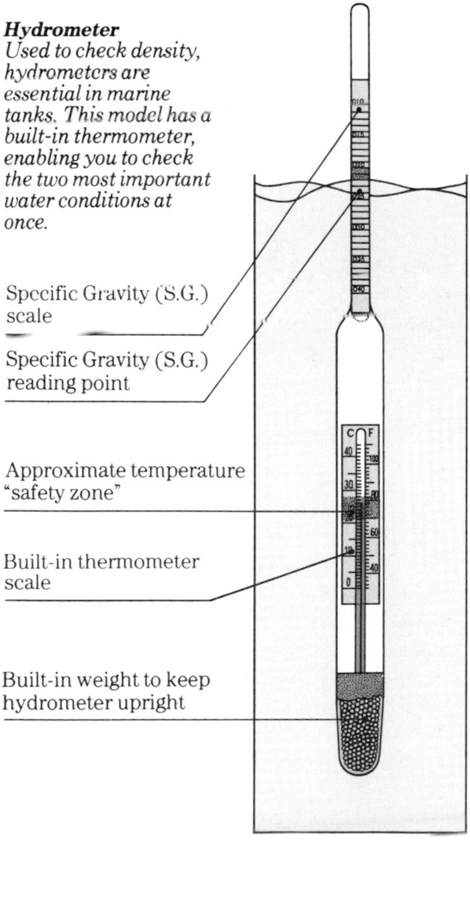

# LIGHTING

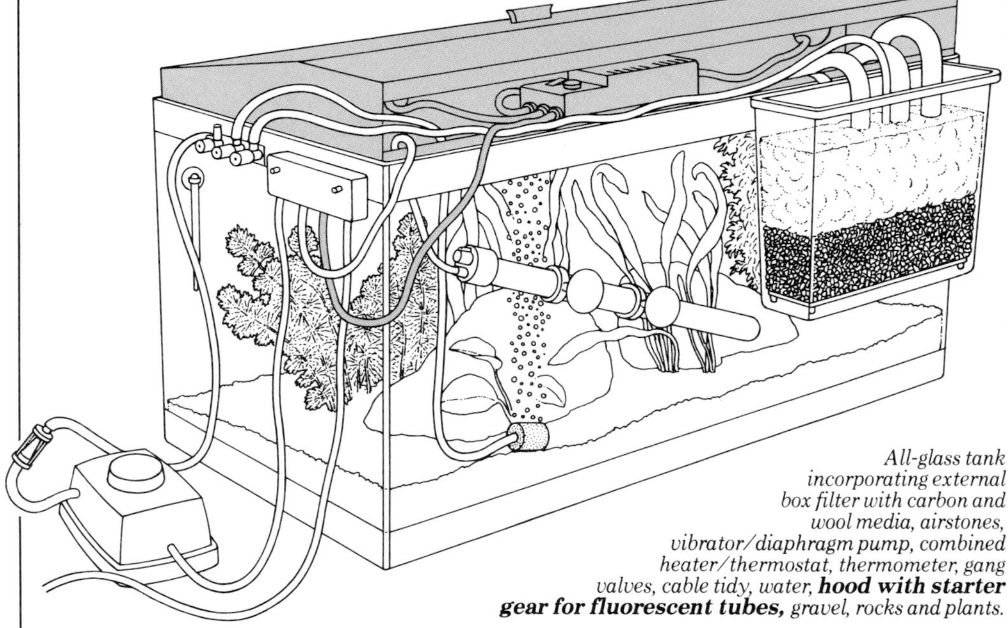

*All-glass tank incorporating external box filter with carbon and wool media, airstones, vibrator/diaphragm pump, combined heater/thermostat, thermometer, gang valves, cable tidy, water,* ***hood with starter gear for fluorescent tubes,*** *gravel, rocks and plants.*

Light is necessary in an aquarium to allow you to see the contents properly, and artificial light is used because natural light is too unpredictable. However, making the aquascape easy to view isn't the most important function of lighting; its prime role is as the stimulus for life, both for fishes and plants. Light is essential for plants to photosynthesize. And this action is beneficial to the fishes – it helps to keep carbon dioxide levels down to a minimum, and releases oxygen into the water. In order to perform this valuable service efficiently, plants require the right amount and strength of illumination.

Plants growing under water need more light than land-cultivated plants because the light's strength is cut down as its rays pass through the water. If sufficient light isn't provided, the plants will die or fail to grow. Tropical types, in particular, need plenty of light as they are used to long, intense periods of sunshine. As a guide, an aquarium with a water depth of up to 38 cms should have at least 60 watts of tungsten light or 20 watts of fluorescent light for each 30 cms of tank length, assuming that the tank is lit for 10–15 hours a day. After 8–10 hours, the plants will have had enough intense light, and the level can then be reduced. Dimmer switches and timers can be used to regulate the duration and intensity of light. Reducing the lighting level in the evening has another bonus: after a day of bright light, the nocturnal species will think that dusk has arrived and come out of their hiding places.

# Installing lighting

To use light to the best effect, you should direct as much of it as possible down into the aquarium, fitting the lamps so as to give you the best view of the fishes. The hood doesn't just serve for keeping the dust out and the fishes in, it is also the ideal place to accommodate the lamps. For information on choosing lamps, see the chart on p. 151.

## Improving lighting efficiency

To make the most use of the light, you must make the hood work as a reflector. To do this, you should paint the inside white. Don't follow the common practice of lining it with metal cooking foil as the foil will block the very necessary ventilation holes in the hood through which unwanted carbon dioxide and heat escape from the tank.

## Positioning lamps

To show your aquarium off to its best advantage mount the lamps in the hood so that the light is reflected slightly towards the rear of the tank. This means that any shadows cast are directly behind the fishes, away from the viewer. A centrally mounted lamp will cast shadows towards the viewer, hiding many of the fishes' natural iridescences.

***Hood with tungsten lamps***
*Cheap and simple to install, bayonet or screw fitting tungsten lamps (below left) have a shorter lifespan than fluorescent tubes. Other disadvantages are an uneven spread of light and a tendency to produce heat.*

***Hood with fluorescent lamps***
*Tubes (below right) aren't always available in the right length for your hood. If this is the case, you will have to use two shorter ones. Waterproof caps protect connections from water or condensation damage. Although they are more expensive than tungsten lamps, tubes last longer, are cooler-running and provide a more even spread of light.*

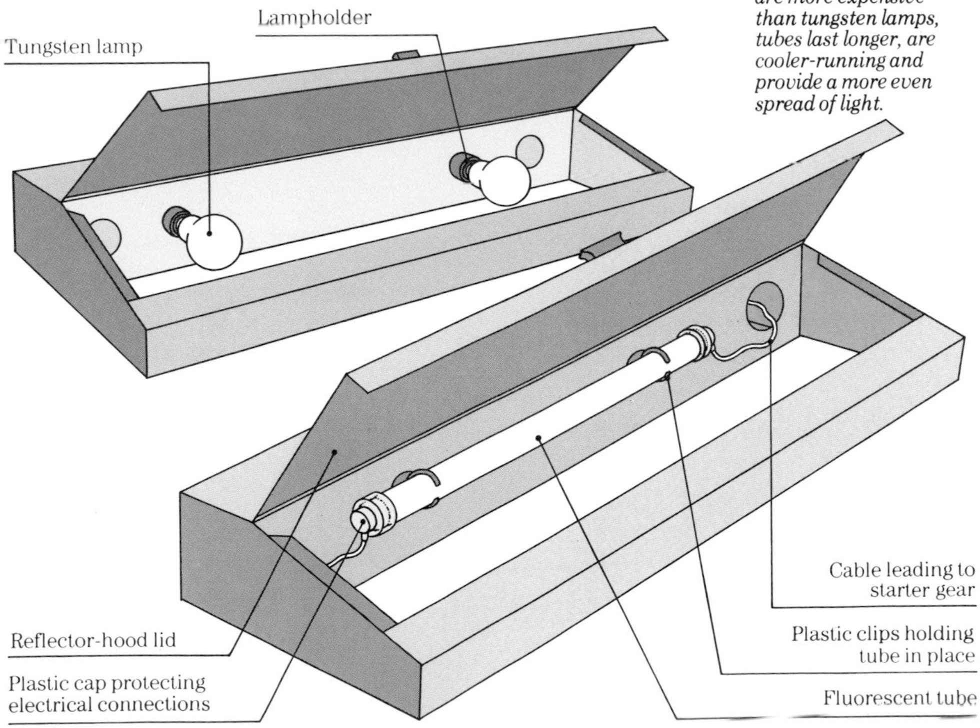

**Positioning the starting gear**

Either a choke or an auto-transformer is needed to run fluorescent lighting. Unlike household tubes, this heavy equipment isn't built-in to the fitting, but sited separately, either in a compartment in the hood, on the outside of the hood or near the aquarium. One danger of having the starting gear in or on the hood is that it makes it heavy and unbalanced, and a moment of clumsiness when lifting it up to feed the fishes could result in the hood falling and cracking the cover-glass. If you are going to use a number of fluorescent tubes, you will need several sets of starting gear, so it is better not to mount the gear on the tank itself.

**Safety factors**

Electricity and water *don't* mix, and aquariums provide plenty of opportunities for this deadly combination, so extra caution is called for. Because they are sited immediately over the water surface, hot lamps are particularly vulnerable to any splashing from fishes, airstones or filters, which could cause an explosion. You *must* take precautions to prevent damage to the fittings by water or condensation. The easiest way to protect the lamps against any form of water damage is by buying waterproofed lamp fittings and connectors and using a cover-glass on the tank. This is fitted above the water level, but below the reflector-hood. All-glass tanks have a narrow glass ledge to sit the cover-glass on. The glass can be in two pieces, so that one half can be slid to one side for feeding or maintenance purposes. And you can cut off a corner of the cover-glass to provide access to airlines or filter tubes.

**The effect of lighting on algal growth**

In general, the average aquarium is underlit, and a doubling of the light in your tank will make an enormous difference, providing you take a few precautions to keep algae at bay. More light usually means more algae, but algae will only grow if the conditions are right.

**COVER-GLASS FUNCTIONS**

- Protects the lamps and electrical fittings from water damage
- Keeps dust out of the aquarium
- Stops children dabbling in the tank
- Cuts down on evaporation losses
- Prevents the more adventurous fishes from jumping out
- Provides a spawning site for some fishes and snails

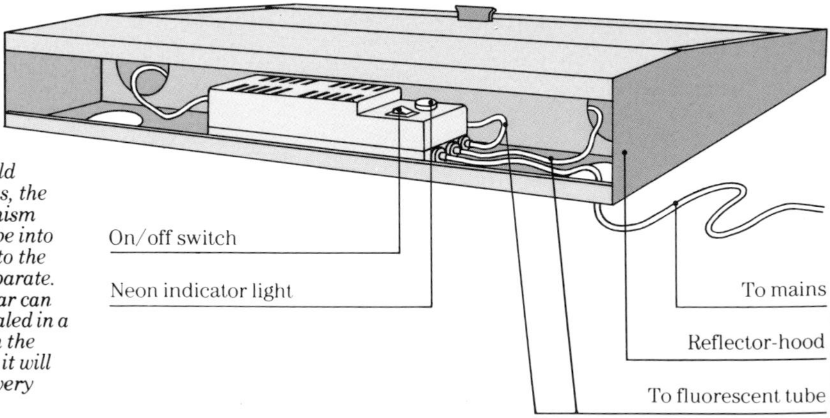

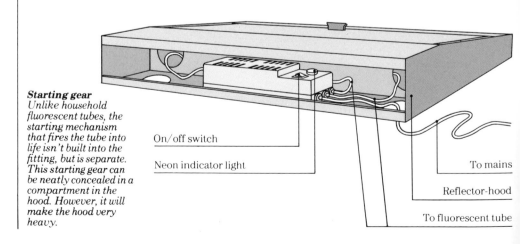

***Starting gear***
*Unlike household fluorescent tubes, the starting mechanism that fires the tube into life isn't built into the fitting, but is separate. This starting gear can be neatly concealed in a compartment in the hood. However, it will make the hood very heavy.*

In the case of light, algae only take hold when there is an excess of light energy left over after the aquarium plants' needs have been met. The majority of fish-keepers advise cutting down on the intensity and/or duration of the lighting to prevent excessive algal growth. But there is a better line of attack: keep more aquarium plants, make use of all the light energy available, and thus deny the algae any source of light.

If you are interested in the plant side of aquarium-owning you could follow the example of some European fishkeepers who specialize in luxuriant plant growth in their aquariums, achieved through very intense lighting. They may have up to 60 watts of fluorescent lighting for every 30 cms of tank length.

**LIGHTING DON'TS**

- Don't spoil the effect of well-installed lighting by having a dirty cover-glass that cuts out a lot of the light
- Don't switch the lamps off or on suddenly as you will stress the fishes. To avoid this, switch off the aquarium lights a few minutes before turning off the room light, and switch the room light on before the aquarium lights
- Don't use lamp fittings and connectors that aren't waterproofed
- Don't line the hood with foil; this will block the ventilation holes
- Don't mount the lamps centrally as they will cast ugly shadows

## TYPES OF LAMP

Basically, there are only two types of lamp to choose from: tungsten and fluorescent. This table outlines the merits and disadvantages of each sort. In the main, your choice will depend on your budget – if you can afford them, buy fluorescent tubes.

| | **Tungsten** | **Fluorescent** |
|---|---|---|
| **Type of lamp** | Individual glass bulbs or filament striplight | Tubular, in varying lengths |
| **Ease of installation** | Simple to install in hood; bayonet or screw fittings | Complicated to install; hood may need cutting. Tube lengths not always compatible with hood length. Requires special starting gear |
| **Lamp cost** | Bulbs are inexpensive | Tubes are more expensive than bulbs |
| **Lamp brightness** | Various brightnesses available in same-size bulbs | Brightness proportional to length |
| **Running cost** | High running costs | Low running costs |
| **Lamp life** | Short lamp life | Long lamp life |
| **Heat factor** | Produces heat | Runs cool |
| **Light output** | Output may be localized and uneven | Even spread of light |
| **Light spectrum** | Light spectrum limited | Various "colours" and light spectrums available |

# 5

# SETTING UP AN AQUARIUM

There is plenty of scope for creativity when setting up an aquarium from scratch – your aim should be to establish a safe, healthy environment for your fishes whilst at the same time creating an attractive focal point in your home. Plants play an important part in this, particularly in freshwater aquariums, as they are sources of food and oxygen for the fishes, they help to keep the tank conditions healthy, and add much to the look of the aquarium. Rocks and corals also make tanks more interesting, and are useful for hiding some of the technical equipment. This chapter shows you how to use these elements to design a well-laid-out aquascape and guides you through the setting-up processes.

# AQUASCAPING

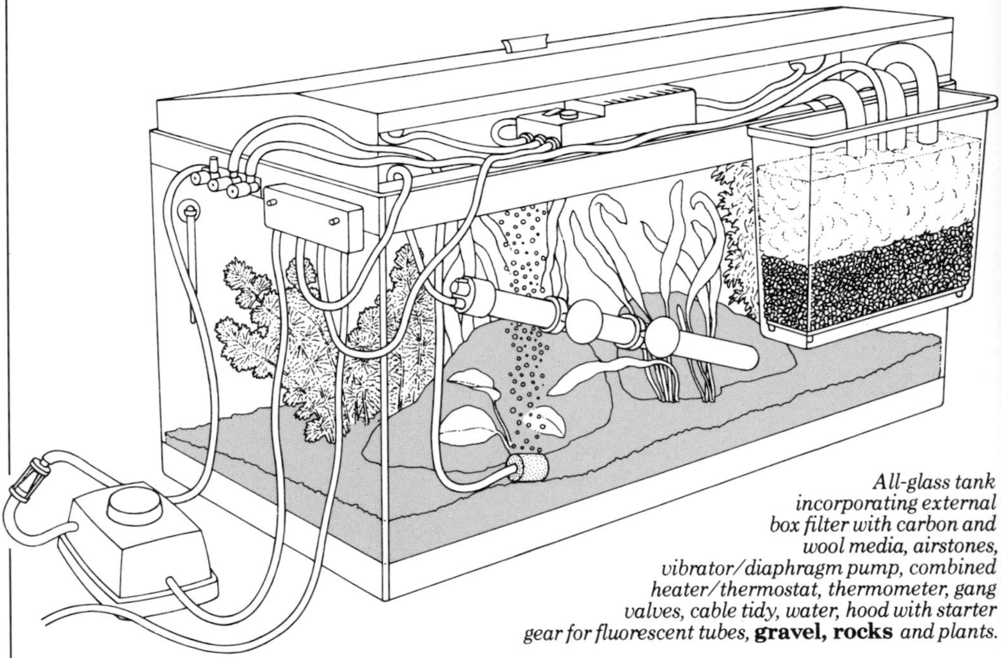

*All-glass tank incorporating external box filter with carbon and wool media, airstones, vibrator/diaphragm pump, combined heater/thermostat, thermometer, gang valves, cable tidy, water, hood with starter gear for fluorescent tubes,* **gravel, rocks** *and plants.*

Aquascaping an aquarium has two purposes: to make the bare tank more attractive for the viewer and to provide several functional benefits for the fishes. Inert materials like rocks and gravel provide comfort and security for the fishes, and may be used as spawning sites or retreats by some species. Moreover, the aquarium plants depend upon them as a growing medium. Also, a deep layer of gravel on the aquarium floor can make up part of a filtration system (see p. 128).

When decorating their tanks, many fishkeepers try to create an aquascape that resembles the environment their fishes came from as closely as possible. This is achieved by adding suitable rocks and other "furnishings" such as plants and logs. For example, a soft-water rain-forest stream (Tetras and *Corydoras* Catfishes) or a hard-water rocky African shoreline (*Julidochromis* or *Lamprologus* Cichlids) can be set up in your aquarium if you do your "research" properly and pay attention to detail. In marine aquariums, where it isn't possible to keep aquatic plants, corals make the tanks much more attractive, and simulate the coral reefs from which many of the fishes originate.

Part of the interest and challenge of aquascaping is that the aquarium isn't static. The carefully arranged plants may grow at different rates (or the fishes may eat them), so you will need to change your design from time to time.

# The base covering

An aquarium should have a base covering for three main reasons: to provide camouflage for the fishes, to serve as a growing medium for the plants, and to act as part of a filtration system. It is extremely important that the base material is the correct size and type to suit the aquarium system.

### The base as camouflage

If the colour of the base covering is the same shade as the fishes' backs the base will act as camouflage. In most cases, the top of a fish is a dark shade which, when seen from above, merges with the river bed, partially camouflaging the fish from predators. (Equally, when viewed from below, the lighter belly merges with the light coming through the water surface.)

In the aquarium, however, freshwater fishes will look more attractive if their tank has a dark-coloured base as the light reflected up from light-coloured material, such as silver sand, will make fishes look paler in colour. This is because the colours are caused by reflection from guanin beneath the scales (see p. 16). But in the marine aquarium, the fishes' brilliant colours are caused by pigmentation, and so are unaffected by the colour of the tank base covering.

### Fishes' use of the base material

Many freshwater fishes make practical use of the the aquarium gravel when breeding; for example, *Nothobranchius rachovi* (Rachov's Nothobranch) excavates depressions in the gravel and deposits its eggs in them. Some marine species like to bury themselves in the gravel at night, whilst others such as *Opistognathus aurifrons* (Yellow-headed Jawfish) make tunnels in it which they use as retreats.

### The base as a growing medium

Although true aquatic plants don't feed through their roots, they need to anchor themselves with their root systems, and many use the gravel bed for this purpose.

### The base as a filter medium

Biological filtration requires a medium on which nitrifying bacteria can multiply (see p. 128). A gravel bed, or the coral sand of a marine tank, is ideal.

### Materials for freshwater aquariums

The most widely used material for covering the base of the tank is gravel. It is available from aquarium dealers, and is safe to use in most freshwater tanks. However, calcareous gravel (containing

## CHOOSING SUITABLE MATERIALS

Any materials used in aquarium decoration must be non-toxic, and mustn't adversely affect the water chemistry. Some substances should only be used in one type of aquarium system – either freshwater or marine.

| Material | Freshwater | Marine |
|---|---|---|
| Gravel | ✓ | ✗ |
| Coral sand | ✗ | ✓ |
| Corals | ✗ | ✓ |
| Shells | ✗ | ✓ |
| Rocks | ✓ | ✓ |
| Wood | ✓ | ✓ |
| Oolite | ✗ | ✓ |

water-hardening materials such as limestone) shouldn't be used in a soft-water tank. To test the calcium content of gravel, add some dilute hydrochloric acid or vinegar to a sample. If the acid fizzes, then water-hardening calcium is present.

Gravel is usually a dark, yellow-brown mottled colour. Although coloured gravels are available, they aren't advisable: the glaring colours will detract from the fishes' coloration, and the dyes used to colour the gravel may escape and affect the water chemistry and the fishes' health.

### Materials for saltwater aquariums

Calcium-rich sands and gravels are quite safe to use in this type of aquarium; in fact they will help to maintain the correct water conditions. Crushed coral sand can also be used, but is expensive. Crushed limestone, crushed sea-shells, oolite and dolomite are also suitable. A cheaper alternative is sand or shingle collected from the seashore.

### Base material size

A particle size of around 3 mm is best. If the base material is too large or too small, it may have harmful effects:

- *Too large:* uneaten food can fall into the material, beyond the reach of the fishes, and cause water pollution.
- *Too large:* as the surface area on which the nitrifying bacteria live will be small, self-cleaning won't be so effective.
- *Too small:* plants will have difficulty in rooting.
- *Too small:* waterflow through the undergravel filter will be impeded.

### Preparing the base covering for use

Whatever material you use, it must be clean and free from foreign bodies. Gravel and most materials are best washed in reasonably small amounts, say, half-a-bucketful at a time. Take the bucket outside and feed a garden hose into it. When the water runs clear from the bucket, the gravel is clean. If you want to use naturally collected gravel, stand a sample in a bucket of water for a few days to see if anything happens. A strong smell emanating from the bucket will indicate that the natural lifeforms in the gravel have died. You can then wash the gravel thoroughly, after which it will be ready for use. Coral sand can be prepared by immersing it in a solution of domestic bleach for a couple of days, and then washing it carefully.

### Installing the base covering

If you intend using an undergravel filter make sure that you put the filter plate in the aquarium before the base covering, and that there is a sufficient quantity of base material. As a rough guide, allow a 10 litre bucketful of base covering for every 900 $cms^2$ of base area.

The base covering should be spread over the tank floor to a good depth, and sloped so that the back is higher than the front (a ratio of 1:5 from front to back will allow good filtration and give the aquarium a sense of perspective).

***Keeping the slope in place*** *Rock or glass strips will hold back the base material and prevent it "levelling out".*

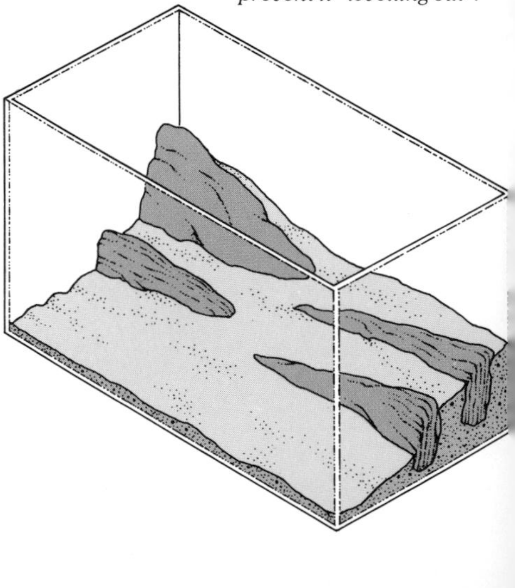

# Adding decorative materials

In addition to plants (see p. 161), rocks, shells and logs can be used to decorate your aquarium, and to hide unsightly aquarium hardware. Make sure any material you add is clean and safe.

**Ways of using rocks**
All rocks should be prepared as for gravel (see p. 156). Although they will look good individually, you can also make rocky outcrops, or even a cliff-face by sticking rocks together with sealant. To prevent large rocks from toppling over, place them directly on the tank floor, but don't obstruct the filter pipes. For the fishes' safety, don't use sharp-edged rocks, and don't design a layout where you can't see or reach parts of the tank.

**Other aquarium decorations**
You can simulate tree roots by using branches. Because these may contain resins that will pollute the tank, you *must* boil the wood repeatedly in changes of fresh water before use. For extra safety, seal with a coat of polyurethane varnish.

Other suitable materials are cork bark, shells and non-toxic replicas of tree roots available from aquarium shops.

## NATURAL FITTINGS

Pieces of branch or cork bark will look like sunken logs if they are positioned correctly. Unfortunately, they have a tendency to float, so you will have to weight or fix them down.

***Fixing in place***
*To prevent the "log" floating, fix it to a slate or plastic base-plate using sealant or a plastic screw.*

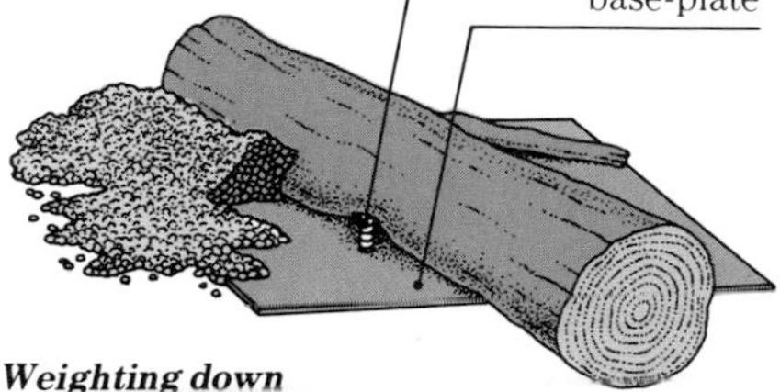


***Weighting down***
*Cork bark can be weighted down by rolling it around a gravel-filled plastic pipe.*

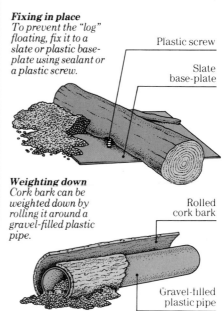

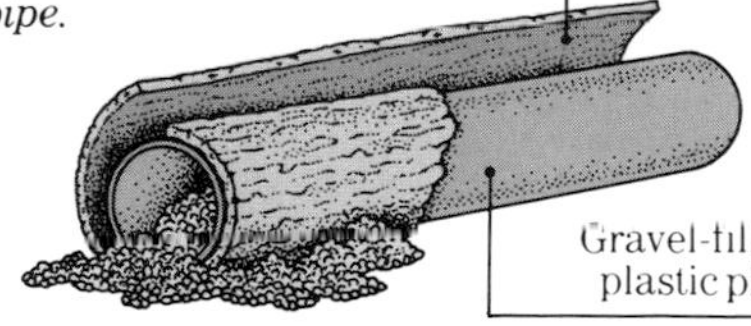


## CHOOSING SUITABLE ROCKS

**FRESHWATER – SOFT AND MEDIUM/HARD**
**Safe**
- Basalt
- Granite
- Quartz
- Slate
- Hard sandstones
- Artificial "rocks" precast from unsealed concrete or plastic

**Unsafe**
- Soluble rocks
- Calcareous rocks
- Shells

**FRESHWATER – HARD**
**Safe**
- Soluble rocks
- Calcareous rocks without metallic-ore veins

**Unsafe**
- Rocks with metallic-ore veins

**MARINE**
**Safe**
- Oolite
- Calcareous rocks without metallic-ore veins
- Shells
- Dead and live corals
- Slate

**Unsafe**
- Rocks with metallic-ore veins
- Soft, soluble sandstones

# PLANTS

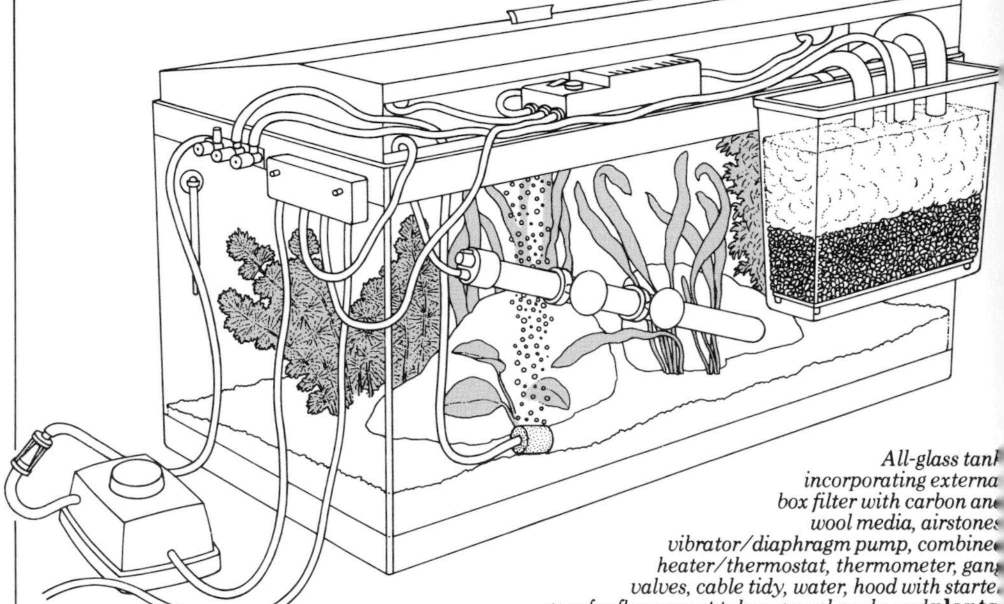

*All-glass tank incorporating external box filter with carbon and wool media, airstones, vibrator/diaphragm pump, combined heater/thermostat, thermometer, gang valves, cable tidy, water, hood with starter gear for fluorescent tubes, gravel, rocks and* **plants**

Aquatic plants are very important in freshwater aquarium-keeping, as they help to keep the water conditions pure.

Plants can do this because, under the influence of light, they create food in their green cells by absorbing carbon dioxide and giving off oxygen; a process known as photosynthesis. This makes plants a very useful tool in coping with the supply of oxygen and the removal of carbon dioxide in the aquarium. However, the process only occurs whilst the aquarium lights are on. When the lights are switched off, the plants reverse these functions, absorbing oxygen and giving off carbon dioxide, just like the fishes. So for a healthy aquarium, you must make sure that your plants receive an adequate amount of light.

Plants are also useful in removing another waste product: nitrate. If you have a biological filter installed in your aquarium, this will break down ammonia (which the fishes excrete) into less-harmful nitrate, which the plants can then use as food.

Plants have other uses too: they provide the fishes with shade and shelter, make useful breeding sites, and form part of vegetarian fishes' diets.

# CHOOSING PLANTS

By comparing the countries of origin of plants and fishes, you can create a definitive collection from practically any location in the world. But bear in mind that many plants have been "exported" to other countries.

**KEY ■ Coldwater ▒ Tropical**
Some genera contain both coldwater and tropical species. But, this does *not* mean that every species within these genera can be used in both systems. Some are coldwater types, whilst others are tropical plants.

| Species | Origin | Tank | Type and use |
|---|---|---|---|
| **Acorus** | Americas | ■ ▒ | Rooting; foreground |
| **Aponogeton** | Africa, Asia | ▒ | Rooting; specimen |
| **Bacopa** | Worldwide | ■ ▒ | Rooting; filler |
| **Cabomba** | Americas | ▒ | Cutting; filler; spawning mop |
| **Ceratophyllum** | Worldwide | ■ ▒ | Cutting; filler; shading; spawning mop |
| **Cryptocoryne** | Asia | ▒ | Rooting; (small) foreground; (large) specimen; spawning mop |
| **Echinodorus** | Americas | ▒ | Rooting; specimen; spawning mop |
| **Egeria** | Americas | ■ ▒ | Cutting; filler; spawning mop |
| **Eleocharis** | Worldwide | ■ ▒ | Rooting; foreground |
| **Elodea** | Americas | ■ ▒ | Cutting; filler; spawning mop |
| **Fontinalis** | N. Hemisphere and S. Africa | ■ | Anchoring; filler; spawning mop |
| **Hydrilla** | Worldwide | ■ ▒ | Cutting; filler |
| **Hygrophila** | Asia | ▒ | Cutting; filler |
| **Lagarosiphon** | Africa | ■ ▒ | Cutting; filler; spawning mop |
| **Lemna** | Worldwide | ■ ▒ | Floating; shading |
| **Limnophila** | Asia, Africa | ▒ | Cutting; filler; spawning mop |
| **Ludwigia** | Americas | ■ ▒ | Cutting; filler |
| **Microsorium** | Asia | ▒ | Anchoring; foreground |
| **Myriophyllum** | Worldwide | ■ ▒ | Cutting; filler; spawning mop |
| **Najas** | Americas | ▒ | Rooting or floating in midwater; filler; spawning mop |
| **Nomaphila** | Asia, Africa | ▒ | Cutting; filler |
| **Pistia** | Worldwide | ▒ | Floating; shading |
| **Potamogeton** | N. Hemisphere | ■ | Cutting; background |
| **Sagittaria** | Worldwide | ■ ▒ | Rooting; background |
| **Vallisneria** | Worldwide | ■ ▒ | Rooting; background |
| **Vesicularia** | Asia | ▒ | Anchoring; filler; spawning mop |

# Cultivating and using plants

If they are to thrive, aquarium plants need just as much attention and care as the fishes. They have three main needs: light, food, and clean water conditions.

### Lighting requirements

In order for a plant to synthesize food from carbon dioxide and the hydrogen component of water, it needs light. However, too much light encourages algal growth, and too little restricts plant growth. Moreover, some plants require less light than others. To accommodate these differing needs, taller plants, which often need more light, can be used to shade the shorter species which thrive in lower light levels. As a guide, an aquarium with water up to 38 cms deep should have at least 60 watts (tungsten) or 20 watts (fluorescent) for each 30 cms of tank length, assuming that it is lit for 10–15 hours each day.

### How do plants obtain their nourishment?

Most aquatic plants obtain food in the form of nitrates and organic matter through their leaves. But a few plants also feed via their root systems. For feeding methods, see right.

### Water requirements

Clean water is essential for successful plant growth for two reasons: any dirt in the water will settle on the surface of the leaves and choke the plants, and dirty water dramatically reduces the amount of energy-giving light reaching the plants.

Some plants are adversely affected by a change in water conditions, dropping their leaves as soon as they are introduced into a new aquarium. Usually this is just a temporary reaction, and the leaves soon regrow. It is important to make water changes gradually over several days (see p. 218) to avoid harming plants.

### The effect of undergravel filtration on plants

Undergravel filtration (see p. 128) is the subject of a longstanding controversy amongst fishkeepers. Some think that it hinders plant growth, arguing that water flowing through the gravel takes away nutrients from the plants' roots too quickly, and that the equalization of water temperature through the whole tank may have a detrimental effect on plants, as many species are used to having their roots in cooler water than their leaves. Conversely, advocates of undergravel filtration insist that as long as the gravel bed is deep enough, has the correct gravel size, and the waterflow isn't too strong, then plant growth won't be affected. In fact, this filtration system helps to turn waste products into food that plants can assimilate.

### Planting techniques

You can either plant the tank "dry" or "wet". Planting dry is easier as the plants won't float upwards, cling to your arms or become tangled together. However, you can't judge how the tank will look until the water is added, and therefore alterations will need to be made. When you plant wet you will see the final result straightaway.

When planting your aquarium, begin by putting tall plants around the back and sides of the tank. Once you are satisfied with the arrangement, fill in with bushy, low-growing and specimen plants by working from the sides towards the centre. Don't push the plants too deeply into the gravel – the crown should be level with or just above the gravel bed.

*Background plants*
A major reason for adding plants to an aquarium is to disguise its artificiality and make it appear larger by hiding the rear and side walls of the tank. But don't hide the glass completely – if small gaps are left, the eye will be deceived into thinking that there is still more space beyond. Tall, fast-growing plants are most suitable for this task and, if kept to the outside edges of the tank, won't obscure shorter plants nor take up any of the fishes' valuable swimming space. These plants will also provide retreats for shy fishes or for harrassed females.

*Space-fillers*
Once the background has been filled, you can use bushy plants to disguise the tank corners, aquarium hardware and any spaces between rocks.

*Foreground plants*
Using low-growing species, it is possible to create the equivalent of a garden lawn in your aquarium. This looks particularly good in front of rocks and caves. Some egglaying fishes use the rigidly held leaves of these plants as spawning sites.

*Specimen plants*
The splendid foliage of these "star" species stands out from other plants. As some species are difficult to keep, you should grow them in individual flower-pots so that they can be looked after more easily.

*Plants for shading*
Rooted species with long leaves which trail along the surface of the water will provide shade, as will plants which float freely on the surface or in midwater.

## NOURISHING YOUR PLANTS

If you want to feed your plants, I would advise the use of liquid or tablet plant foods specially formulated for aquarium use (garden plant foods aren't entirely safe). These can be put into the gravel near to the plants' roots. Alternatively, you can buy nutrient-rich, preformed "plugs" or rooting fibres to put in individual pots.

**Plant plug**
*Specimen plants grow more luxuriantly if given individual attention. They can be planted in nutritious plugs which are then buried in the gravel.*

**Plant pot**
*Another way to encourage faster initial growth, particularly with cuttings, is to use miniature, open-sided pots in conjunction with nutrient-enriched, rooting fibres. The pot holds the fibres together around the stem, so that the plant gets adequate food, but can be moved easily to other positions in the tank.*

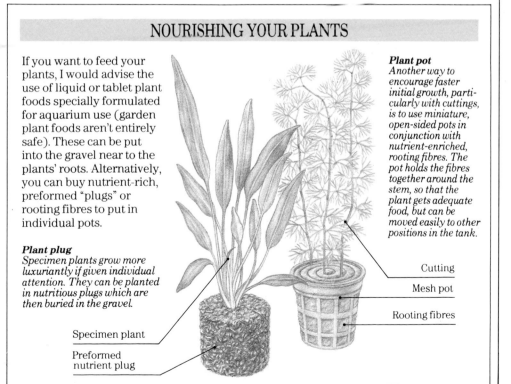

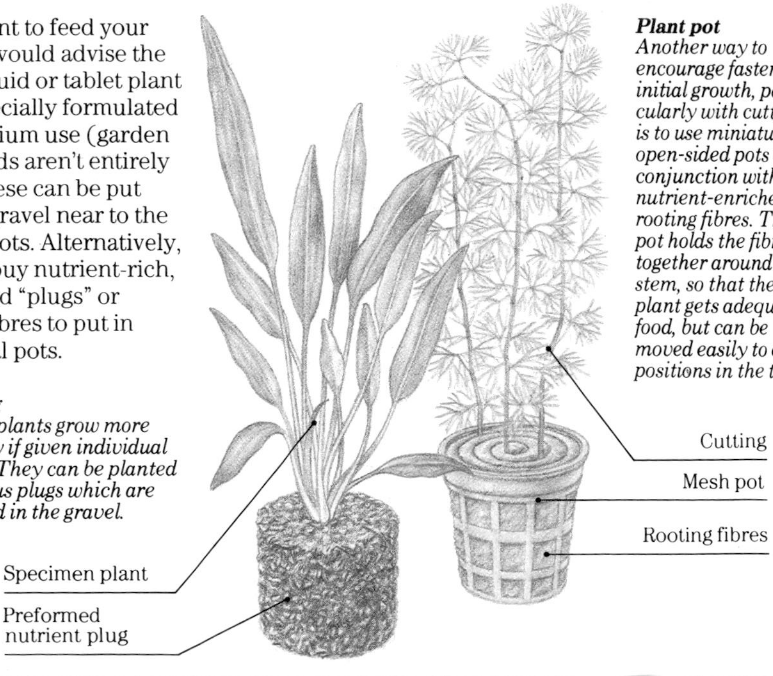

# Types of plants

Aquarium plants can be divided into three distinct groups: rooted plants, floating plants and cuttings.

### Rooted plants

These plants are rooted in the gravel, although they don't draw the majority of their nourishment through their roots as land plants do. True aquatic plants feed mostly through their leaves, and have roots mainly for anchorage in the gravel or to cling onto rocks or logs.

*Reproduction*

Most rooted plants reproduce vegetatively, sending out runners from the main stem which re-root to form young plants. But a few develop "daughter" plants directly on the adult plant's leaf surfaces. You can also increase rooted plants by dividing and replanting the rootstock.

### Floating plants

Plants that float in the water aren't anchored by their roots at all. In the aquarium their growth can be rampant, and you will therefore have to thin them out continually to allow light through to the plants below, or keep vegetarian fishes to eat the surplus. Floating plants have several uses in the aquarium: they bring shade; they provide nest-building material for spawning fishes; and their long, trailing roots make a safe refuge for newly born fry. These plants are liable to suffer scorch damage from lamps so always use a cover-glass to protect them.

*Reproduction*

Floating plants reproduce prodigously — in fact, much too quickly for the fish-keeper's liking. They do so by division, through daughter plantlets or by developing from severed pieces.

### ARTIFICIAL PLANTS

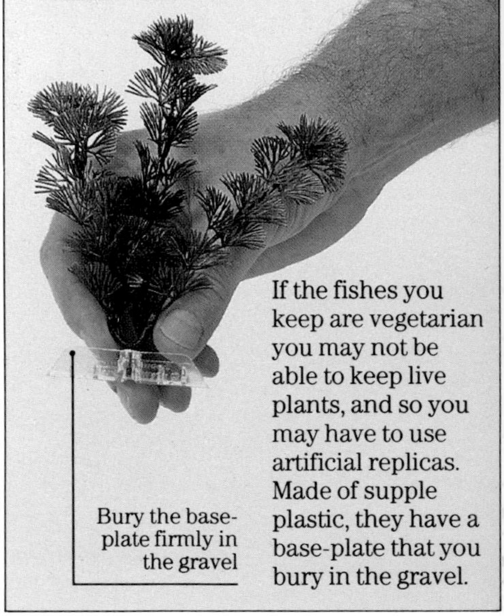
Bury the base-plate firmly in the gravel

If the fishes you keep are vegetarian you may not be able to keep live plants, and so you may have to use artificial replicas. Made of supple plastic, they have a base-plate that you bury in the gravel.

### Cuttings

These plants have been grouped artificially, purely for the benefit of fish-keepers, as many will either root in the gravel or grow floating free. Because they tend to grow very quickly they make good tank fillers.

*Reproduction*

To obtain extra plants take cuttings from the parent plant's top and re-root them in the gravel. The effect of this pruning is that the parent plant becomes bushier and throws out further side-shoots. Some species are so fast-growing that a single leaf will send out roots if floated in water for a short period. Severed cuttings can be bunched and weighted down by tying them together with thin lead strips until rooting occurs, but take care not to crush the stems when attaching the lead.

## PLANT GUIDE

Large aquatic dealers and garden centres will almost certainly stock a wide range of plants suitable for aquarium use. You may also collect plants from local ponds, but be especially careful to clean them thoroughly to avoid introducing disease into your tank. As plants will flourish and multiply at different rates, don't stock your aquarium to the limit right from the outset. A rough guide is to allow one plant for every 25 $cms^2$ of base area. For example, in a tank that is 60 cm long by 30 cm wide, you will need 72 plants.

### KEY

■ **Coldwater**

▒ **Tropical**

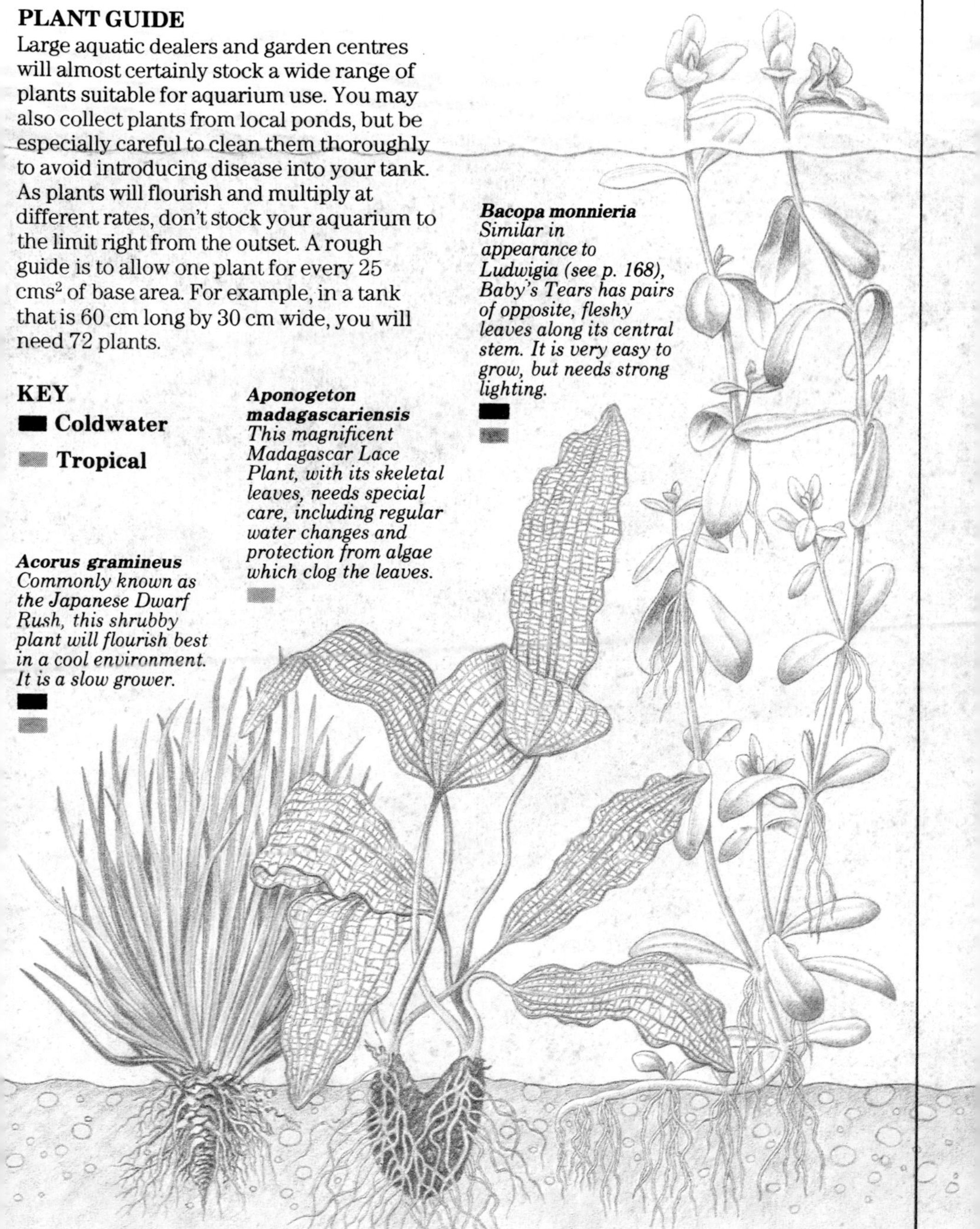

***Acorus gramineus***
*Commonly known as the Japanese Dwarf Rush, this shrubby plant will flourish best in a cool environment. It is a slow grower.*
■
▒

***Aponogeton madagascariensis***
*This magnificent Madagascar Lace Plant, with its skeletal leaves, needs special care, including regular water changes and protection from algae which clog the leaves.*
▒

***Bacopa monnieria***
*Similar in appearance to Ludwigia (see p. 168), Baby's Tears has pairs of opposite, fleshy leaves along its central stem. It is very easy to grow, but needs strong lighting.*
■
▒

**Ceratophyllum demersum**
*This species, known to fishkeepers as Hornwort, has very stiff leaves which are tough enough to withstand attempts by fishes to eat them. It can be propagated by taking cuttings.*

**Cabomba aquatica**
*Although it isn't easy to cultivate, Fanwort is worth growing as a space-filler because of its feathery leaves. Good conditions are important as the leaves can soon become choked with detritus and algae.*

**Cryptocoryne blassii**
*Like other* Cryptocoryne *species, Water Trumpet is often used as a foreground or specimen plant. However, it prefers low lighting conditions and so is best used in a deep tank or under tall plants whose leaves trail on the surface and shade out the available light.*

**Egeria densa**
*Previously known as* Elodea densa*, this fast and densely growing coldwater plant requires regular pruning. Commonly called Water-weed, it can be used in bunches as an egg-trap in a breeding tank.*

**Echinodorus grandiflorus**
*The main feature of this particular species of the* Echinodorus *family, known as Amazon Sword plant, is its large, heart-shaped leaves. It likes soft or medium-hard water, and good lighting.*

**Eleocharis acicularis**
*Hairgrass is an apt name for this plant, as its leaves are very tall, narrow and needle-like in appearance. When planted against a rocky background they look particularly attractive. In order to produce the best growth, good lighting is needed.*

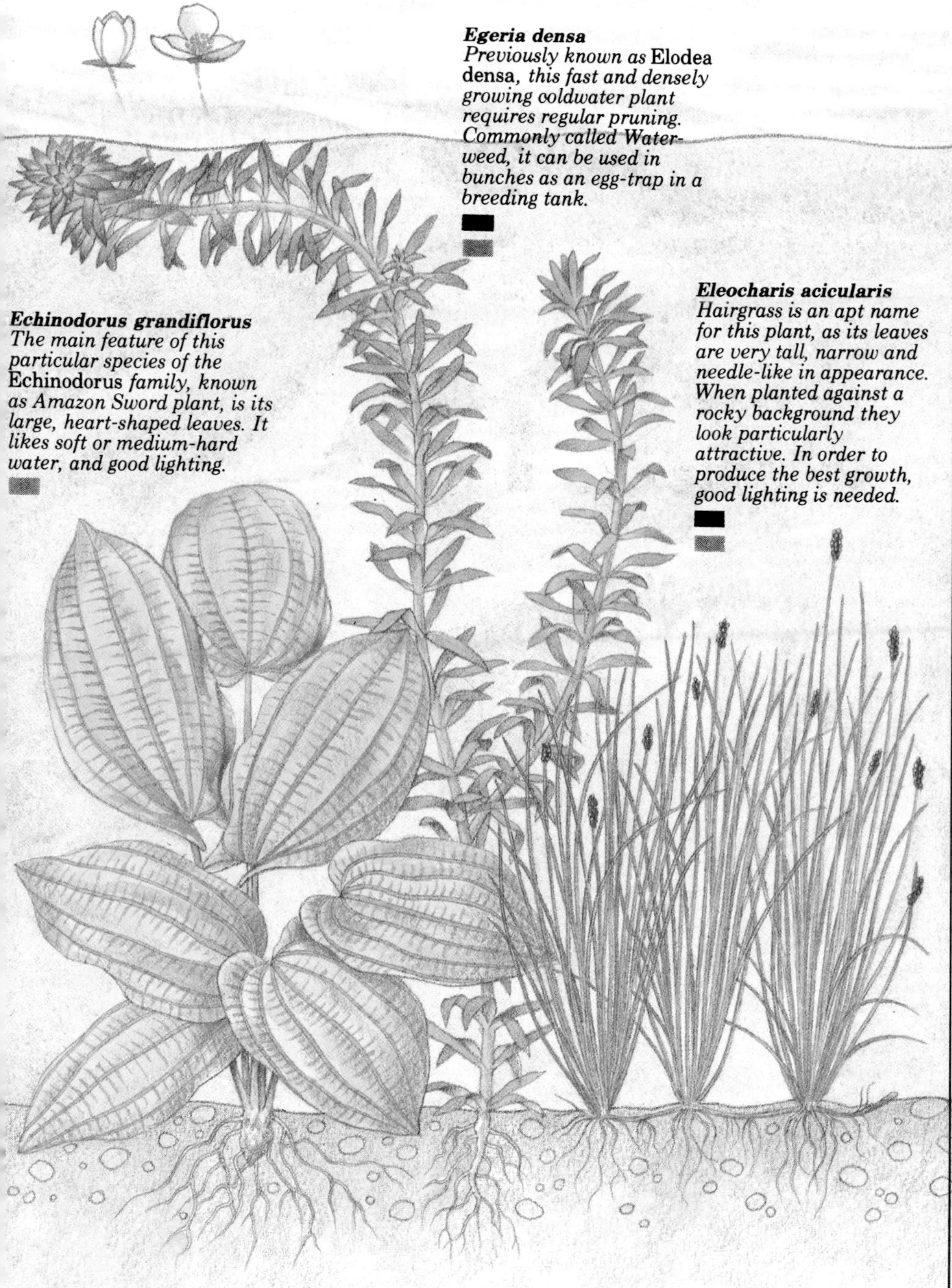

**Elodea canadensis**
*Commonly called Water Thyme or Canadian Pondweed, this aquatic plant is very fast-growing. Although it can be grown in both coldwater and tropical aquariums, it tends to become more spindly in warmer water.*

**Fontinalis antipyretica**
*An excellent spawning medium for coldwater fishes, Willow Moss is a bushy fine-leaved plant which attaches itself to the surface of rocks.*

**Hydrilla verticillata**
*This hardy plant, which doesn't have a common name, has many narrow leaves on its long stems. It is a very fast grower, and may therefore require occasional pruning.*

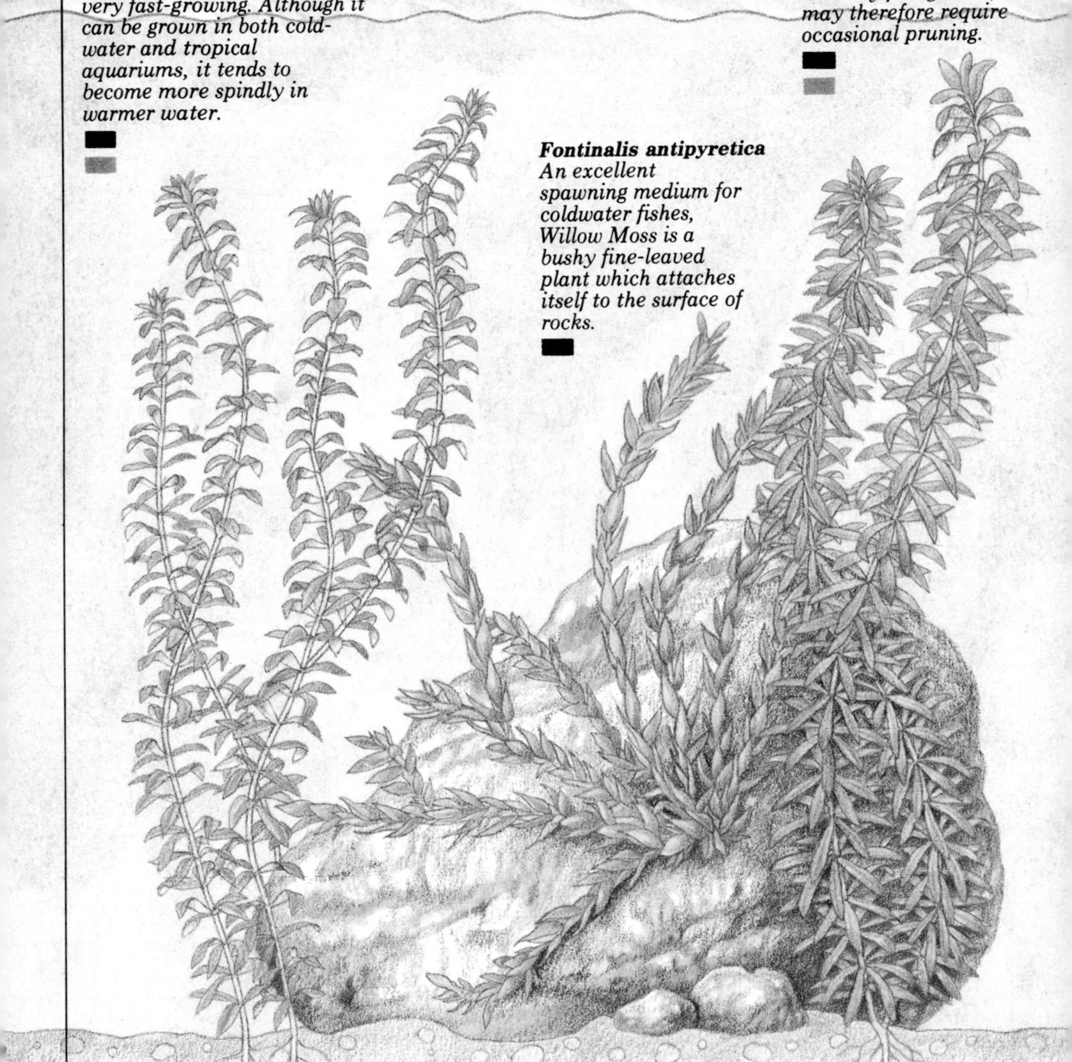

***Hygrophila difformis***
*Water Wisteria (also known as* Synnema triflorum*) has green, ragged-edged leaves, the undersides of which may appear quite white. It is a hardy species.*

***Lagarosiphon madagasericuse***
*This plant is often sold under the name* Elodea crispa. *It likes bright lighting conditions and medium-hard water.*

***Hygrophila polysperma***
*A fast-growing plant, Water Star has narrow oval-shaped leaves, and can become quite bushy. It will flourish if kept in strong light and prefers soft water.*

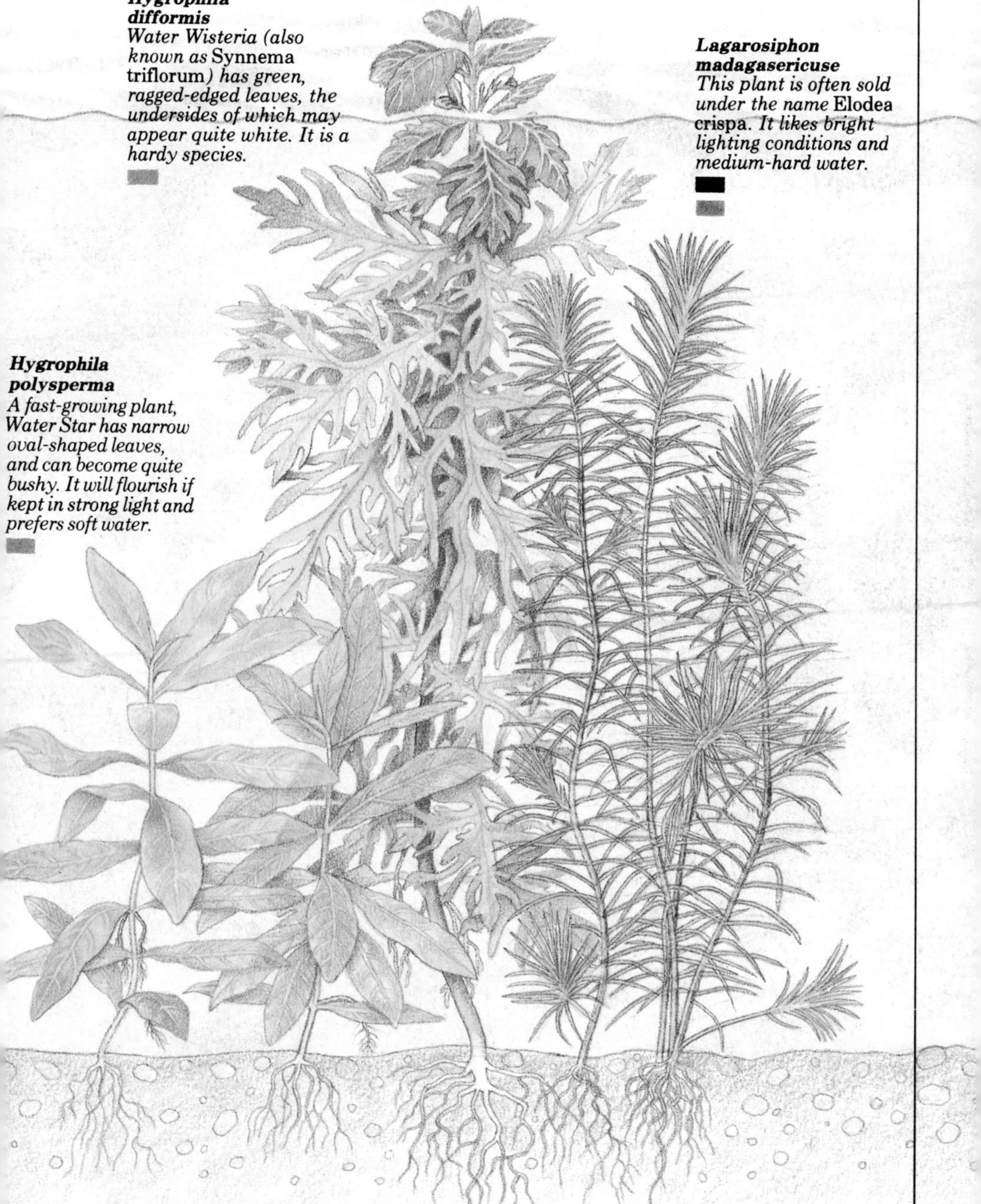

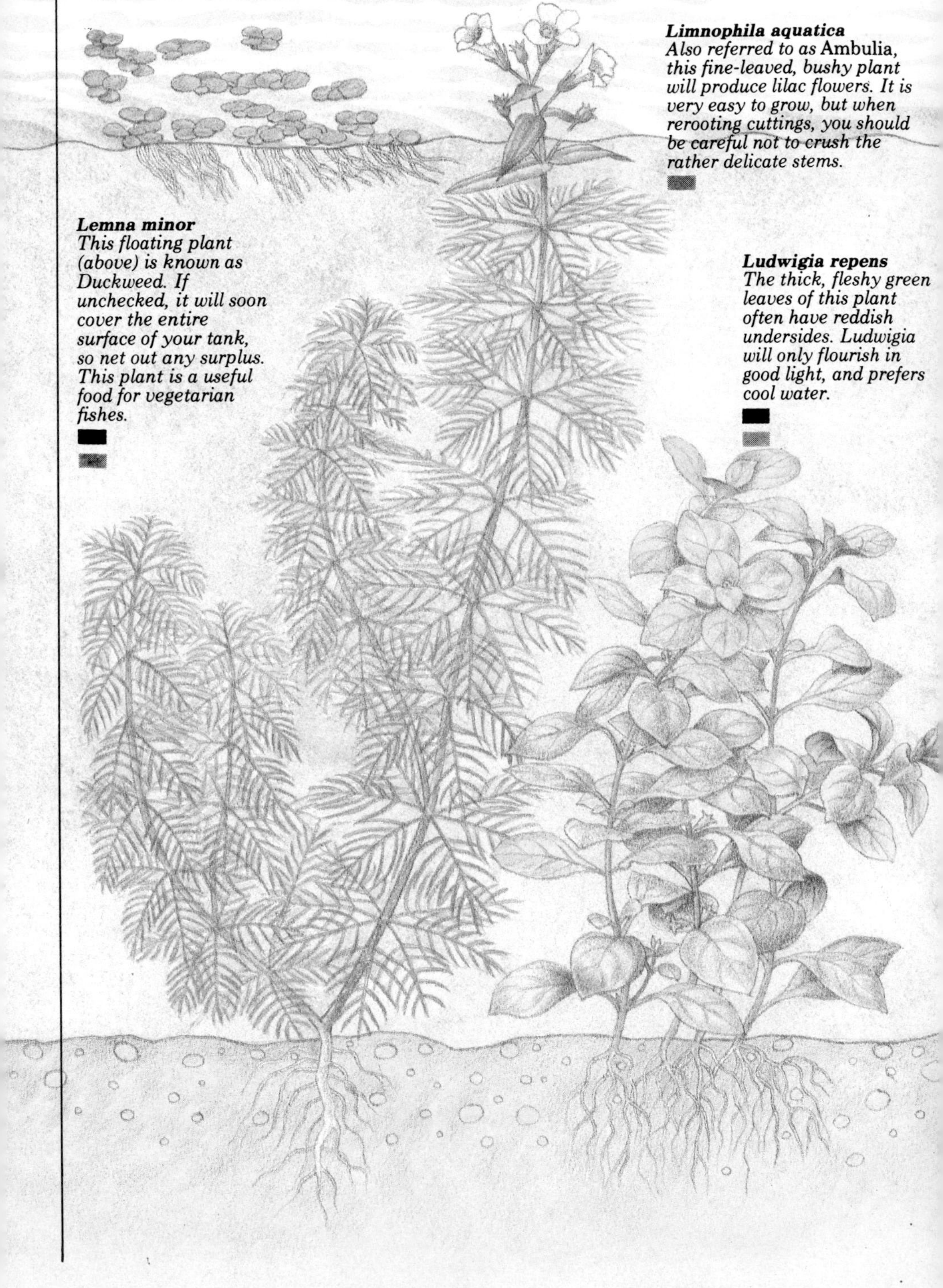

***Limnophila aquatica***
*Also referred to as* Ambulia, *this fine-leaved, bushy plant will produce lilac flowers. It is very easy to grow, but when rerooting cuttings, you should be careful not to crush the rather delicate stems.*

***Lemna minor***
*This floating plant (above) is known as Duckweed. If unchecked, it will soon cover the entire surface of your tank, so net out any surplus. This plant is a useful food for vegetarian fishes.*

***Ludwigia repens***
*The thick, fleshy green leaves of this plant often have reddish undersides. Ludwigia will only flourish in good light, and prefers cool water.*

***Myriophyllum aquaticum***
*Similar in appearance to* Cabomba *(see p. 164), the fine leaves of the Water Milfoil need protection from suspended dirt in the water, provided by good filtration.*

***Microsorium pteropus***
*Java Fern has fairly large, tapering leaves. Its long, hairlike roots don't extend into the gravel but anchor the plant to rocks or logs in the aquarium. It requires only moderate lighting, and is a slow grower.*

***Najas guadelupensis***
*This rather delicate-looking plant has deceptively sturdy stems. It is hardy, and may either root or remain suspended in midwater.*

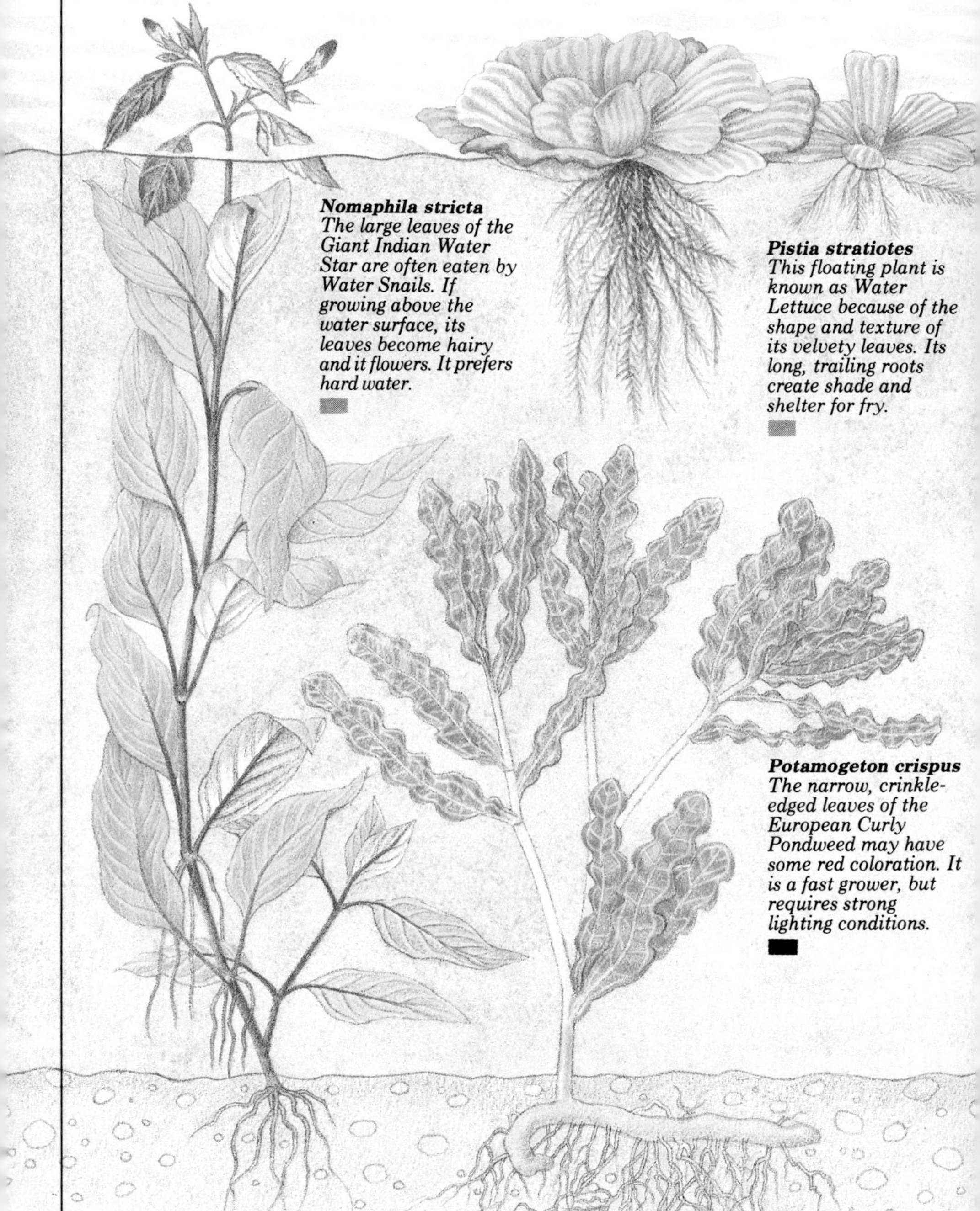

***Nomaphila stricta***
*The large leaves of the Giant Indian Water Star are often eaten by Water Snails. If growing above the water surface, its leaves become hairy and it flowers. It prefers hard water.*

***Pistia stratiotes***
*This floating plant is known as Water Lettuce because of the shape and texture of its velvety leaves. Its long, trailing roots create shade and shelter for fry.*

***Potamogeton crispus***
*The narrow, crinkle-edged leaves of the European Curly Pondweed may have some red coloration. It is a fast grower, but requires strong lighting conditions.*

**Vallisneria gigantea**
*A grass-like species, Giant Eel Grass is quite easy to grow, but likes bright lighting conditions. It reproduces by sending out runners, and is a fast-growing plant.*

**Vesicularia dubyana**
*Java Moss anchors itself on rocks, and is an excellent spawning medium for tropical fishes. Good water conditions are important to prevent this feathery leaved plant becoming clogged with detritus or algae.*

**Sagittaria graminea**
*A large aquarium plant, this species is commonly known as Arrowhead. To produce the best growth, the aquarium water should be slightly alkalinc.*

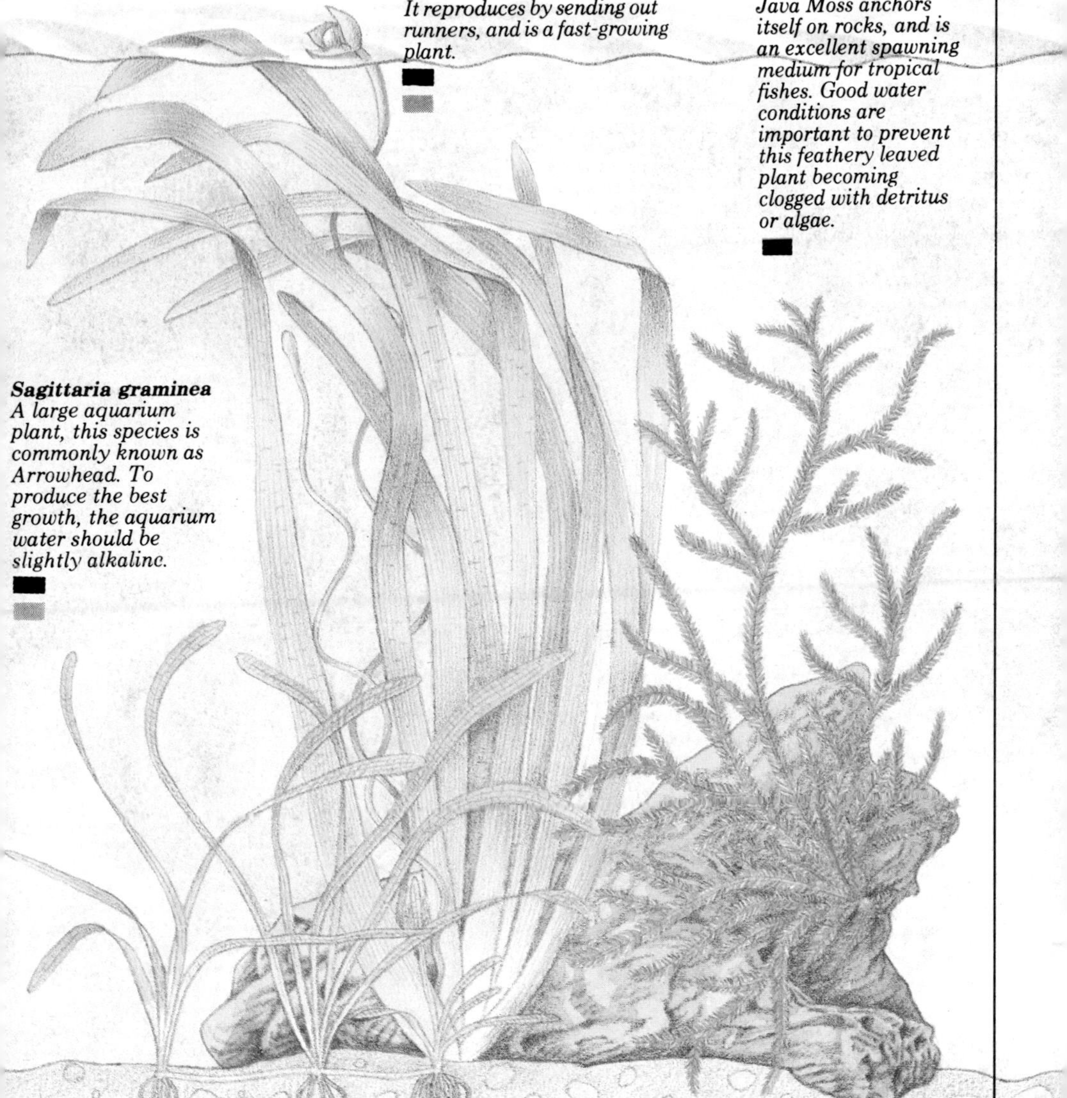

# SETTING UP CASE STUDIES

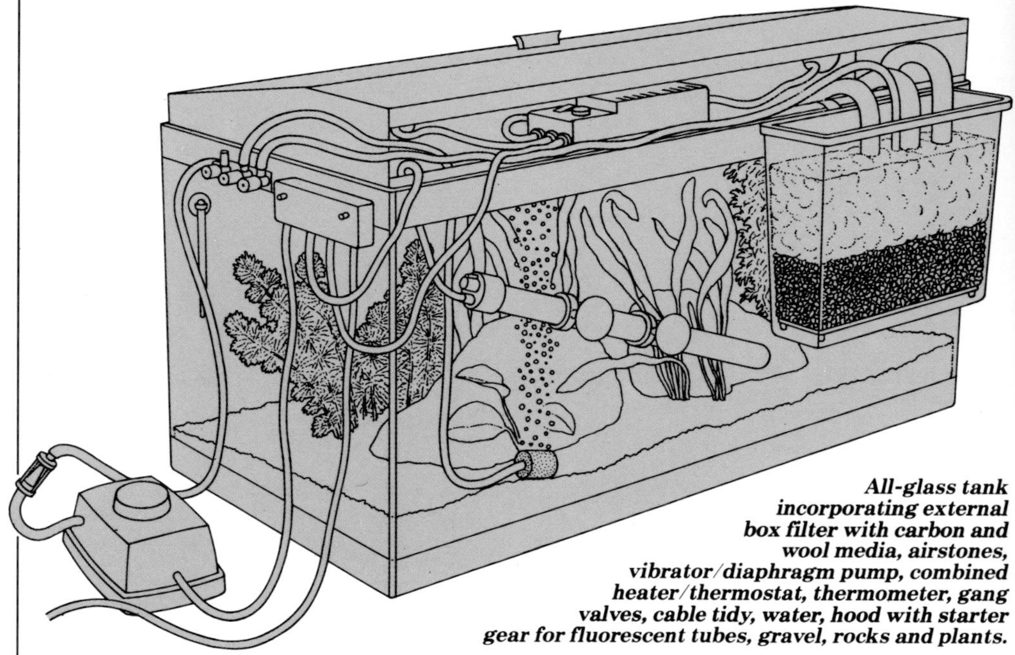

***All-glass tank incorporating external box filter with carbon and wool media, airstones, vibrator/diaphragm pump, combined heater/thermostat, thermometer, gang valves, cable tidy, water, hood with starter gear for fluorescent tubes, gravel, rocks and plants.***

The key to setting up an aquarium successfully is to plan everything thoroughly beforehand. The first step is to choose a system (see pp. 24–5), and to amass all the necessary equipment for it. This will include the correct size of tank (see pp. 115–16), a suitable filter (see pp. 124–9), an airpump (p. 123), a heater and thermostat for tropical systems (see pp. 130–6), lighting (see pp. 148–51), and a hydrometer for marine systems (see p. 147).

If you wish to recreate a natural habitat as closely as possible (see pp. 138–43), choose a suitable geographic location from an area of the world that interests you and find out which species are native to it. Before you buy, you *must* check that all your chosen fishes are compatible (see *Species Guide* pp. 35–111). This is important because fast-moving or territorial fishes can intimidate smaller, more nervous species, and may cause them physical harm. Bear in mind that fishes also have differing needs, and if, for example, they are herbivorous there is little point spending money on expensive plants that will soon get eaten. Not only do the fishes need to be compatible, but everything that goes into the tank including rocks (see p. 157) and plants (see p. 159) must be suited to the particular system you have chosen.

Although you will undoubtedly make adjustments as you set the tank up, planning the aquarium on paper before you begin, including the order of planting, will save you time on the day. Don't be disheartened if your finished tank doesn't look "natural"; it will take a few weeks to settle down and will change as the plants establish themselves.

# Tank designs

The tank layouts shown below are intended to give you basic ideas on which to build up your own individual furnished aquarium. There are three designs for tropical freshwater tanks: one for soft water, one for soft/medium-hard water, and one for hard water. Each is designed to suit the needs of fishes that require those water conditions. Also shown are designs for coldwater freshwater, tropical marine and coldwater marine set-ups.

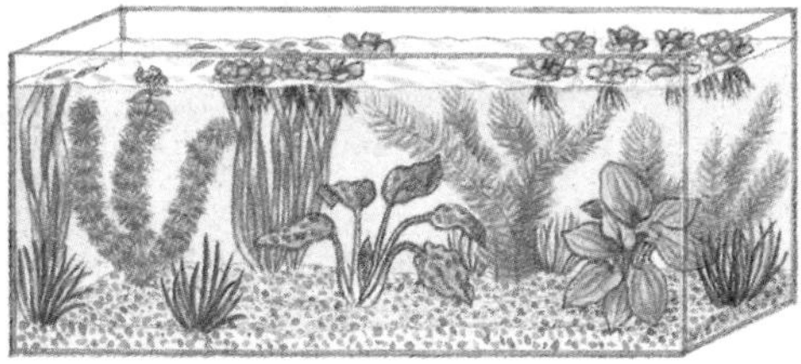

**TROPICAL FRESHWATER**
**Soft/medium-hard**

- *A.madagascariensis* (p. 163)
- *C.blasii* (p. 164)
- *E.acicularis* (p. 165)
- *E.grandifloris* (p. 165)
- *L.aquatica* (p. 168)
- *P.stratiotes* (p. 170)
- *S.graminea* (p. 171)
- Lime-free gravel

**COLDWATER FRESHWATER**

- *B.monnieria* (p. 163)
- *L.repens* (p. 168)
- *V. gigantea* (p. 171)
- Suitable rocks (p. 157)
- Any gravel

**TROPICAL FRESHWATER Soft**

- *C.demersum* (p. 164)
- *L.aquatica* (p. 168)
- *N. guadelupensis* (p. 169)
- Petrified wood
- Bamboo canes
- Willow/alder roots
- Lime-free gravel

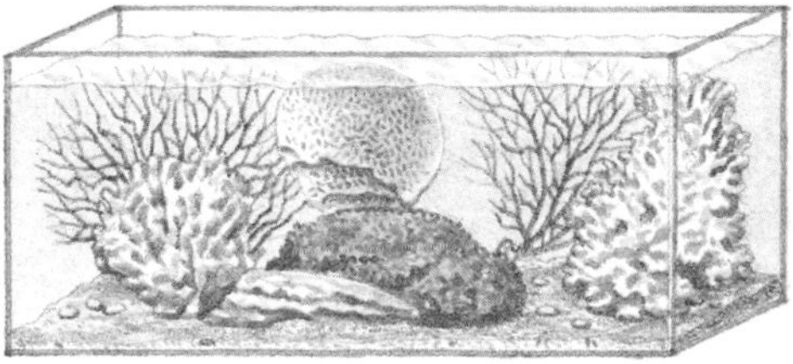

**TROPICAL MARINE**

- Coral heads
- Sea fans
- Living rock or live corals
- Crushed coral sand

**TROPICAL FRESHWATER Hard**

- Slate
- Calcareous rocks
- Ordinary gravel

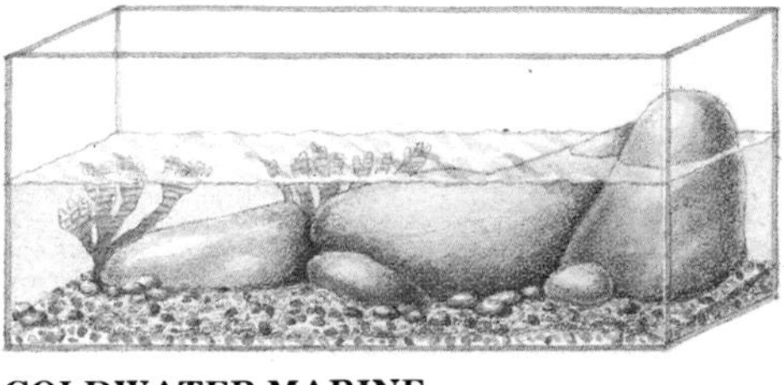

**COLDWATER MARINE**

- Seaweed *(Taonia atomaria)*
- Broken seashells
- Well-worn rocks
- Shingle
- Calcareous gravel

# A tropical freshwater set-up

The tropical freshwater aquarium is probably the best system for a beginner as it is compact, relatively inexpensive, and requires less technical knowledge than other systems.

Once you have decided on a tropical freshwater tank, chosen your site and bought your equipment, your next step is to set up the tank in position at its intended site. Working "on-site" is essential because moving a heavy gravel and rock-filled tank can be very difficult, if not impossible. Stand the tank on a sheet of polystyrene (this absorbs irregularities in the surface below), and make sure that you can get to all four sides easily. When you have the tank in position get the ingredients ready:

- Washed gravel
- Washed plants grouped into planting order
- Tools – pliers, screwdriver, planting stick, sharp knife, scissors and hosepipe
- Technical equipment – filter, airlift, heater, thermostat, airpump, gang-valve, check-valve, airstone, cable-tidy, reflector-hood, lamps
- "Decorating" materials – washed rocks
- Cover-glass
- Fishes

**Warning**: Water and electricity *don't* mix, so never connect up the wires from any electrical equipment to the main power supply until you have filled the aquarium.

**1 Fitting an under-gravel filter** Using a craft knife, trim the airlift tube to suit the water depth of the tank, and slot it into the airlift hole in the filter plate. Then, with the airlift positioned in one rear corner, lower the plate into the tank, making sure that it covers all of the tank base, and that it lies perfectly flat.

**2 Adding the gravel** Use gravel with a particle size of 3–5 mm to allow good water circulation and to enable the plants to root. For the undergravel filter to work successfully, you must cover it with gravel to a depth of *at least* 5 cms at the front of the tank, rising to double that at the back.

**3 Installing the airpump** Attach the pump's outlet to a gang-valve for separate control of the air supply to each piece of equipment. Include a check-valve in the airline to prevent back-siphoning of water. Place the airstone at the back of the tank.

**4 Fitting the heater/ thermostat** Mount a combined heater/ thermostat diagonally on the back glass, keeping the heater clear of the gravel.

**5 Adding rocks** Make sure that the rocks are suitable (see p. 157). Bed large rocks firmly down into the gravel so that they can't topple over. Where possible, use rocks to hide aquarium hardware. If you decide to use an internal rather than an undergravel filter, make sure that the rocks don't impede the waterflow to and from it.

**6 Arranging rocks** Group the rocks so that they look natural, keeping any strata lines running continuously from rock to rock. If you want a "cave" in your tank, you can make one by sticking several pieces of rock together with silicone sealant. Bury small, flat rocks horizontally at intervals in the gravel to maintain its slope.

**7 Filling the tank** There are two important points to remember. Firstly, you should direct the hosepipe over a rock to disperse the water so that it won't spoil the gravel, and secondly you shouldn't fill the tank completely; if you do it will overflow when you put your hands in to add the plants.

**8 Putting in the plants** Using tall specimens, start at the back and sides, then fill in the corners with bushier species. Spread the roots out in the gravel, making sure that you don't bury the crown (the junction between stem and roots). Cuttings may need weighting down with lead. Next, top up the tank with water until the water-line is hidden by the tank rim.

**9 Siting the thermometer** Put the thermometer where it can easily be seen without being too obtrusive. The model shown is stuck to the aquarium glass by means of its rubber sucker. However, it can also be floated upright in the water.

**10 Putting the cover-glass into position** Cover-glasses are either one or two pieces of plain, clear glass or a single pre-formed plastic sheet, as shown here. This type has cut-outs for cables and for feeding access.

**11 Fitting the "cable-tidy"** This tidy is a safe, neat method of connecting and switching the electrical supply to the aquarium equipment. There are separate switches for lighting and airpump circuits (the heating circuit is always "on"). Cables are wired up via a "maze" to prevent disconnection.

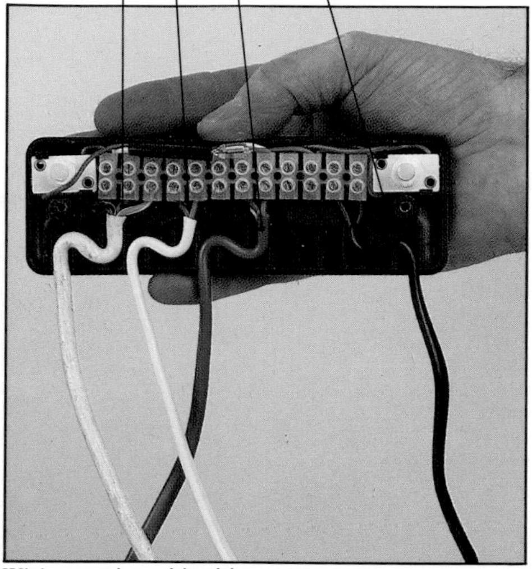

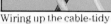

Wiring up the cable-tidy

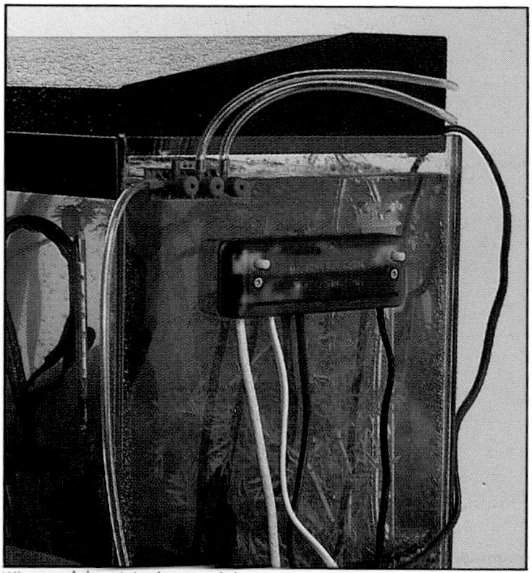
The cable-tidy in position

**12 Setting up the hood and lighting** This is the last stage before you put in the fishes. Some reflector-hoods are designed to house the starting gear for the fluorescent tubes as well as the lights themselves. This will make the hood heavy, so be careful when raising or lowering it, especially if you have a glass cover-glass. In general, fluorescent tubes are used, but you can remove their mounting clips and fit tungsten lamps instead. Next, switch on the heating and the airpump.

**13 Adding the fishes** Float the bag containing your new specimens in the water for 10–15 minutes to equalize the two water temperatures (see p. 29), then release the fishes.

*Vallisneria spiralis* Fast-growing Common Eel-grass is a useful background plant
Male *Brachydanio rerio* (Zebra Danio) see p. 41
Male *Xiphophorus* hybrid (Swordtail) see p. 68
Female *Xiphophorus* hybrid (Swordtail) see p. 68
Male *Poecilia reticulata* (Guppy) see p. 67
Male *Rasbora heteromorpha* (Harlequin Fish) see p. 42
Thermometer (see Step 9)
Loop the airlines to raise them above the tank and prevent water siphoning back into the pump
Gang-valve for airlines (see Step 3)
Cable-tidy (see Step 11)
*Hygrophila difformis* (Water wisteria) see p.167
*Aponogeton madagascariensis* (Madagascar Lace Plant) see p. 163
An airstone in the air-lift tube will improve the rate of waterflow
*Corydoras julii* (Leopard Catfish) see p. 74
*Cryptocoryne blassii* (Water trumpet) see p.164
Male *Xiphophorus variatus* hybrid (Variatus Platy) see p. 69
*Corydoras aenus* (Bronze Catfish) see p. 74
Airline check-valve (see Step 3)
Airpump (see Step 3)

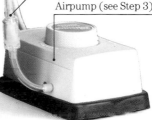

# A coldwater freshwater set-up

The coldwater freshwater system doesn't require as much technology as the tropical freshwater type since no heating is needed. However, keeping this type of aquarium healthy is more difficult, and therefore demands more understanding and experience on your part. Also, because coldwater fishes consume more oxygen than tropical types, a larger tank is necessary. Before you make your choice of system read *Choosing fishes* (see pp. 22–31).

Once you have decided on a coldwater freshwater tank, chosen your site and bought your equipment, your next step is to set up the tank in position at its intended site. Working "on-site" is essential because moving a heavy gravel and rock-filled tank can be very difficult, if not impossible. Stand the tank on a sheet of polystyrene to absorb any irregularities in the surface below. Make sure that you can get to all four sides of the tank easily. When you have the tank in position, you should get all the ingredients ready:

- Washed gravel
- Washed plants grouped into planting order
- Tools – pliers, screwdriver, planting stick, sharp knife, scissors and hosepipe
- Technical equipment – power filter, spray bar, reflector-hood, fluorescent tubes, starting gear
- "Decorating" materials – washed rocks, pebbles
- Cover-glass
- Fishes

**Warning**: Water and electricity *don't* mix, so never connect up the wires from any electrical equipment to the main power supply until you have filled the aquarium.

**1 Putting in the gravel** Use gravel with a particle size of 3–5 mm to allow good water circulation and to enable the plants to root. The gravel should be 5 cm deep at the front, rising in a slope to about 7 cm at the back. To keep the slope in place, bury small, flat rocks horizontally in it at intervals.

**2 Installing the filter** Stand the power filter next to the tank. Place the inlet tube low down near the gravel so that it picks up any dirt. Then fix the spray bar near the top of the tank, so that it will return the clean water over the entire water surface.

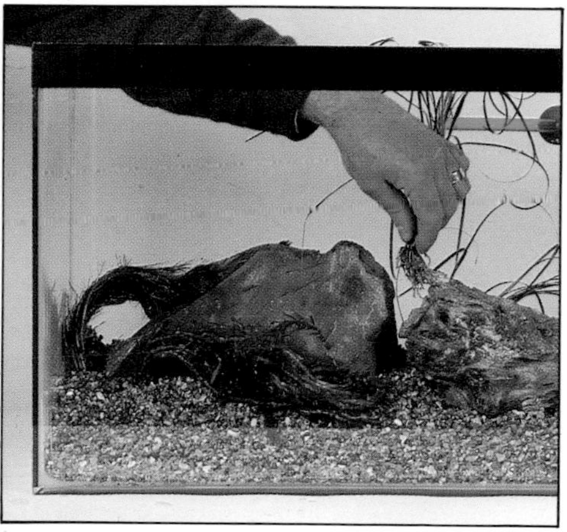

**3 Adding the rocks and plants** Place the rocks in position, bedding them down firmly into the gravel. If possible, arrange them so that they hide the filter tubes. Next, add the plants, making sure that you don't bury the crown (the junction between stem and roots). If you prefer, you can add plants *after* filling the tank.

**4 Adding the water** Using a hosepipe, add water until the tank is two-thirds full and you can see how the plants look. Adjust them, then finish filling the tank.

**5 Priming the filter system** Remove the return hose at the filter end and suck at the filter outlet in order to start the water flowing. Once water emerges from the outlet, fit the return hose back into the filter, then switch the filter on. If water doesn't emerge from the spray bar, there may be an airlock in the filter body, in which case you will have to shake the filter to dispel it.

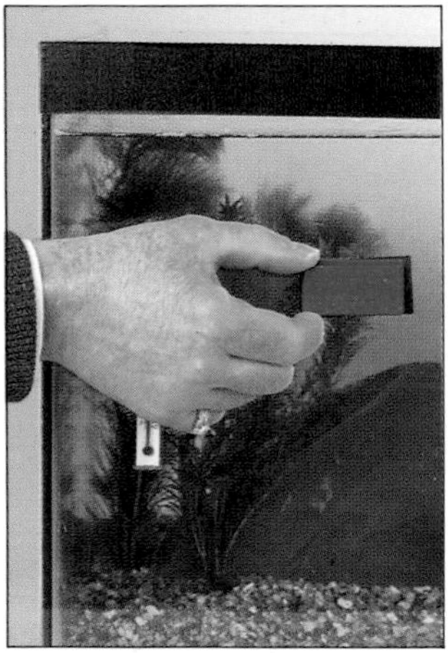

**6 Adding the thermometer and cleaning the glass** Stick the thermometer in position on the glass. As you filled the tank, gassy bubbles may have accumulated on the front glass. If you want to clear these off quickly, you can wipe the glass with a magnetic algae scraper.

**7 Adding the fishes in bags** Fishes are usually supplied in clear plastic bags of water. Before releasing the fishes, float the bags containing them in the aquarium for 15—30 minutes to equalize the temperature of the bag with that of the tank water.

**8 Introducing the tank water** Once the bags have floated in the tank for 15–30 minutes, gently introduce some of the tank water into them. This will help to acclimatize the fishes to the water.

**9 Freeing the fishes** Once the temperatures of the two waters have equalized, you can release the fishes. *Don't* tip them out; tilt the bags gently, allowing the fishes to swim out in their own time.

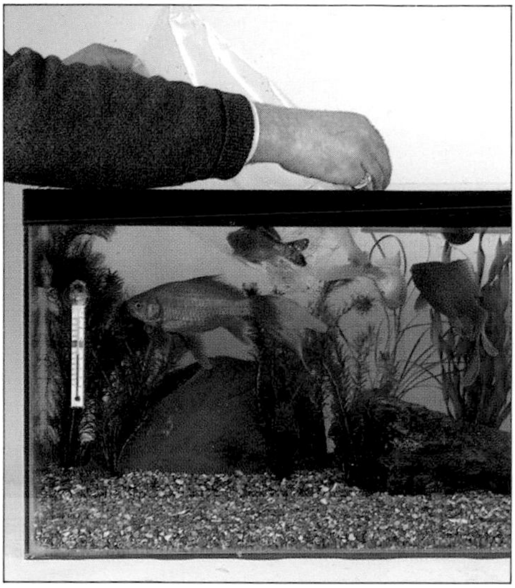

**10 Setting up the hood and lighting** Some reflector-hoods are designed to house the starting gear for fluorescent tubes, as well as holding the lights themselves. The gear is located in a built-in compartment at the rear of the hood. This type of hood will be heavy, so be careful when raising or lowering it, especially if you have a glass cover-glass.

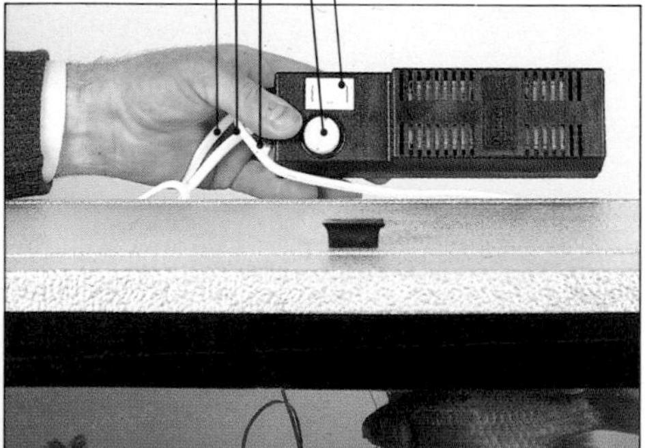

Positioning the starting gear behind the hood

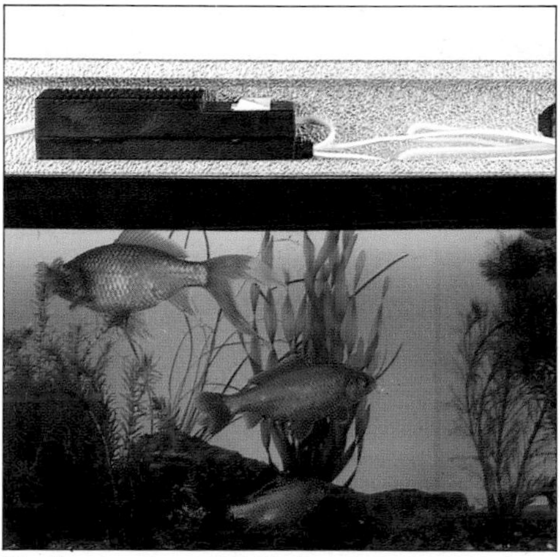
Starting gear in position in its compartment

**11 The finished tank** It will take several weeks for the fishes to "settle in", and for their environment to become established, with algae and beneficial bacteria at the right levels. As this happens, the tank will begin to look more natural.

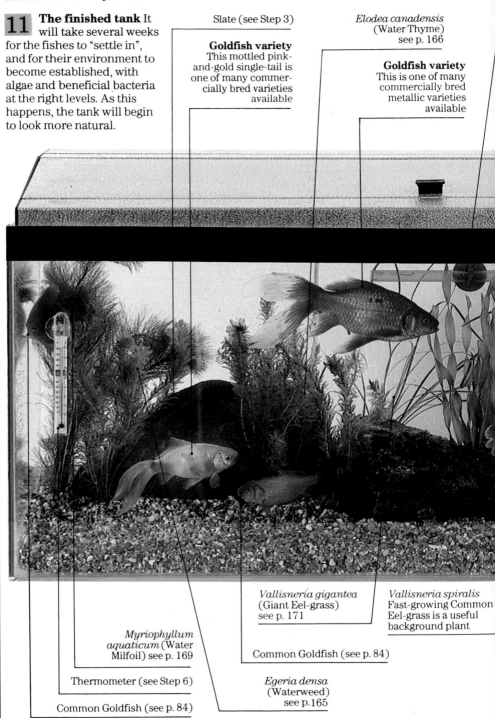

Suckers holding spraybar (see Step 2)
Spraybar (see Step 2)
Reflector-hood (see Step 10)
*Elodea canadensis* (Water Thyme) see p. 166
Inlet tube from tank to filter (see Step 2)
Return tube from filter to tank
Top-plate release clips
Filter motor
Slate (see Step 3)
Activated carbon (see Step 2)
Power filter body (see Step 2)
Artificial rock (see Step 3)
Common Goldfish (see p. 84)
Filter wool (see Step 2)
Common Goldfish (see p. 84)
3-mm particle size gravel (see Step 1)

# A tropical marine set-up

Like its freshwater counterpart (see p. 174), the tropical marine system requires heating technology, but unlike the freshwater system a marine set-up isn't recommended for beginners. Experience is demanded because the water conditions have to be kept within very narrow tolerances; you will have to prepare the salt water to the correct chemical requirement and monitor nitrite levels regularly. Before you choose a system read the *Choosing fishes* chapter, pp. 24–31.

Once you have decided on a tropical marine tank, chosen your site and bought your equipment, your next step is to set up the all-glass tank in position at its intended site. Working "on-site" is essential because moving a heavy, sand and rock-filled tank can be very difficult, if not impossible. Stand the tank on a sheet of polystyrene to absorb any irregularities in the surface below. Make sure that you can get to all four sides of the tank easily and that you can work round it comfortably. When you have the tank in position, prepare the ingredients:

- Washed coral sand
- Synthetic saltwater ("sea-mix")
- Tools – pliers, screwdriver, sharp knife, scissors, plastic bucket
- Technical equipment – undergravel filter plates (two will probably be needed), airlift, electric impeller, filter wool, heater/thermostat unit, plastic mounting clips, hydrometer, reflector-hood, fluorescent tube
- "Decorating" materials – washed rocks, coral heads
- Cover-glass
- Starter fishes – with a marine system start with a few hardy specimens and build up to the full complement of fishes as the tank becomes established

**Warning**: Electricity and water *don't* mix, so never connect up the wires from any electrical equipment to the main power supply until you have filled the aquarium.

**1 Fitting an under-gravel filter** In a large tank like this one you will probably need two filter plates. Trim two airlift tubes to length with a craft knife, then slot one into each hole at the rear of the filter plates. Next, lower the plates into the tank (they should cover the entire base). To improve the rate of water flow in the tank, fit an electric impeller to the top of each airlift.

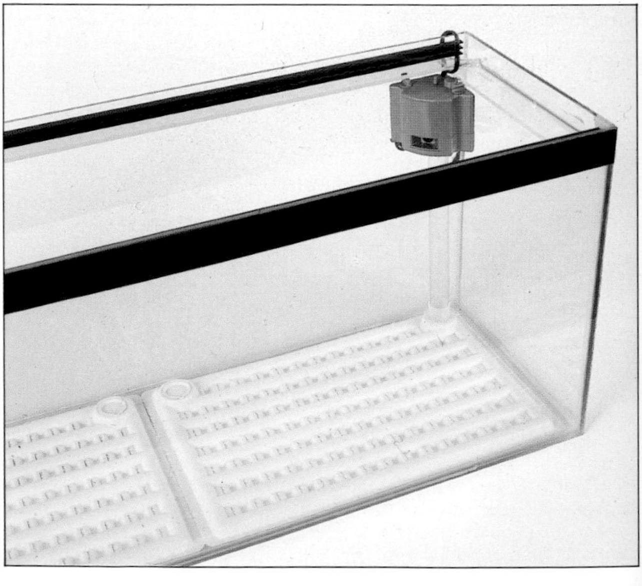

**2 Adding filter wool** Put a layer of filter wool over the entire surface of the filter. This prevents any fine particles of sand from clogging up the slots in the plates and thus impeding water circulation.

**3 Adding the base covering** For the undergravel filter to work well, you must cover it with a base material up to a depth of *at least* 5 cms at the front of the tank, rising to 7—10 cms at the rear.

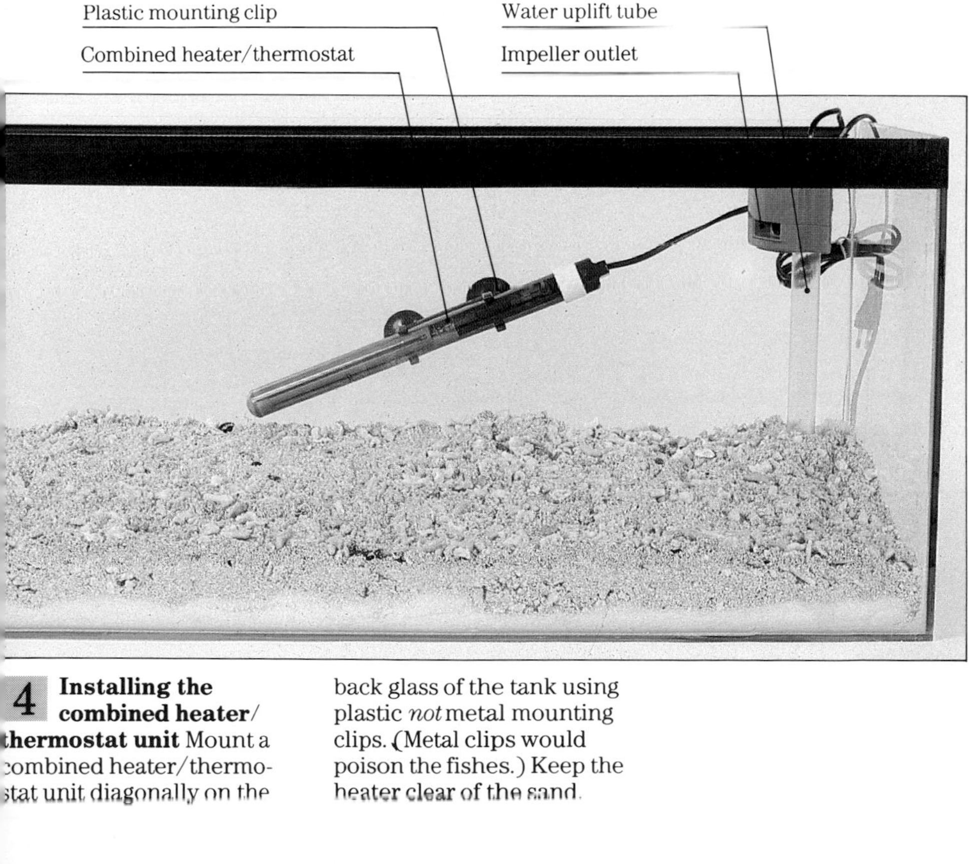


**4 Installing the combined heater/thermostat unit** Mount a combined heater/thermostat unit diagonally on the back glass of the tank using plastic *not* metal mounting clips. (Metal clips would poison the fishes.) Keep the heater clear of the sand.

**5 Adding rocks and corals** Make sure that any rocks you choose match the water conditions in the aquarium (see p. 157). Bed the rocks and corals firmly down so that they can't topple over if the fishes start digging in the coral sand.

**6 Mixing the synthetic saltwater** Follow the "sea-mix" manufacturer's instructions. This will usually involve filling a plastic bucket with tap water and adding the entire packet of mix. Aerate the mixture vigorously for 2–3 hours before you start adding it to the tank.

**7 Adding the sea mix to the tank** Pour the mix into the tank until the water level is 2–5 cms from the top. Set aside the rest of the mix. Switch on the filtration and heating systems. Add the floating hydrometer, then measure the density (see above right).

## MEASURING DENSITY

The "saltiness" or strength of seawater is measured in terms of density or "Specific Gravity" (see p. 147). To check the density of a sea-mix, use a hydrometer.

**Checking and stabilizing density in a new aquarium**
Once the temperature of the water in the tank has stabilized, measure the Specific Gravity (S.G.) on the hydrometer. To read it, check the scale where the surface line of the water crosses it (see p. 147). A coloured zone indicates the approximately correct S.G. value of around 1.020–1.023. If the reading is correct, add the remaining sea-mix to the tank. If it is too high, top up the tank with fresh water to reduce the reading to its correct level; if it is too low, top up with the rest of the mix, then add more synthetic saltwater until the reading is correct.

**Regular density checks**
Once established, you will have to check the density in your tank every 2–3 weeks.

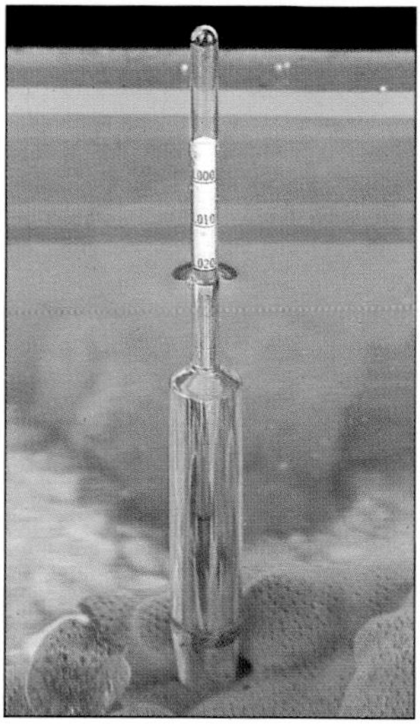

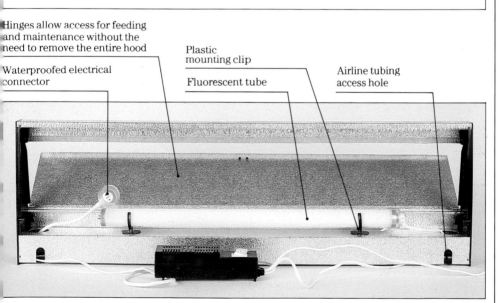

**8** **Setting up the hood and lighting** In a marine tank you must have a cover-glass to prevent condensation dripping off the hood into the water and polluting it with metals. Fix the fluorescent tubes into the hood, making sure that the mounting clips are plastic and that the electrical connectors are waterproofed. Some reflector-hoods house the starting gear for the tubes (see p. 183). This type of arrangement will make the hood heavy, so you must take care when raising or lowering it, especially if you have a glass cover-glass.

**9 Adding the thermometer and the fishes** Stick the thermometer to the inside of the front glass with its rubber sucker. Next, float the bag containing your new specimens in the water for 10–15 minutes to equalize the two water temperatures (see p. 29), then release them.

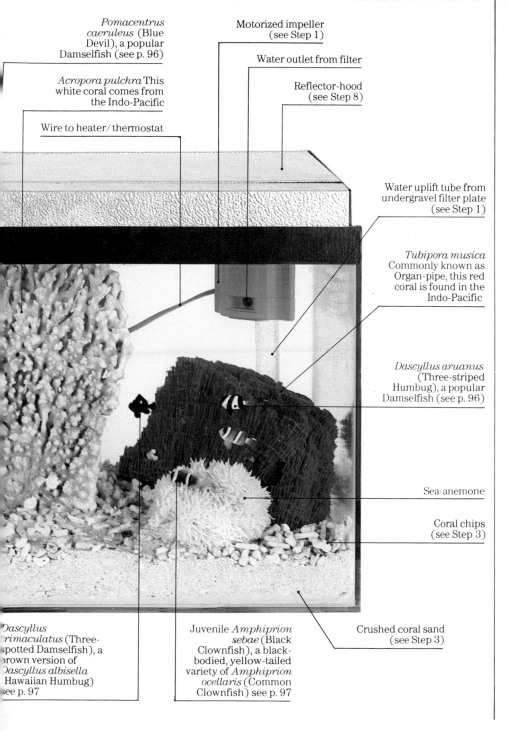

*Pomacentrus caeruleus* (Blue Devil), a popular Damselfish (see p. 96)
*Acropora pulchra* This white coral comes from the Indo-Pacific
Wire to heater/thermostat
Motorized impeller (see Step 1)
Water outlet from filter
Reflector-hood (see Step 8)
Water uplift tube from undergravel filter plate (see Step 1)
*Tubipora musica* Commonly known as Organ-pipe, this red coral is found in the Indo-Pacific
*Dascyllus aruanus* (Three-striped Humbug), a popular Damselfish (see p. 96)
Sea-anemone
Coral chips (see Step 3)
Crushed coral sand (see Step 3)
*Dascyllus trimaculatus* (Three-spotted Damselfish), a brown version of *Dascyllus albisella* (Hawaiian Humbug) see p. 97
Juvenile *Amphiprion sebae* (Black Clownfish), a black-bodied, yellow-tailed variety of *Amphiprion ocellaris* (Common Clownfish) see p. 97

# A coldwater marine set-up

This system can be set up with a very small financial outlay – the fishes can be collected free, and you can use second-hand or home-made equipment. Choose the site for the tank (make sure you can get to all four sides). Next, get the ingredients ready:

- Washed gravel
- Synthetic saltwater ("sea-mix")
- Tools – pliers, screwdriver, sharp knife, hacksaw, scissors, plastic bucket
- Technical equipment – tank sealant, corrugated plastic sheet and piping for DIY undergravel filter, airstone, check-valve, airpump, plastic guttering with two endcaps, lamp fittings and white paint for DIY reflector-hood, lamps
- "Decorating" materials – collected rocks, pebbles and shells
- Cover-glass
- Collected fishes and invertebrates

**COMMON MARINES**

The following species are often found on North Atlantic shores:

- *Blennius gattorugine* (Tompot Blenny) see p. 110
- *Blennius pholis* (Blenny or Shanny)
- *Gobius scorpoides* (Spotted Goby)
- *Taelia felina* (Dahlia Anemone) see p. 110
- *Actinia equina* (Beadlet Anemone)
- *Asterias rubens* (Common starfish) see p. 111
- *Serpula* (Tubeworm) see p. 111
- *Leander serratus* (Common Prawn)

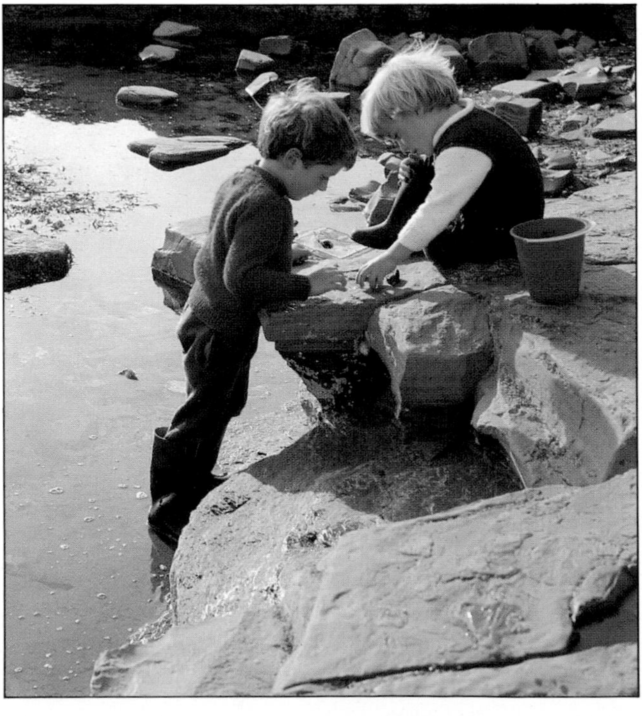

**Collecting at rockpools**

Look carefully: fast-swimming fishes and shrimps are often cleverly camouflaged, whilst closed-up sea-anemones can be difficult to spot. Take safety precautions:

- Seaweed-covered rocks surrounding pools can be slippery at low tide, so don't move too rapidly.
- Always keep one eye on the tide; it may come in more quickly than you anticipate.

There are also ecological rules you should follow:

- Don't collect more specimens than you need.
- If you lift out stones to get at fishes underneath, replace them afterwards as they provide hiding places for fishes brought in by the next high tide.

**1 Resealing a second-hand tank** Working in a well-ventilated area, apply sealant in an unbroken line and smooth with a finger.

**2 Making an under-gravel filter** Using a hacksaw, cut corrugated plastic sheet to fit the tank base. Next, cut slits across the ridges. Turn the sheet over so that the slits are at the bottom, and make a 25 mm hole in one rear corner to take the uplift tube.

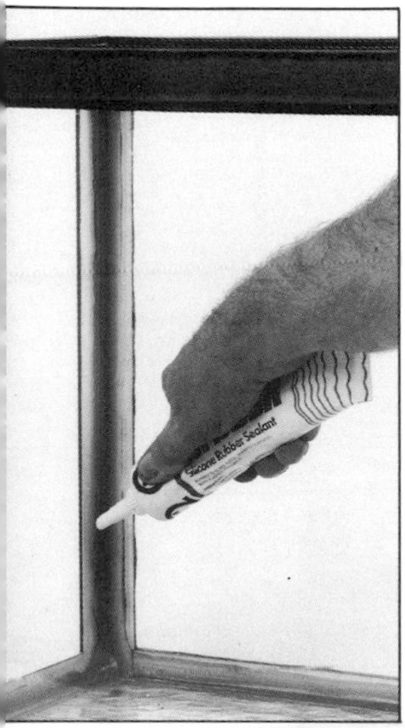

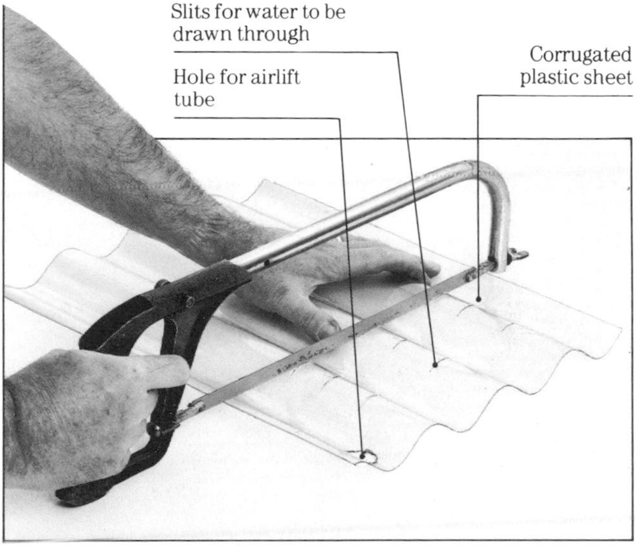


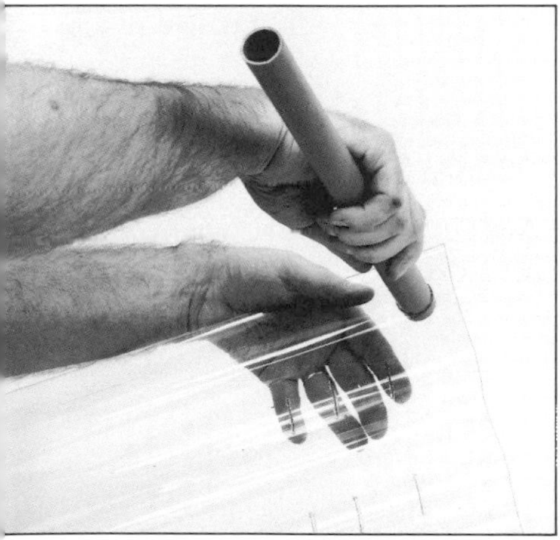

**3 Making an airlift tube** Cut a length of 25 mm plastic pipe to slightly less than the water depth of the tank, and fit it in the hole in the corrugated plastic, fixing it in place with aquarium sealant. (The slits should be in the bottom "troughs", with the tube pointing upwards.) If you wish, fit a plastic elbow to the top of the airlift to direct the filter outflow across the water surface.

**4 Adding the base** Cover the filter plate with 3-mm gravel, topped with shingle. For the undergravel filter to work well, the base material must be *at least* 5 cms deep at the front, rising to double that at the back.

**5 Adding rocks** Arrange suitable rocks (see p. 157) on top of the shingle, bedding them down firmly.

**6 Installing the airstone and airpump** For a high water flow through the filter use an airstone in the airlift, powered by an airpump. As the pump is next to the tank, install a check-valve to protect it from back-siphoned water.

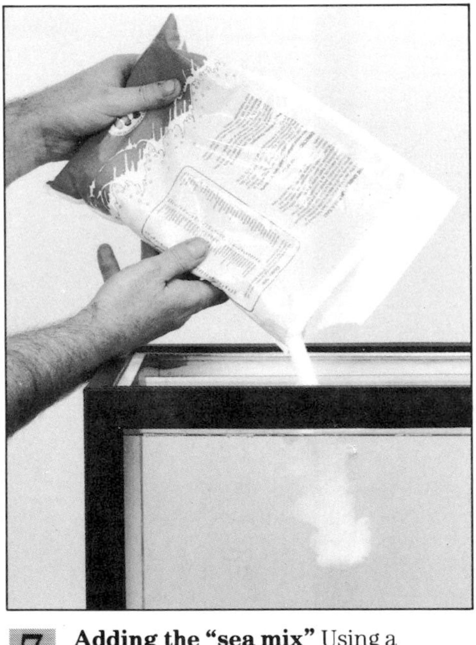

**7 Adding the "sea mix"** Using a hosepipe, fill the tank with tap water, then add the mix and turn on the filter to help the salt dissolve. Check the Specific Gravity (see p. 189) – it should read around 1.023. Aerate the mix vigorously for 24 hours.

Snap-to bucke

Plastic film-covered bucket

**8 Collecting and transporting the fishes** The next stage is to collect your fishes from the seashore (see p. 192). Take two buckets with you – you can't put fishes and invertebrates such as sea-anemones in the same bucket because the fishes will probably be stung to death. If sea-anemones are still clinging to their home rocks, don't pile up the rocks as they might topple over in transit and crush the anemones. To make the buckets leakproof and escapeproof for the return journey, use a snap-on lid or simply cover with plastic film.

**9 Making the reflector-hood** Before you add your finds to the tank, buy or make a reflector-hood. You can make your own by painting the inside of a tank-size length of plastic guttering white or silver.

**10 Fitting the lamp holders to the hood** To add lamp housings to a DIY hood, mount two lamp fittings in standard guttering endcaps, and fit these onto each end of the length of guttering.

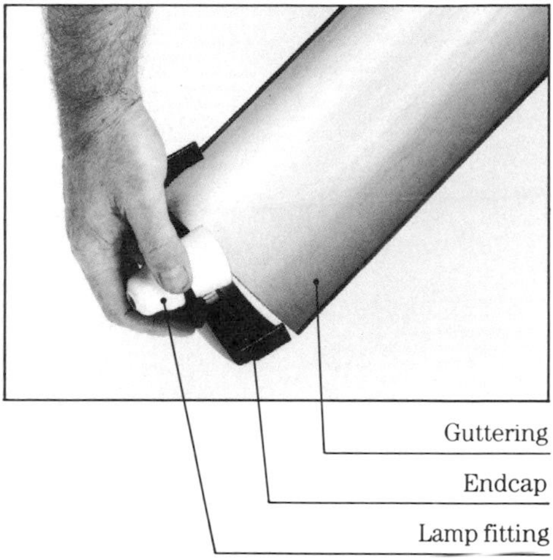

**11 Installing a cover-glass and lamps** Add a cover-glass to the tank; this will protect the lamps from spray (evaporation will only be a problem during warm summer months). To allow the reflector to sit flat on the cover-glass, use slim, candle-shaped lamps or a slim fluorescent tube. You may need to drill small ventilation holes along the top of the reflector-hood to allow the heat from the lamps to escape.

## 12 Adding the livestock

Using a net or plastic bag (see p. 215), carefully transfer the collected fishes and invertebrates one by one from their buckets (see p. 194) to the tank.

*Patella vulgata* (Common limpet) This gastropod is found clinging to rocks in the N. Sea, Atlantic and English Channel

*Actinia equina* (Beadlet anemone) These deep crimson or green sea-anemones share the habitats of the Dahlia Anemone (see p. 110)

*Patella vulgata* (Common limpet)

Fine shingle (see Step 4)

*Leander serratus* (Common Prawn)
*Actinia equina* (Beadlet anemone)
*Buccinum undatum* (Common Whelk) Empty Whelk shells can be found on most European beaches
*Blennius gattorugine* (Tompot Blenny) see p. 110
*Littorina littorea* (Edible Periwinkle) These molluscs are found in the Atlantic, N. Sea, English Channel and Mediterranean. They feed on organic detritus and may help to keep the tank clean
Reflector-hood (see Step 9)
Guttering endcaps (see Step 10)
Airlift tube (see Step 3)
Lamp holders (see Step 10)
Well-rounded, calcareous rocks collected from the beach. Avoid any containing metallic-ore streaks or patches of oil
*Leander serratus* (Common Prawn) Found in rock pools and shallow water in the Atlantic, N. Sea, English Channel and Mediterranean
*Actinia equina* (Beadlet anemone)
*Actinia equina* (Beadlet anemone)
*Ensis ensis* (Razorshell) Discarded shells are readily available at the seashore
Airpump regulator (see Step 6)
Airpump (see Step 6)

# 6

# FEEDING

A hungry aquarium fish can't obtain food by itself unless it turns its attentions to the smaller inmates of the tank Therefore, it is very important that you understand how to provide an adequate food supply for your fishes, as well as what to feed them, since they are totally dependent on you. In fact, a poor diet can endanger fishes' lives.

However, this doesn't mean that their dietary needs are difficult to provide – manufacturers market a wide range of well-balanced, easy-to-use fish foods.

# The basic diet

You must make sure that the type, form and amount of food you provide is right for your particular specimens. If you add too much food, or the wrong sort, the fishes will ignore it, and it will rot, upsetting conditions. Your aim should be to match, as closely as possible, the diet that a fish would eat in the wild.

**The natural eating habits of fishes**

In the wild, fishes' food may be brought to them from outside their watery environment – insects, seeds or fruits that fall into the water. Or it may occur naturally in the water; such food ranges from aquatic worms, crustaceans and insect larvae, to green plant matter, algae, coral heads, molluscs and small fishes.

*Seasonal variations*

Fishes' food supply is affected by the seasons. Insect life, for example, becomes very plentiful when the summer rains cause rivers to flood, and as a result, insects become an important component of fishes' diet at this time. Because plenty of food is available, the breeding season of many species coincides with this type of natural event.

*Feeding level*

Depending on the anatomy of its mouth (see p. 15), a fish may feed from any of three localities. Some fishes feed on the surface of the water, snapping up floating insects; others feed at mid-water, consuming algae or food carried

## VITAMINS

If you give your fishes a balanced diet that includes both processed and natural foods there is no need for you to worry about their vitamin intake. Vitamins are just as important for fishes as they are for humans. In fact, fishes probably require larger amounts of vitamins in proportion to their body size to assist their less efficient biochemical processes.

A balanced diet will contain all the major vitamins:

**A** Crustaceans, egg yolk, green foods

**B1, B2, B6, B12** Algae, green foods, fish meat, cow's liver, beef, eggs, yeast

**C** Green foods, algae, cow's liver, fish eggs

**D** Earthworms, mealworms, algae, seaweeds, snails, water fleas, shrimps

**E** Algae, green foods, egg yolk

**H** Egg yolk, liver, yeast

**K** Cow's liver, green foods, water fleas

| Type |
|---|
| **PROTEINS** **Manufactured** (pellet, flake, tablet and granular) |
| **Meat** (red) |
| **Meat** (white) |
| **Fish** |
| **Insects and crustaceans** |
| **Worms** |
| **FILLERS** **Vegetable matter** |
| **Starchy foods** (potato and cereals) |

along by water currents; and a third type forage on the river or lake floor, digging small animals out of the mud.

**Substitute food for aquarium fishes**
As the majority of tropical fishes are kept outside their native countries, access to their natural foods is impossible, and so you will have to feed them acceptable substitutes. Even for non-tropical species, feeding a totally natural diet isn't usually practical. Foods that are suitable for feeding to aquarium fishes fall into one of three groups: manufactured foods (see pp. 204–5), live foods obtained from nature or cultured (see pp. 206–8), and household scraps (see p. 208).

**Nutritional requirements**
Fishes require proteins, fats, carbohydrates, minerals, vitamins and water, as do other vertebrates, including humans. Proteins provide the necessary material with which a fish builds up muscle, cells and tissue. Carbohydrates provide energy, whilst vitamins and minerals help to build up a fish's health, give it resistance to disease, and strengthen its bones.

*Providing nutrients*
Vitamins and minerals are often added to manufactured, freeze-dried and frozen foods during the preparation process, and, of course, all live or natural foods contain them. Protein and carbohydrate content figures are often quoted on packets of manufactured food, but these can be misleading. The best way to give your fishes a good level of protein, carbohydrate, vitamins and minerals is to feed them a varied diet of good-quality foods.

## THE MAJOR FOODS

| Preparation | Value | Comments |
|---|---|---|
| Ready-to-serve | Formulated to provide a balanced diet, often geared to certain types of fishes | Convenient to use |
| A thin sliver of raw meat should be suspended on a length of cotton | High in protein | Amounts given should be very small |
| Must be cooked, otherwise serve like red meat | High in protein | Amounts given should be very small |
| Small live fishes can be fed. Slivers of raw fish meat are also appreciated | High in protein | Some fishes are cannibals |
| Buy freeze-dried, or catch or culture them | A source of protein | Wild-caught types may carry disease |
| Buy live or freeze-dried, or catch or culture them | High in protein | Wild-caught types may carry disease |
| Chop up large pieces finely | Provides vitamins | Some fishes are vegetarian |
| Chop up large pieces finely | Provide carbohydrate | Particularly useful for vegetarian fishes |

# Choosing pre-packed foods

Fortunately for fishes, things have progressed a long way since the days when most people thought that they could live on an exclusive diet of "ants' eggs". There has been extensive research into the dietary needs of fishes, and today many large concerns supply a wide range of well-balanced, processed foods. Many packed foods are multi-purpose, and can be fed to all types of fishes. Foods for particular species only will specify this on the packet.

## MANUFACTURED FOOD TYPES

Foods made for aquarium fishes are produced in several forms — pellet, flake, tablet or granular. Each type is designed to behave differently in water, so as to suit the feeding actions of different fishes. These foods are made from many ingredients — cereals, meat, fish, and vegetables, plus added vitamins and minerals.

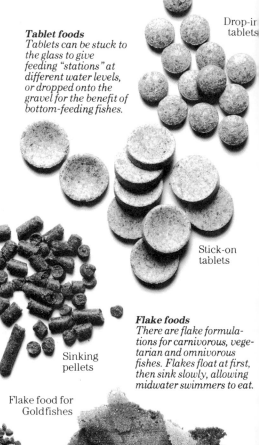

***Tablet foods***
*Tablets can be stuck to the glass to give feeding "stations" at different water levels, or dropped onto the gravel for the benefit of bottom-feeding fishes.*

Drop-in tablets

Stick-on tablets

***Pellet foods***
*Some pellets float, and are aimed at surface-feeding fishes. Other types sink, and are designed for fishes that live at lower levels.*

Floating pellets

Sinking pellets

Tropical freshwater pellets

***Flake foods***
*There are flake formulations for carnivorous, vegetarian and omnivorous fishes. Flakes float at first, then sink slowly, allowing midwater swimmers to eat.*

Flake food for Goldfishes

***Granular foods***
*Because the grains sink very fast; granular foods are very good for bottom-dwellers. These fishes may be nocturnal, so serve granular products last thing at night.*

Granular food

## FREEZE-DRIED FOODS

Freshly caught natural foods that have been dried or frozen for storage can be fed to fishes at a later date, with no degradation of their food values. Like tablet foods, many freeze-dried types can be stuck on the glass or floated in the aquarium. In order to feed them to bottom-dwelling fishes, you will have to put them in a suitable weighted container.

***Tubifex worms***
*These nutritious frozen worms are available either bulk-packed or formed into handy cubes that are sufficient for one feeding.*

***Water fleas***
*The popular* Daphnia *(see p. 204) is available in freeze-dried form.*

Tubifex cubes

Water flea

Black Mosquito larvae

***Black mosquito larvae***
*This food is good for fishes you intend to breed or show, as it improves condition.*

***Bloodworms***
*These "worms" are actually midge larvae, and are very good fish food.*

Krill

Bloodworms

***Krill***
*Because of their size, these shrimps should be broken up if they are fed to smaller fishes.*

***Pacific shrimp***
*This crustacean is very similar to Krill, and can be fed to all types of fishes.*

Brine shrimp

Pacific shrimp

***Brine shrimp***
*These very small crustaceans, properly known as* Artemia salina, *are suitable for all types of fishes, even young fry (see p. 209).*

# Collecting live foods

Fishes derive a good deal of nutritional benefit from natural, live food. They also enjoy hunting and catching for themselves live insects that you introduce into their aquarium. These foods can be found in gardens, ponds, rivers and even rainbutts.

You must inspect collected foods extremely carefully as it is very easy to introduce dangerous predators into your tank. And it is very important to clean Tubifex worms thoroughly before use, and keep them under running water or in a container in the refrigerator until needed. After feeding live foods, you should remove all uneaten matter immediately as these animals will eventually die and cause pollution.

### WATERBORNE FOODS

A pond is the best source of fresh aquatic food. Try to find one with no fishes in it because this lessens the chance of introducing fish diseases into your aquarium. You should use a fine net, and have some snap-top containers ready to store your catch.

***Mosquito and gnat larvae***
*Found in ponds and garden rainbutts during the summer, these larvae hang by their tails at the water surface.*

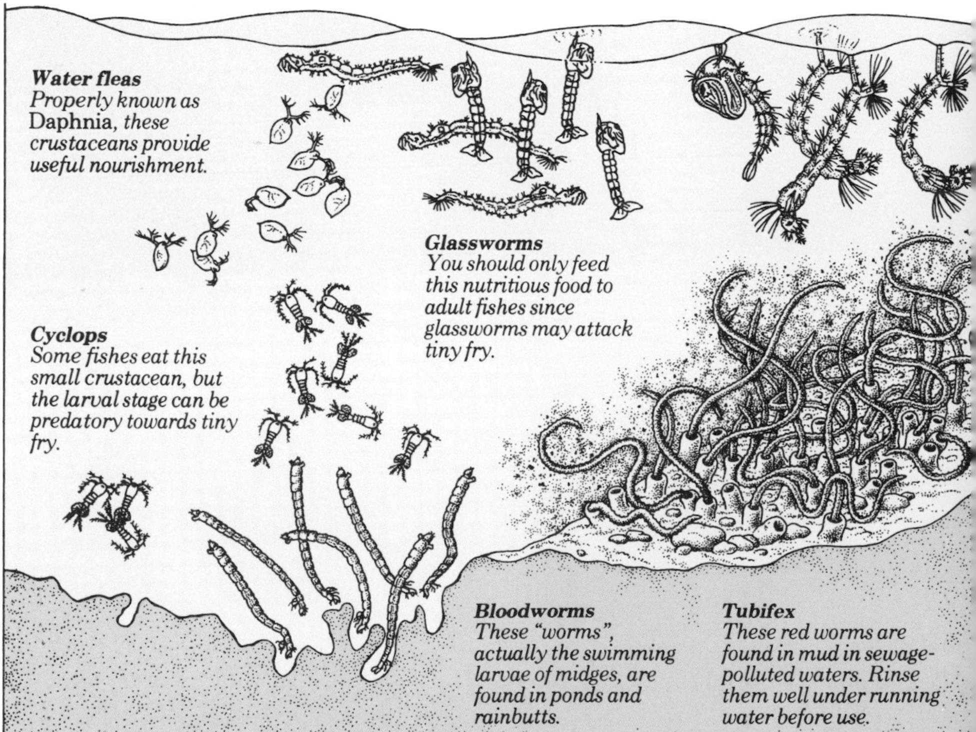

***Water fleas***
*Properly known as* Daphnia, *these crustaceans provide useful nourishment.*

***Glassworms***
*You should only feed this nutritious food to adult fishes since glassworms may attack tiny fry.*

***Cyclops***
*Some fishes eat this small crustacean, but the larval stage can be predatory towards tiny fry.*

***Bloodworms***
*These "worms", actually the swimming larvae of midges, are found in ponds and rainbutts.*

***Tubifex***
*These red worms are found in mud in sewage-polluted waters. Rinse them well under running water before use.*

## TERRESTRIAL FOODS

A range of fish-suitable foods can be found on land. Your own garden or a convenient meadow or river bank are good sources. The compost heap is a particularly fruitful site. You will need to take along some containers with aerated lids to store your catch. Don't use worms or any other live food that you have found in a part of the garden that has been treated with weedkillers or other garden chemicals. And chop up, shred and rinse the larger specimens before using them as food.

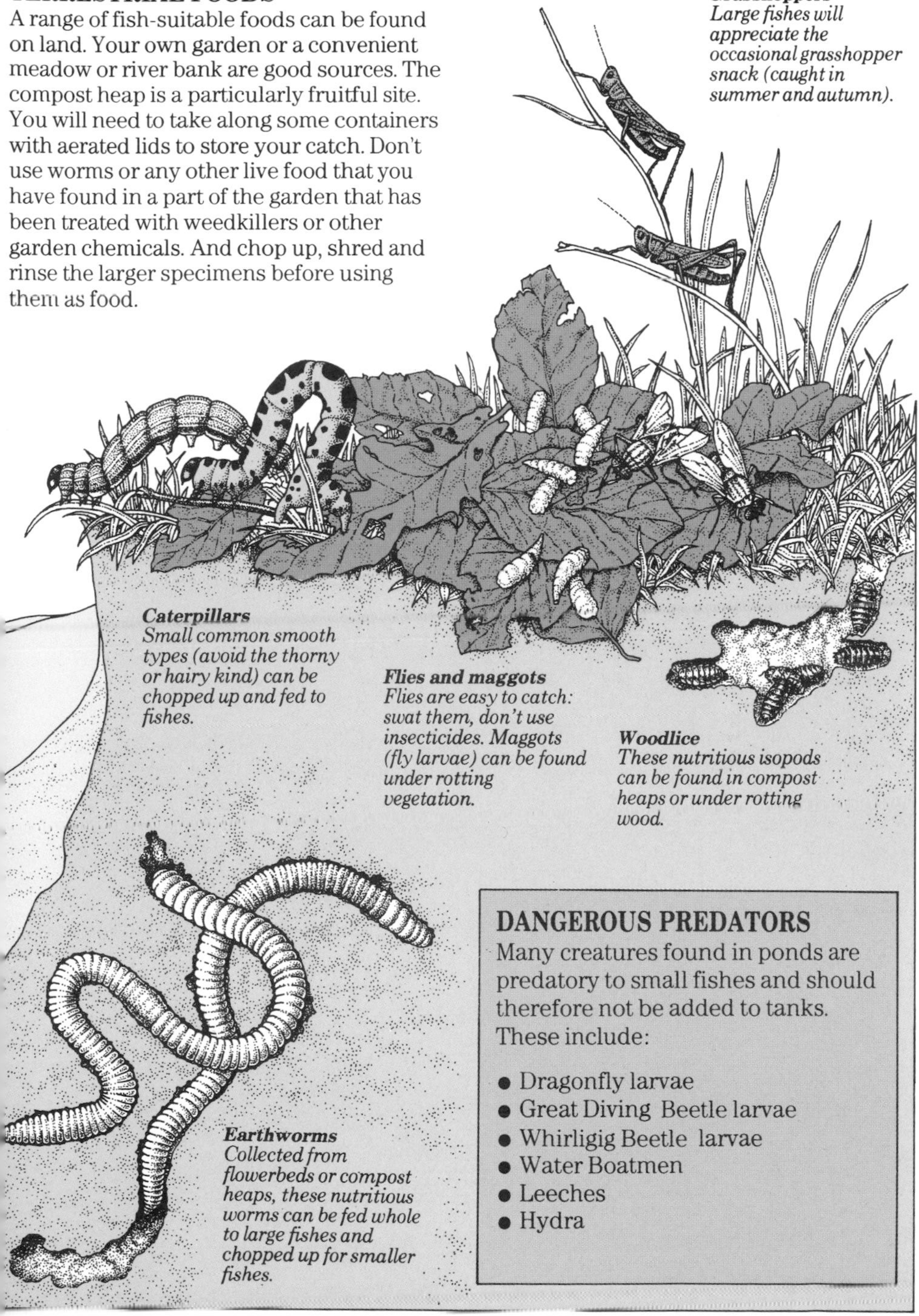

***Grasshoppers***
*Large fishes will appreciate the occasional grasshopper snack (caught in summer and autumn).*

***Caterpillars***
*Small common smooth types (avoid the thorny or hairy kind) can be chopped up and fed to fishes.*

***Flies and maggots***
*Flies are easy to catch: swat them, don't use insecticides. Maggots (fly larvae) can be found under rotting vegetation.*

***Woodlice***
*These nutritious isopods can be found in compost heaps or under rotting wood.*

***Earthworms***
*Collected from flowerbeds or compost heaps, these nutritious worms can be fed whole to large fishes and chopped up for smaller fishes.*

### DANGEROUS PREDATORS

Many creatures found in ponds are predatory to small fishes and should therefore not be added to tanks. These include:

- Dragonfly larvae
- Great Diving Beetle larvae
- Whirligig Beetle larvae
- Water Boatmen
- Leeches
- Hydra

# Other foods

Not all fish foods need come out of a packet or pond; several other types of food are suitable. In fact, a broad diet is very healthy for your fishes. For example, many fishes need green vegetable matter in their diet, and you can provide this by feeding algae (grown in their tank or scraped from another tank) or household scraps (see right). You can also devise your own recipes, mixing ingredients with gelatine in a blender.

### Feeding animal pet foods

Some fishkeepers economize by feeding cat or dog foods to fishes. But until you have gained experience in feeding manufactured fish foods, you shouldn't experiment with other pet products. Some animal pet foods are high in fat, and may pollute the tank.

### HOUSEHOLD SCRAPS

Many items we eat are acceptable to fishes in suitably scaled-down servings.

- Slivers of raw, lean meat (beef, heart, liver)
- Slivers of cooked (cold) white meat such as chicken
- Crumbled cheese
- Finely chopped cooked potato
- Slivers of raw fish meat
- Fish roe
- Slivers of shellfish meat
- Chopped or shredded lettuce
- Chopped spinach leaves
- Tinned peas
- Wheatgerm
- Oatflakes

## CULTURING LIVE FOODS

Small worms and shrimps can be cultured to provide a disease-free, year-round supply of live food. To start a worm culture obtain a sample (see p. 205), then follow the instructions below. They should breed rapidly, providing a continuous supply. A shrimp culture uses shop-bought eggs, which are hatched out.

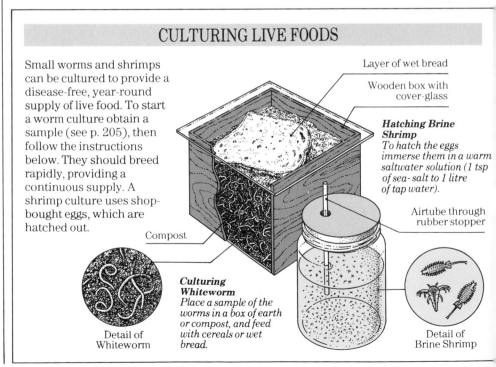

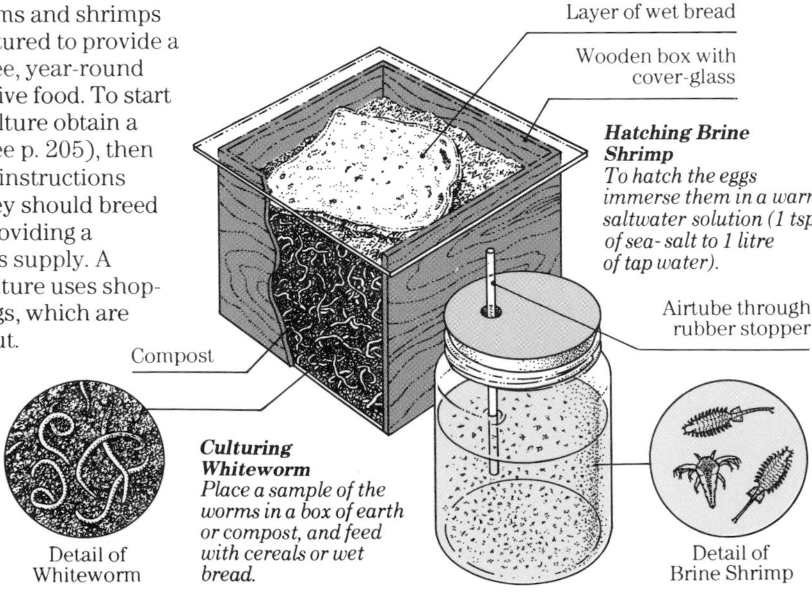


***Hatching Brine Shrimp***
*To hatch the eggs immerse them in a warm saltwater solution (1 tsp of sea-salt to 1 litre of tap water).*

***Culturing Whiteworm***
*Place a sample of the worms in a box of earth or compost, and feed with cereals or wet bread.*

# Feeding methods

One of the secrets of good fishkeeping is to feed sparingly. Fishes will stop eating when they've had enough, and any uneaten food will begin to decompose in the tank and eventually cause pollution problems.

### Judging the correct amount of food

You shouldn't give fishes any more food than they will eat in a few minutes. Although it is very tempting to give them more, especially if they are up at the glass and seem to be begging for food, you mustn't give in. In fact, it is a good idea to keep your fish slightly hungry, so that they will devour every scrap of food and thus keep the tank clean. A good method for measuring out the correct amount of food for one meal is to give only as much as you can pinch between your thumb and your middle fingertips. If you have a community tank, you will find that it is impractical to make up a meal containing small amounts of different foods for each species, so the best thing to do is to feed tablets at one feed, flake at another, live food at another, and so on.

**FEEDING DON'TS**

- Don't overfeed; make sure that you know the fishes' natural food requirements
- Don't stick to a set diet or to one type of manufactured food only
- Don't feed wild-caught aquatic live foods to fishes without screening for predatory insect larvae first (see p. 205)
- Don't neglect nocturnal species; feed some food just after "lights out"
- Don't give fishes Whiteworms too often as they are very fatty
- Don't forget to clean Tubifex worms before using them as food

### Frequency of feeding

Feeding fishes "little and often" is preferable to providing them with one large daily meal. Add small amounts of food two or three times a day, and try to vary each meal so that your fishes don't get bored with their diet. If possible, give some live food as a regular part of your feeding programme; your fishes will appreciate it, and they will repay you by looking their best.

If you share your aquarium with the rest of the family, make sure that you have set feeding times, or that everyone else knows when the fishes have just been fed, so that you don't over-feed between you.

### Coping with feeding problems

Sometimes, your fishes will refuse to eat a nutritious item that they haven't encountered before. If this happens, you will have to educate them into accepting the new food. The best way to do this is to keep them hungry for a day or two, and then try them with the new food gradually, instead of switching over to it suddenly. You should watch the fishes closely to check that they are actually eating the food, not just tasting it and spitting it out again.

### Vacation feeding

Unlike other pet-owners, fishkeepers don't have a problem at vacation time; as long as fishes have been well-fed in the period immediately prior to the holiday, they will be able to withstand a two-week fast without any ill-effects. This is a cheaper solution than buying an expensive automatic feeder, and safer than handing over responsibility to a neighbour who is new to aquariums. If you decide to trust a neighbour, you should make up several small packets of fish food in the correct amount for one serving, and tell your helper when to feed each "packet meal".

**Ensuring thorough feeding**

Some fishes will miss out on a meal because they can't reach food at a particular level in the water. However, if you plan carefully, you can make sure that all your fishes are fed. Various devices are available to help you, including a worm feeder (shown below) for keeping worms at the top of the water, and weighted perforated baskets that are used for sinking freeze-dried food to the tank floor.

## FEEDING FISHES AT DIFFERENT LEVELS

Some fishes' swimming style means that they can't reach food at every level in the water. Top-swimmers often miss worm foods, because these normally sink quickly; similarly, bottom-dwelling species can lose out on freeze-dried foods, as these tend to float. Take these factors into account when selecting foods.

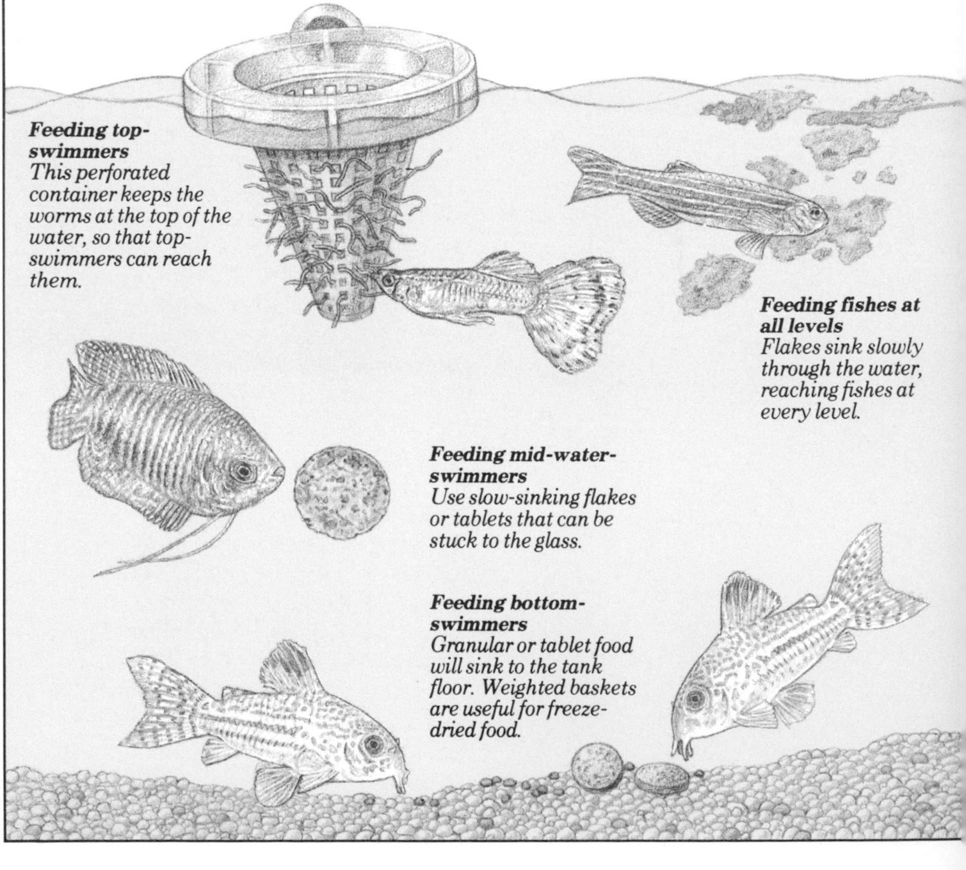

**Feeding top-swimmers**
*This perforated container keeps the worms at the top of the water, so that top-swimmers can reach them.*
**Feeding fishes at all levels**
*Flakes sink slowly through the water, reaching fishes at every level.*
**Feeding mid-water-swimmers**
*Use slow-sinking flakes or tablets that can be stuck to the glass.*
**Feeding bottom-swimmers**
*Granular or tablet food will sink to the tank floor. Weighted baskets are useful for freeze-dried food.*

# Feeding young fishes

Most fish fry require the tiniest morsels of food at first. Fry of egglaying species need smaller food than fry from livebearers.

### Live food for fry

One of the best live foods to start tiny fry on is *Infusoria* (or "green water"). There are two ways of obtaining this: you can collect it from an established pond using a fine muslin net, making sure that you search the catch for predatory larvae before feeding it to your fry, or you can culture it using special tablets. These tablets are more successful than the old-fashioned method of pouring boiling water on to something like potato peelings, or hay, and then seeing what develops. Other suitable foods are Micro-eels (cultured in a syrup solution) and Microworms (cultured in a box, see p. 206). Starter cultures can be obtained from a dealer or a fellow aquarist.

Fry are also able to eat another highly nutritious food: the newly hatched young of the Brine Shrimp, *Artemia salina*. The Brine Shrimp lays eggs which can be stored in a dry state for many months, even years. When these are placed in a saltwater solution (1 tsp of sea-salt to a litre of tap water), they hatch in two days.

With all cultured foods, a succession of cultures is necessary to maintain a good supply for your fishes.

### Manufactured fry food

Proprietary brands of fry food are available either in a liquid or a powder form. They are made in two basic formulations: one sort suits the dietary needs of egglaying fish fry, and the other type is designed for livebearing fish fry which need a higher proportion of vegetable matter.

### When to start feeding fry

Livebearer fry can take powdered food immediately, and they also relish Brine Shrimp. It is difficult to judge when to begin to feed fry from egglaying species. If you start too early, they won't need it, and the tank may become polluted; but if you leave it too late, the fry will starve. Watch the fry closely, and as soon as they have finished absorbing the nourishment from their attached yolk sacs, start giving food. Make sure that you time your food cultures carefully so that some is ready at the right moment.

### Feeding frequency

Young fishes should have a supply of food available at all times. The easiest way to provide this is to set up a header tank of *Infusoria* (this is simply a container and siphon tube arrangement that uses gravity to deliver the culture). To do this, take a glass jar and fill it with the culture, then set up the jar so that its base is higher than the top of the tank. Run a length of airline tubing from the bottom of the jar into the tank, and fix a clamp just above the end that enters the tank so that the culture drips in slowly.

You can make sure that the fry are eating enough by encouraging feeding activity; the simplest way to do this is to leave the light over the fry tank on.

### Introducing the adult diet

After a few days, you can start to feed fry with mashed-up dried foods. And as the fry grow, gradually increase the size of the food particles, estimating what the young fishes will be able to get in their mouths. You should also give regular partial water changes.

# 7

# HEALTH CARE

When you first selected your fishes you will have followed the advice in *Choosing fishes* (see p.24), and made every effort to choose strong, healthy specimens. However, isolation in the "safety" of the aquarium won't protect your fishes from disease. The two main causes of ill-health in fishes are stress and bad aquarium management; the first section of this chapter – *Preventing Disease* – provides information on avoiding these. If, despite proper care, disease strikes, the second section of this chapter – *Ailments and Disorders* – covers the diagnosis and care of sick fishes.

# What is a healthy fish?

Fish-watching is the first step in health care. A regular observation routine will help you in two main ways: you will know straightaway if any fishes are sick, and you will be aware of any tasks that need to be done to maintain the aquarium in good condition. For examples of the difference between healthy and unhealthy fishes see pp. 26–7 and 260–1.

### Observing fishes

Start by checking regularly that all your fishes are present. You can carry out this "roll call" quite easily at feeding time, and you will soon notice any absentees. Don't panic the first time you can't see all your fishes at once. If you have stocked the aquarium with fishes that occupy all levels in the tank, the nocturnal specimens may be hiding away during the daytime. If you do miss seeing a fish for 2–3 days, then you should investigate. It may have died, got trapped behind a rock, been eaten by a larger fish or even jumped out of the tank. Whatever the cause for its disappearance, you must find it as its corpse may pollute the tank.

### Looking out for bad habits

Watch out for any aggressive or undesirable traits in your fishes, for example bullying or fin-nipping. Often, an anti-social fish can be cured by a brief spell in "solitary" – a small jar floated in one corner of the tank. However, some fishes behave this way out of loneliness or boredom when deprived of the companionship of their own species. If this is the case, try providing it with enough companions before you sentence it to solitary confinement.

## HOW TO USE THIS CHAPTER

**Diagnosis**
To find out what your fish's symptoms mean, you should start by looking at the *Diagnosis charts*:

- Visual diagnosis of health problems (see p. 221)
- Skin problems (see pp. 222–3)
- Swimming problems (see pp. 224–5)

These charts will lead you to the appropriate text box in the *Ailments* section (see pp. 226–35).

**Prevention**
If you wish to keep your fishes healthy, read *Preventing Disease* (see pp. 213–9). Information on looking after plants is given on p. 219.

**See also:**

- Eye problems (see p. 228)
- Poisoning (see p. 214)
- Caring for sick fishes (see pp. 234–5)

*Parasites*
Anchor Worm (see p. 236)
Coral Fish Disease (see p. 231)
Fish Louse (see p. 232)
Freshwater White Spot (see p. 230)
Gill Flukes (see p. 231)
Marine White Spot (see p. 230)
Skin Flukes (see p. 231)
Slime Diseases (see p. 232)
Velvet Disease (see p. 231)

*Diseases*
Finrot (see p. 233)
Mouth Fungus (see p. 233)
Septicaemia (see p. 229)
Tuberculosis (see p. 229)
Fungus (see p. 233)

# PREVENTING DISEASE

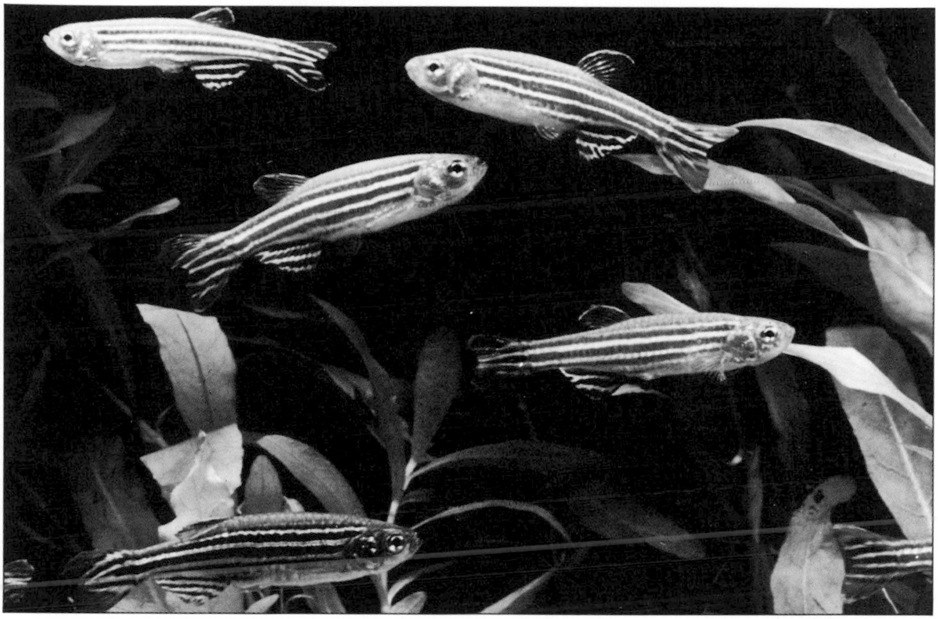

As diseases aren't always easy to cure, prevention is the key factor in fish health care. If your fish is already ill you should consult the *Diagnosis* (see pp. 220–5) and *Ailments* (see pp. 226–35) sections.

The good health of your fishes is very dependent on their environment – poor water conditions, stress or incorrect feeding can lead to illness. The fishes have no control over their conditions, and it is therefore up to you to manage the aquarium correctly. This involves various important procedures to prevent the introduction of disease: quarantining new specimens, handling and feeding the fishes correctly, preventing the introduction of poisons or predators, keeping the heating, lighting, air supply and filtration systems functioning properly, and last, but not least, ensuring that the water is clean, fresh and correctly balanced for the fishes in it. And since plants are a vital part of a freshwater aquarium's "eco-system", caring for them is important too.

# Maintaining a healthy aquarium

To keep your fishes healthy, you must make sure that they aren't subject to stress, and that their aquarium is clean and its conditions stable.

### Quarantine

All fishes, wherever their place of origin, will have travelled some distance to reach the aquatic dealer, and so it is important that they are given time to regain their strength before their final journey to your aquarium. This interim "stopover" takes place in the dealer's tanks. During this period the fishes should be checked for any outbreak of latent disease.

If you are setting up an aquarium for the first time, then all your newly bought fishes will undergo an additional communal quarantine period in the first few weeks in your aquarium. After this, *without exception*, new additions must be quarantined in a separate tank for a few weeks before they join the main collection.

### Reducing stress

Stress is caused by bad handling or inconsiderate treatment, so it is very important to develop a gentle technique when handling fishes (see *Using a net*, opposite), and to follow the correct procedures when moving them between tanks. A fish under stress may well fall victim to a disease to which it is normally immune when it is handled and treated correctly.

### Moving fishes safely

Any net used to catch or transfer a sick fish shouldn't be used in the main aquarium again until it has been disinfected – rinse in a strong proprietary aquarium disinfectant. It is good policy to keep a net associated with one tank only, as a net used in an infected aquarium may spread the disease to another tank.

For advice on transporting fishes and transferring them to a new aquarium without harming them see the *Choosing fishes* chapter, p. 29.

### Environmental effects on health

Various factors under your control can influence the well-being of your fishes. To keep them healthy, you must take these into account.

*Food*

Overfeeding can lead to excess algal growth (see p. 219) and, eventually, pollution of the aquarium, so make sure that you give your fishes the correct amount of food (see p. 207).

A common cause of low resistance to disease is lack of vitamins (see p. 200) due to feeding an incorrect or too-narrow diet. Even if you can only feed

### PREVENTING POISONING

The best way to prevent your fishes being poisoned is to exercise strict control over everything that is put into the aquarium:

- Don't let any metal get into the water, especially in marine aquariums where even condensation dripping back from a metal hood will contaminate the water.
- Cover any metal fittings (for example, external thermostat clips) with airline tubing to prevent contact with the water.
- Don't have metal clips on internal thermostat/heater units.

dried manufactured foods, it is a good idea to use different brands in rotation. This ensures that your fishes get all the vitamins they need, and has the added bonus of preventing them becoming bored with a fixed diet (this is an unsatisfactory situation because it might lead the fishes to leave food, which could then pollute the tank). When adding freshly caught foods to the aquarium, screen them first for dangerous predators (see p. 205).

*Water changes*
Surprisingly, a water change (see p. 218) isn't stressful; many fishes seem to enjoy the addition of fresh water to the tank. There are two schools of thought about the temperature of the new water. Many fishkeepers are quite scrupulous about getting the temperature of the new water as near as possible to that of the existing tank water, whilst others are quite happy to add cold water straight from the tap, saying that their fishes enjoy the stimulation. It may be prudent to steer a middle course, giving the fishes a "cold shower" in summer, but avoiding the addition of very cold water directly to the tank in winter, when the temperature differences between the two bodies of water may be greater.

*Light changes*
Since all aquariums are lit for several hours a day (see p. 148), the transition from light to dark (and vice versa) can be stressful for the fishes. The effect of a sudden change can be severe – the fishes may even dash about in the aquarium, hitting the sides of the tank. You can lessen the shock of a sudden light change by following this procedure: always switch off the aquarium lighting before the room light, and, on

## USING A NET CORRECTLY

Netting can be a terrifying experience for fishes, but you can help by handling the net with care, and using a net of a suitable size for the fish in question. An alternative to using a net is to use a plastic bag. There are two main advantages to this: this type of bag is virtually invisible underwater, so the fish is easier to catch, and the soft plastic does less damage to fins than a coarse-meshed net.

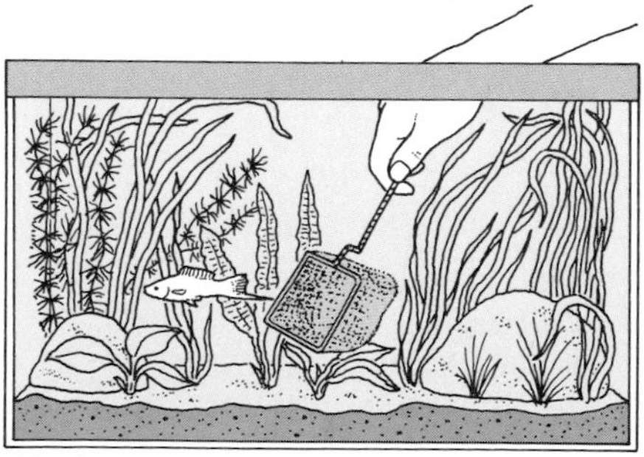

***Single net method***
*When using just use one net, keep it constantly on the move. The net shown here has a special handle for reaching corners or crevices.*

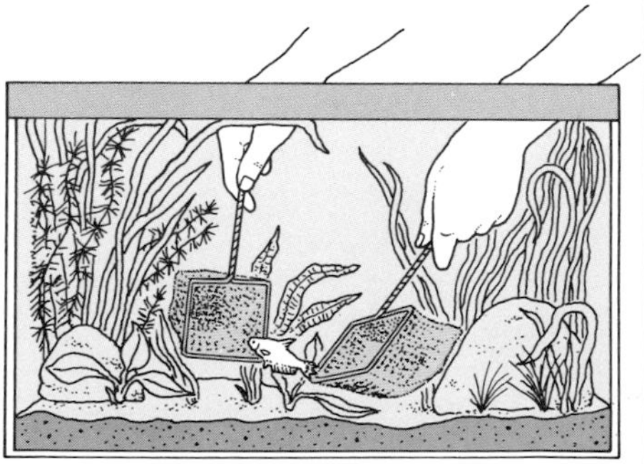

***Two-net method***
*To speed up the catching process, use two nets. Keep one moving, and hold the other stationary in the fish's path.*

dark mornings, always switch on the room light before you switch on the aquarium lighting.

*Vibration*
Fishes are likely to be shocked and stressed by any knocking on the front glass, so site the tank in a position where it can't be bumped accidentally, and discourage children from tapping on the glass. Similarly, any violent vibrations from slamming doors should be avoided. And siting your television or hi-fi loudspeakers very close to your aquarium is a bad idea.

**Aquarium care and maintenance**
A clean, stable environment is essential to fishes' well-being.

*Maintaining the correct temperature*
In tropical aquariums, the temperature should be self-adjusting, but it is worth keeping an eye on the thermometer in case of a mechanical failure. In the case of total heater failure or a power cut, heat loss only occurs gradually, and the larger the tank the slower the loss rate will be. In such circumstances, you can protect the aquarium by covering it with thick layers of newspaper or a blanket.

There is almost a reverse problem with coldwater aquariums during the summer months, where a rise in temperature may occur. Increased aeration will help in maintaining oxygen levels, but in very hot weather you may have to resort to floating ice-cubes in a plastic bag in the water.

*Keeping the air supply pure*
Because the airpump pumps air from the room into the tank, any atmospheric pollution such as cigarette smoke, aerosol polish or paint vapour will enter the aquarium and affect the fishes. And there is also a danger of polluting the fishes' atmosphere by placing your hands in the tank as you might transfer particles of dirt, strong soap or disinfectant into the water.

However, white coats and surgical masks aren't called for – all that is needed is commonsense. If the room is likely to be affected by any strong smells or vapours, then simply open a window. And make sure that your hands are clean and unscented before putting them into a tank.

*Maintaining the water quality*
It is vitally important to fishes' health that the aquarium water is kept clean and fresh. Your filter should keep the water in good condition, but it will only work properly if it is cleaned regularly. The filter medium must also be cleaned (very often it can be re-used after a thorough washing), and the tubes to and from the tank kept free of algae.

Most power filters have their motors "sealed for life", but some of the older types may require lubrication.

The timing of filter-cleaning depends on the type and number of the fishes and the size and efficiency of the filter. Small filters in a tank that contains heavy-feeding fishes will become blocked very quickly, whilst a large filter fitted to a tank that contains a few small fishes may work for a long while without needing attention.

Undergravel filters require very little maintenance. From time to time you should attach a siphon tube to the up-tube to remove any sludge from beneath the filter plate. Also, rake over the gravel occasionally to prevent it from packing down too tightly.

*Maintaining the aquarium lighting*
Before replacing or adjusting lamps you *must* switch off the electricity supply. To make sure that your plants and

fishes thrive, replace any faulty lamps and keep the reflector clean. If a fluorescent tube fails to light, check all the connections before buying and fitting a replacement. Install a cover-glass to reduce the danger of condensation damage to the lamps. The amount of light required by the plants will be affected if the cover-glass isn't spotless (it quickly becomes dirty, and covered in algae and water splash-marks).

**Timetable for keeping your fishes' environment healthy**

For your fishes to thrive and your tank to look attractive, it is important to carry out certain routine tasks. Some need to be done daily or weekly, whilst others only require periodic attention.

*Daily*

- Check that all the fishes are present and in good condition.
- Make sure that the water temperature is correct in tropical aquariums, and that the temperature in coldwater tanks isn't too high during summer months.
- Test the Specific Gravity and Nitrite levels in tropical marine aquariums.

*Weekly*

- Remove any dead plant leaves.
- Prune back any fast-growing plants and replant cuttings.
- Siphon off any unsightly detritus from the tank floor.

*Periodically*

- Clean and/or replace filter medium.
- Change 20–5 percent of the aquarium water (see p. 218). For marine tanks, you must prepare the necessary quantity of synthetic salt-water mix in advance.
- Clean algae from the front glass and the cover-glass (see p. 219).
- Rake over the surface of the base covering, particularly in tanks with undergravel filters.
- Check the pH value in marine aquariums (see p. 145). In freshwater systems, it will only be necessary to check it if you are trying to keep the water conditions within narrow parameters for a particular reason such as when breeding difficult species.
- Inspect airstones for blockages, especially in marine tanks.
- Clean filter pad and air-valves in the airpumps (see below).

## PUMP MAINTENANCE

Since the pump (see p. 123) provides the vital supply of fresh oxygen to your fishes, you should give it regular maintenance checks. You *must* disconnect the electricity supply before removing the pump cover. Clean the air-filter pad and the air valves within the diaphragm assembly block once every three months. (The pad is generally fitted in a recess underneath the pump body, and the air valves are within the body of the pump.)

If your pump requires lubrication make sure that you oil it regularly, and fit an oil-filter in the airline to prevent oil entering the aquarium.

If a vibrator airpump starts to make a clattering noise, its diaphragm has split and needs replacing. Remove the arm that holds the diaphragm in place, then unclip the diaphragm from the valve block. Whilst the pump is dismantled, take the opportunity to clean the small, flexible input and output valves in the valve block. Next, fit the new diaphragm (obtainable from your aquatic dealer), and reassemble the pump.

- Check all hose connections, especially on power filters.
- Replace faulty lamps as required, and renew fluorescent tubes after long periods of service.

### Making water changes

Although an efficient filter will do much to maintain the water quality (see p. 216), you can also help to keep your aquarium clean by carrying out regular, partial water changes. In a freshwater aquarium replacement of about 20 percent of the water every 3–4 weeks is recommended, whilst in a marine tank changing 25 percent of the water every 2–3 weeks is advisable. Either siphon out some old water and add new manually, or use an automatic device attached to two hosepipes (one connected to the taps, the other to the drain). This action reduces (by simple removal and further dilution) the amount of dissolved waste materials in the water. When making water changes in a marine aquarium you must use the recommended density of "sea-mix" in the new water in order to maintain the correct density. Take care not to stress the fishes during this (see p. 214).

*Replacing "lost" water*
It is quite normal for the water level in any aquarium to drop. This isn't necessarily a sign of a leak, but often merely of evaporation, particularly if there is no cover-glass. All such water losses must be replaced with fresh water, even in a saltwater aquarium, as sea salt isn't lost during evaporation.

*Balancing the water correctly*
Freshwater must be of the correct pH (see p. 145) – neither too acid nor too alkaline – and saltwater should be of the correct density and pH (see p. 189). An imbalance in water conditions can seriously affect fishes' health.

### METHODS OF CHANGING WATER

Regular, partial water changes will remove dissolved wastes from the tank without disturbing your fishes. You can carry out a water change with a siphon tube and buckets or a purpose-made device.

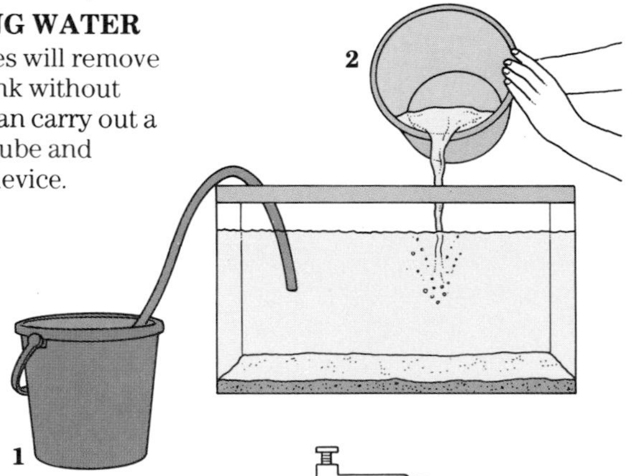


***Stage one***
*Siphon about 20 percent (fresh) or 25 percent (salt) of the aquarium water into a bucket.*

***Stage two***
*Refill the tank with fresh water or sea mix at the same temperature.*

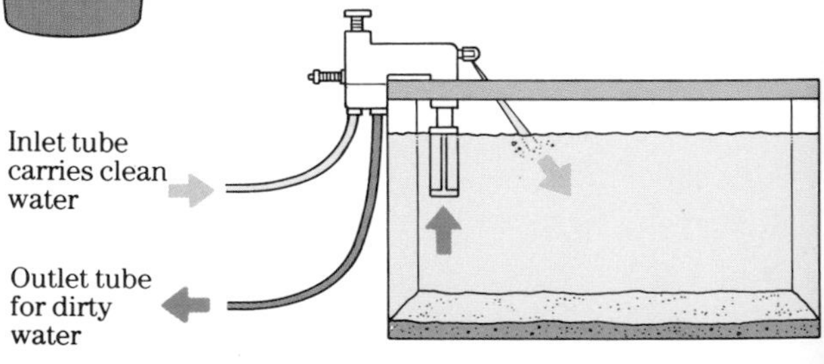


***A water changer***
*This device automatically drains off dirty water and replaces it with fresh. Attach the inlet tube to a tap, and position the outlet tube over a sink.*

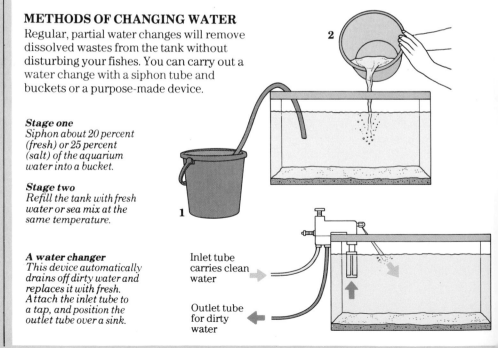

**Health tips**

Although you can't protect your fishes from all ailments, there are a number of measures you can take to keep them in good health:

- Buy only good-quality, compatible fishes (see p. 24).
- Quarantine all new acquisitions (see p. 214).
- Avoid stressing the fishes (see p. 214).
- Don't overfeed – if feeding household scraps or other "non-aquatic" types of food, remove any uneaten portions immediately the fishes lose interest in them.
- Remove sick fishes to a separate tank for treatment (see p. 235).
- Disinfect nets used to transfer sick fishes.
- Avoid transferring water from the hospital tank to the main aquarium.
- Don't let any metal get into the water (see p. 214).

## ALGAE CONTROL

A slight growth of algae is desirable in marine aquaria, and in freshwater tanks will provide green food for vegetarian fishes. If the growth is too rampant, remove it with a scraper.

## PLANT CARE ROUTINE

New plants should be carefully examined and washed before use, as shown right.

Every month check all existing plants, remove all dead plants and leaves (check the tank floor for any that have fallen there), and re-root any specimens that have floated loose.

If you have any bushy specimens that have grown excessively trim them back, taking cuttings for other tanks. Also, sever the runners from grassy plants such as *Vallisneria* or *Sagittaria*, and replant the young shoots.

Keep a close eye on floating plants as they can grow extremely fast, taking over the entire surface of the water so that any submerged plants are deprived of light and die back.

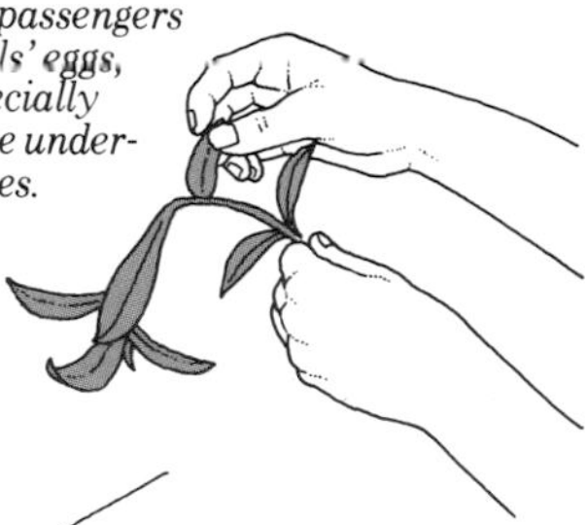

**Stage one**
*Inspect new plants for unwelcome passengers such as snails' eggs, looking especially closely at the under-sides of leaves.*

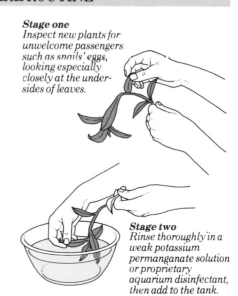

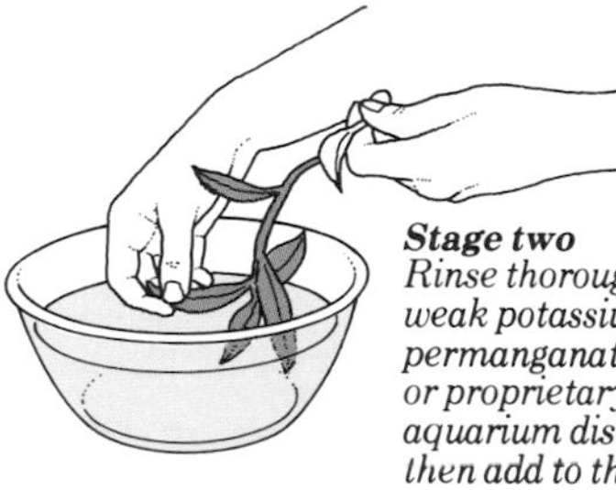

**Stage two**
*Rinse thoroughly in a weak potassium permanganate solution or proprietary aquarium disinfectant, then add to the tank.*

# DIAGNOSIS CHARTS

It is extremely important to detect disease in your fishes as early as possible, since prompt treatment may be able to save the afflicted fish, and equally importantly, prevent the disease spreading to other fishes in the same aquarium. Unfortunately, however, some fish ailments are almost impossible to diagnose since they produce no visible signs. In such cases, the fish will seem perfectly healthy one day yet dead the next, and the only way to ascertain the cause of death is to have a post-mortem carried out. Nevertheless, the majority of fish ailments have characteristic, externally visible signs which you should learn to recognize.

The ability to recognize signs of illness depends upon regular observation of your fishes (see p. 212) so that you are familiar with the normal physical appearance and behaviour of each species. This will prevent unnecessary worry over fishes which have individual characteristics that if seen on other species would normally be signs of disease. For example, the scales of a healthy male *Pachypanchax playfairi* (see p. 72) stand up during spawning, yet for different species this is often a sign of dropsy (see p. 228) as in the case of the fish above.

You should also bear in mind that some species are particularly susceptible to certain diseases, so if you keep such fishes be constantly on the look-out for the relevant signs. For example, *Etroplus maculatus* (see p. 56) is prone to fungal disease (see p. 233), whilst *Holacanthus tricolor* (see p. 99) tends to suffer from skin diseases (see p. 227).

# Making a visual diagnosis

When inspecting your fishes, look for any physical changes – the appearance of abnormal growths, changes to the shape of the fish's body, scales or body-covering or damage to its fins. Also check for the presence of parasites, and see if any fish is having difficulty in swimming. However, not all changes are due to health problems as some species alter colour during the breeding period.

**Gill Flukes** *In less advanced cases (see p. 231), damage to the gills isn't always seen, but the fish's gills will be coated in mucus.*

**White Spot** *This disease (see p. 230) is caused by a parasite, and is easily detected — tiny white spots cover the fish's body and fins.*

**Dropsy** *With its swollen abdomen and its scales standing out, this fish is obviously suffering from acute dropsy (see p. 228).*

**Anchor Worm** *Although this parasite (see p. 232) buries its head in the fish's body, its rear quarters and egg sacs are visible, as on this Moor.*

**Pop-eye** *Septicaemia or tuberculosis (see p. 229) are the most likely causes of this fish's protruding eyes (see p. 228).*

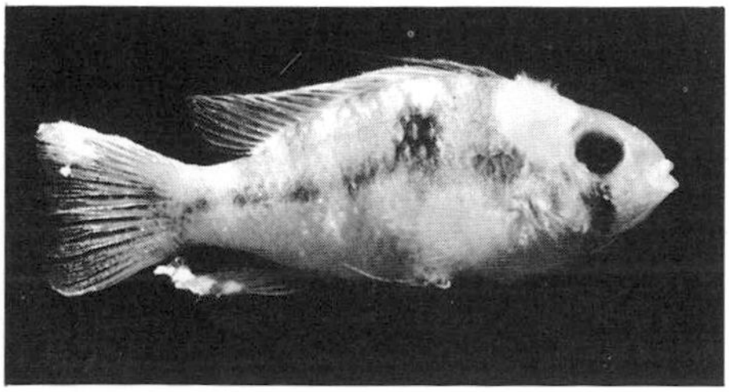

**Fungus** *This fish is showing the classic sign of fungal attack (see p. 233) — tufts of dirty, cotton-wool-like growths on the fins and body.*

**Swim-bladder disorder** *It is easy to see that this Molly has a problem with its swim-bladder (see p. 228) as it is swimming upside-down.*

# Skin problems

A fish that shows signs of a skin disorder is probably suffering from a curable parasitic disease. If you are in any doubt about your fish's health consult an experienced fishkeeper or vet immediately.

---

**START HERE**

Has the fish visible spots on its skin?

YES → Are the spots distinct and separate?

YES → **White Spot Disease** Your fish is probably suffering from *Ichthyophthirius* (see p. 230).

NO

Are the spots numerous and dusty?

YES → Is the fish a freshwater type?

YES → **Velvet Disease** Your fish is probably suffering from *Oodinium limneticum* (see p. 231).

NO

**Coral Fish Disease** Your fish is probably suffering from *Oodinium ocellatum* (see p. 231).

NO

Are parts of its skin excessively slimy?

YES → Is the slime cloudy and is it covering most of the fish's body?

YES → **Parasitic infestation** Your fish is probably affected by one of three parasites – *Costia*, *Chilodonella* or *Trichodina* (see p. 232).

NO

Is the slime cloudy and is it covering the eye only?

YES → **Eye disease** Your fish is probably suffering from an eye disease (see p. 228).

NO

*Go to next page*

*Go to next page*

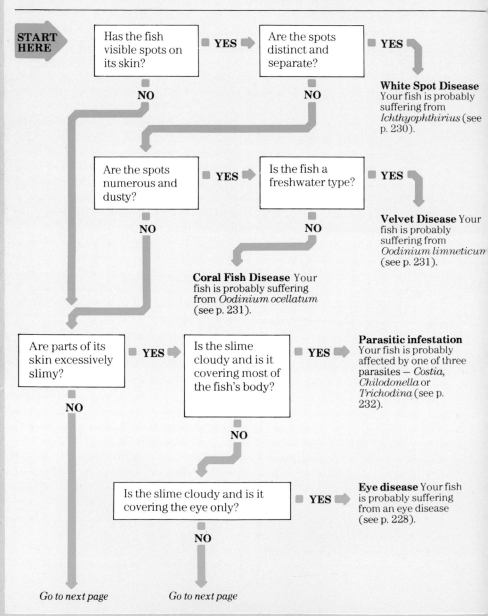

Is the slime around the fish's mouth, and does it consist of very short, fine, tufty filaments?

YES → **Mouth Fungus** Your fish is probably suffering from Mouth Fungus (see p. 233).

NO

Is the slime on the fish's body, and is it tufty and filamentous like cotton wool?

YES → **Fungus** Your fish is probably suffering from Fungus (see p. 233).

NO

Can you see a parasite?

YES → Is it worm-shaped?

YES → **Anchor Worm** Your fish is probably suffering from *Lernaea* (see p. 232).

NO → Is it round in shape?

YES → **Fish Louse** Your fish is probably suffering from *Argulus* (see p. 232).

NO

NO

Can you see an ulcer?

YES → **Ulcer disease** This condition is usually caused by bacteria, and may indicate *Septicaemia* (see p. 229).

NO

Are the fins reddened?

YES → Does it appear dull?

YES → **Septicaemia** This bacterial condition (see p. 229) is a very serious disease.

NO → Is it a long-finned cold-water fish?

YES → **Stress** This type of fish is very susceptible to sudden changes in conditions (see p. 214).

NO

**f you can't make a iagnosis from this chart, onsult a vet who specializes n fish diseases**

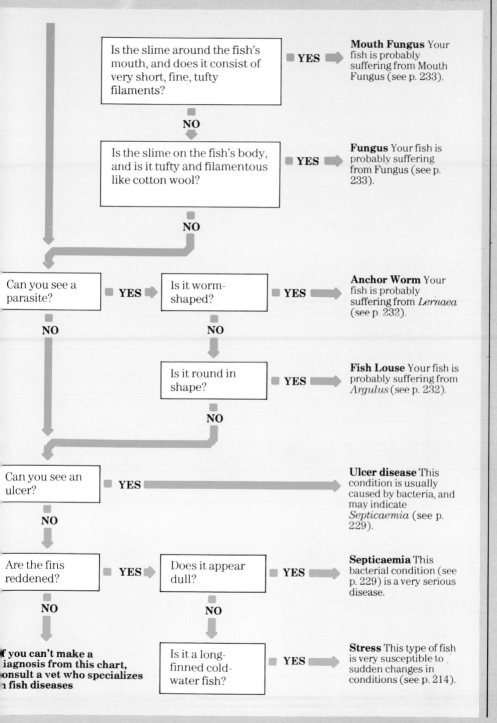

# Swimming problems

If your fish seems to be having difficulty in swimming, this may be a sign of a health problem or a reaction to adverse water conditions. If you are in any doubt about the fish's health, consult an experienced fishkeeper or a vet immediately.

---

**START HERE**

**Is the fish's equilibrium abnormal?**
- **YES** → **Swim-bladder disorder** Your fish probably has a swim-bladder disorder (see p. 228).
- **NO** → Is it "hanging" just below the surface?

**Is it "hanging" just below the surface?**
- **YES** → Is it breathing at its normal rate?
- **NO** → Is it breathing rapidly, and are its gills pale and held open?

**Is it breathing at its normal rate?**
- **YES** → **Lack of oxygen in the water** Your fish is probably suffering as a result of low oxygen levels in the water (see p. 216).
- **NO** → Does it remain still, even when touched?

**Does it remain still, even when touched?**
- **YES** → **Poor condition** Your fish may be affected by too low a pH in the tank (see p. 145). Alternatively, it may be suffering from poisoning (see p. 214) or a terminal disease (see p. 228).
- **NO** → Is it breathing rapidly, and are its gills pale and held open?

**Is it breathing rapidly, and are its gills pale and held open?**
- **YES** → **Gill disease** Your fish is probably suffering from a parasitic infestation such as Coral Fish Disease (see p. 231) or gill flukes (see p. 231).
- **NO** → *Go to next page*

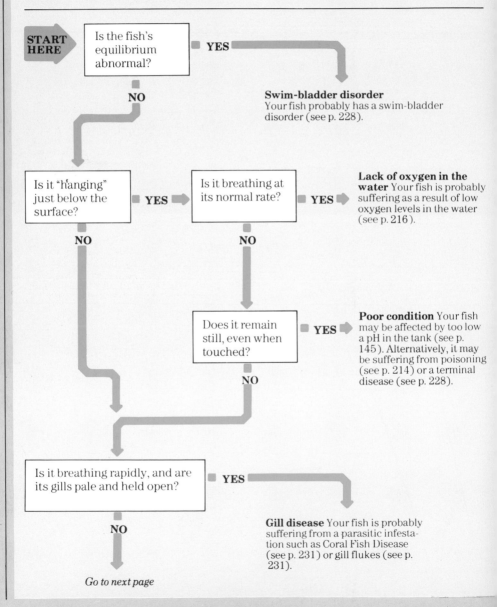

Is it unnaturally restless?

YES → Does it undulate rapidly, but without moving forward at all?

YES → **Shimmying** Your fish probably has the "shimmies" (see p. 228).

NO → Is it a newly introduced fish that dashes about, tries to jump out of the tank or hangs at the surface?

YES → **Poor tank conditions** Your fish is probably affected by too low a temperature (see p. 216) or too low a pH in the tank (see p. 145).

NO → Is it generally restless, twitching convulsively, and are its fins spread and its mouth open?

YES → **Severe disease or poisoning** Your fish is probably suffering from a severe disease (see p. 228) or poisoning (see p. 214).

NO → Does it rub itself against rocks or leaves?

YES → **Flukes** Your fish is probably suffering from gill or skin flukes (see p. 231).

NO

NO

**If you can't make a diagnosis from this chart, consult a vet who specializes in fish diseases**

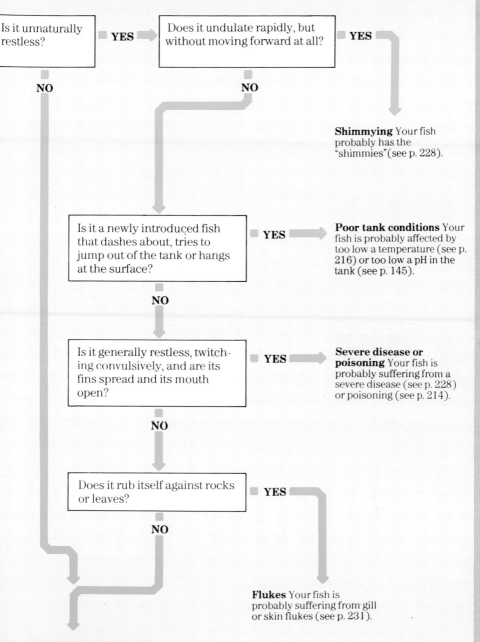

# AILMENTS AND DISORDERS

Most of the ailments that affect aquarium fishes are easy to recognize. These diseases are usually parasitic or bacterial in origin, and the symptoms are visible externally. In most cases, they can be treated quite successfully. There are also a number of internal disorders which affect fishes, and like their counterparts in mammals, these can be very difficult to treat. Moreover, fishes tend not to show signs of such diseases until their condition is too far advanced for treatment to work.

Depending on the type of disease, a sick fish should either be removed from the tank and treated in isolation, or kept with the other fishes and treatment given collectively. Outbreaks of contagious diseases are best dealt with *en masse*, but fishes affected with large, visible adult parasites such as Anchor Worm are best treated individually. A fish affected by parasite larvae should be treated with the rest of the fishes, even if they are apparently healthy.

Very few vets specialize in fish disease, and consulting one will be expensive (more than the price of a single tropical fish, but less than the cost of replacing an individual Koi or a tankful of tropical specimens). If your local vet doesn't usually deal with fishes he or she will put you in touch with the nearest vet who does. When you contact the vet have the following information on your system ready: tank size, filtration, water chemistry, temperature, food and treatment given.

**Warning** : Some treatments are solutions of dangerous chemicals. Treat them *all* as medicines, and keep them where children can't gain access.

# How disease affects fishes

Externally visible skin infections are easily seen and treated, but internal disorders, which are often indiscernible until well-advanced, are very difficult to remedy.

## Parasites

Fishes often find themselves acting as hosts to some very unwelcome guests. Most parasites (see p. 212) feed on the mucus layer on the skin and gills, larger ones penetrate the tissues and feed on blood and tissue fluids. In general, the microscopic parasites are easy to treat using proprietary remedies added to the water, although a few have a cyclical lifestyle where they are resistant to treatment at some stages of the cycle. Most large adult parasites are removed by hand, but larvae have to be treated with chemicals added to the water.

## Bacteria and fungae

Poor tank conditions – whether caused by dead plant matter, faeces or uneaten food – encourage the growth of bacteria or fungae. The best defence against attack is to keep your fishes in a clean, well-oxygenated environment and make sure that you feed them correctly. Some bacteria (see p. 212) are treated with proprietary remedies, others with general antibacterial preparations.

## Skin disorders

The most common problem in fishes, skin disorders are generally caused by bacteria or parasites.

*What are the signs?*

There are several signs of skin problems: □ the fish tries to scratch on rocks and the gravel base □ its colours look "washed out" and dull □ its eyes look dull and cloudy. If left untreated, further signs may appear: □ ragged and frayed-looking fins □ lack of appetite □ grey coloration due to excess mucus production □ areas of reddish inflammation.

Bacterial skin infections have additional signs, including: □ distinct white patches, sometimes with fine whitish filaments visible □ erosion (fin rot or ulcers). Very severely affected fishes may display: □ wheals or ulcers □ haemorrhages around the eyes □ reddened fins □ bulging eyes □ scales standing out.

## Gill disorders

Like our respiratory disorders, infections of the fish's gills (the organs through which it breathes) may be mild or very severe.

*What are the signs?*

Gill trouble isn't always immediately obvious: □ the fish may go off its food □ it may hang at the water surface, airstone or filter uplift (where dissolved oxygen levels are highest) □ the gill covers may move more rapidly than normal □ the gills inside are often visible; they may be swollen and greyer than normal □ light or dark marks (individual parasites) may appear on the gills □ large, eroded areas where tissue has been eaten away may be visible □ mucus may be seen hanging from the gills.

After a time, the gills actually thicken up in response to the chronic irritation, and this slowly suffocates the fish, so that it eventually dies. The cause of the condition is likely to be slime disease (see p. 232), gill flukes (see p. 231) or bacterial damage.

**Disorders of the eyes**
Many diseases that affect a fish's skin can attack its eyes too. Careless handling combined with poor water conditions often causes minor damage which allows bacterial or fungal infections to become established.

*What are the signs?*
There are three main signs which indicate an eye problem: ☐ dirty cotton-wool-like growths – a sign of fungus (see p. 232) ☐ cloudiness – this may be due to a change in the water conditions (see p. 215), a bacterial infection (see p. 212), a slime disease (see p. 232) or skin flukes (see p. 232) ☐ protruding eyes – known as *exophthalmia* or "pop-eye", a condition usually caused by septicaemia or tuberculosis (see p. 229).

**Dropsy**
The term "dropsy" normally means fluid in the abdomen, but in fishkeeping it includes all diseases which make the abdomen swell abnormally. These diseases often can't be distinguished from each other without a post-mortem.

*Chronic dropsy*
A swollen abdomen that develops slowly is generally due to cancer in internal organs or the presence of large parasites (the effect of their size alone, or because they cause liver or kidney damage or peritonitis). In such cases, the dropsy won't spread to other fishes as long as you removed the affected fish from the tank when it became ill. To get an accurate diagnosis (for the sake of your other fishes) take the fish to the vet for a post-mortem.

*Acute dropsy*
Sudden swelling causes the affected fish's scales to stand out. The most common cause of this problem is bacterial septicaemia (see p. 229). Veterinary treatment of the affected fish and its contacts will be necessary. The vet will give antibiotics.

*Malawi bloat*
This condition affects Cichlids, producing a dropsy the cause of which isn't yet fully understood, but may be due to bacteria or a semi-cancerous condition of the stomach. Remove the affected fish from the tank and take it to the vet for a post-mortem (see p. 235).

**Wasting**
Slow weight loss that results in an unbalanced look (because the fish has a normal-size head on a reduced body) can be due to one of two reasons: harassment of an individual fish preventing access to food, or, more commonly, a disease such as tuberculosis (see p. 229).

**Swim-bladder problems**
An abnormal swimming pattern or difficulty in maintaining equilibrium may be due to: ☐ a congenitally deformed swim-bladder ☐ cancer or tuberculosis in organs adjacent to the swim-bladder ☐ constipation ☐ poor nutrition (see p. 201) ☐ chilling (see p. 215) ☐ serious parasitic infestations (see p. 212) ☐ serious bacterial infections (see p. 212).

If you suspect swim-bladder problems in a fish you should first check and treat it for other diseases. If you have eliminated other causes, make sure that you are giving the right food (see p. 200) and make sure that the fish is passing faeces normally. If in any doubt, feed live food as roughage (see pp. 204–5). Finally, check that the temperature is stable and at the right level for your fishes.

# INTERNAL DISORDERS

Because fishes tend to hide signs of these diseases, diagnosis is often too late to help the affected fish, but may help other specimens in the tank. Bacterial septicaemias are a common cause of internal disease, damaging tissues and killing the fish unless treated very promptly. Wasting diseases such as tuberculosis (see p. 229) don't respond well to treatment, but their diagnosis is important for the sake of the other fishes and for human health reasons.

## SEPTICAEMIA

This condition can follow on from skin infections such as finrot (see p. 232), or may occur independently as a result of dirty conditions. Bacteria enter the blood stream and circulate through the tissues causing inflammation and damage. Blood vessel and heart tissue damage results in leakage of fluids into the abdomen, producing dropsy (see p. 228). Inflamed blood vessels in the skin and at fin bases stand out.

**Things to look for**

- Reddening at the bases of the fins
- Small haemorrhages around the eyes
- Very dull, listless behaviour
- Lack of appetite

**Treatment**

Seek veterinary guidance. The vet will prescribe antibiotics. Check the aquarium for the cause and eliminate it.

## TUBERCULOSIS

A fairly infectious bacterial disease, tuberculosis is becoming increasingly common. Affected specimens *must* be removed from the aquarium immediately so that other fishes aren't infected. A tubercular fish usually feeds normally, but loses weight as its internal organs become damaged. Some fishes develop nodules under the skin which eventually ulcerate, in others nodules develop behind the eye, causing "pop-eye" (see p. 221).

The bacteria that causes the disease prefers cooler temperatures than most bacteria that infect humans. However, fish tuberculosis can affect people, usually in the form of an infected nodule on the skin, but there is a small chance that it will cause a serious internal infection. Once diagnosed in one of your fishes, strict hygienic precautions should be observed. A definite diagnosis is only possible by a post-mortem.

**Things to look for**

- Appears dull in colour
- Weight loss
- Folded fins
- Ulcerous skin wounds

**Treatment**

Seek veterinary advice. Affected fishes should be removed and euthanized (see p. 234). *Don't* allow them to die in the tank as the other inhabitants will eat them and become infected too. The tubercular fish's contacts should be treated: move them to a separate hospital tank, and disinfect the original aquarium. If other fishes succumb, don't introduce any new specimens, euthanize all affected fishes, then clean, disinfect and re-stock the aquarium.

# EXTERNAL DISORDERS

Several common health problems, mostly caused by parasites (see p. 212), affect fishes' skin and gills. In general, these disorders are easy to diagnose and treat. Large parasites are easily visible, and smaller ones can be seen under a magnifying glass. Information on the nature of fishes' skin is given in the *Anatomy* chapter (see p. 16).

Parasites are often introduced into the aquarium by newly bought fishes, so a suitable quarantine period (see p. 214) is an important preventative measure.

## FRESHWATER WHITE SPOT DISEASE *(Ichthyophthirius)*

Virtually all freshwater aquariums are affected by this well-known disease at sometime. The parasites are visible to the naked eye as white spots up to 1 mm in diameter on the fish's skin. *Ichthyophthirius* has a cyclical lifestyle – it leaves the fishes to reproduce. When it does, it creates a large hole in the skin which is open to secondary bacterial or fungal infection. The parasite sinks to the tank floor and secretes a jelly-like covering around itself – the cyst. Inside this cyst it divides into many young. Then the cyst bursts and the free-swimming young seek a new host. As with *Cryptocaryon* (see below), the stage within the skin and the cyst stage are both resistant to treatment.

**Things to look for**

- The fish's skin and fins are covered in tiny white spots
- A badly affected fish may make rapid gill movements

**Treatment**

When the parasite is in the fish's skin and at the cyst stage it can't be treated. To eliminate *Ichthyophthirius* add a proprietary remedy to the aquarium before the cyst forms or after the cyst bursts. Remove plants and the activated carbon from filters.

## MARINE WHITE SPOT DISEASE *(Cryptocaryon)*

Only marine fishes are affected by this parasite, the equivalent of the very common *Ichthyophthirius* (see above) found in fresh-water fishes. *Cryptocaryon* has a cyclical lifestyle, and leaves the fish to reproduce. Because it actually buries itself in the skin, it punches a hole to leave when it reaches maturity.

**Things to look for**

- The fish's body and fins are covered in tiny (1 mm diameter) white spots
- The skin is slimy
- The fish makes rapid gill movements
- The eyes look cloudy

**Treatment**

When the parasite is in the fish's skin and at the cyst stage it is resistant to treatment. *Cryptocaryon* can only be eliminated by adding a proprietary remedy to the aquarium before the parasite forms a cyst or when the cyst releases the new adults which infest the fish. Before you add the remedy, read the manufacturer's instructions carefully – you will probably have to remove plants, the activated carbon from filters, and invertebrates from the tank first.

## VELVET DISEASE *(Oodinium limneticum)*

This tiny parasite is seen as a fine "gold-dust" on the skin. It penetrates skin cells with "roots" through which it feeds. Like plants, it contains a pigment which uses light as an alternative energy source.

**Things to look for**

- The fish's skin and fins are covered in a fine velvety, gold sheen that is said to resemble gold-dust
- The fish makes rapid gill movements

**Treatment**

Treat the aquarium with a proprietary remedy. This may also advise shading the tank to rob the parasite of light energy, and removing activated carbon from filters.

## CORAL FISH DISEASE *(Oodinium ocellatum)*

The marine equivalent of Velvet, this parasite lives on the fishes' gills.

**Things to look for**

- Rapid gill movements
- Flared gills
- Dusty skin

**Treatment**

Treat the aquarium with a proprietary remedy, removing invertebrates first.

Copper sulphate treatment may work, but to use it safely you require a very sophisticated test kit. It should therefore only be used by experienced fish-keepers.

## GILL FLUKES *(Dactylogyrus)*

These tiny, worm-like flukes are barely visible to the naked eye. They infect the fish's gill membranes.

**Things to look for**

- The gills move rapidly
- The fish pants at the water surface
- The gills are covered in mucus, and parts are eaten away
- The fish may scrape itself against objects

**Treatment**

The flukes lay eggs which fall to the tank bottom and are resistant to treatment. Attack is two-fold: add a proprietary remedy to a hospital tank and move all the fishes to it, and give affected individuals short-term formalin baths (see p. 234). Treat the hospital tank weekly until all the larvae are killed. Clean out the original aquarium, then return the fishes.

## SKIN FLUKES *(Gyrodactylus)*

Related to *Dactylogyrus*, these skin parasites have a different life-cycle, without resistant stages.

**Things to look for**

- The fish scrapes itself against objects
- Colours fade as the fish becomes mucus-covered
- The skin reddens in places
- The fins become ragged

**Treatment**

Add a proprietary remedy to the aquarium, and give affected fishes a short-term formalin bath (see p. 234).

## SLIME DISEASES *Chilodonella, Costia (Ichtybodo), Cyclochaeta (Trichodina)*

Three protozoan parasites – *Chilodonella, Costia (Ichtybodo)* and *Cyclochaeta (Trichodina)* – which mainly affect the skin cause very similar signs on affected fishes.

**Things to look for**

- Dulling of colours due to over-production of mucus
- Fraying of the fins
- Weakness
- Gill damage
- Death

**Treatment**
If you notice the disease before it has spread to the affected fish's gills, treat the aquarium with a proprietary remedy. If this doesn't work, the parasite is probably *Chilodonella*, and you should treat the fish with a short-term formalin bath (see p. 234). If the fish's gills *are* affected, give it a short-term salt bath (see p. 234), followed by a short-term formalin bath (see p. 234) if the parasite proves resistant to salt.

## ANCHOR WORM *(Lernaea)*

This crustacean buries its anchor-shaped head in the fish's tissues, trailing its two egg sacs behind it. It is clearly visible on the fish, and when it drops off it may leave an unpleasant ulcer.

**Things to look for**

- The fish scrapes itself against objects
- Whitish-green threads hang out of the fish's skin, with an inflamed area at their point of attachment

**Treatment**
Remove the fish from the tank (it can survive for 1-2 minutes), hold it in a wet cloth and pull the parasite off with tweezers. Treat the wound with a cotton bud dipped in an antibacterial preparation such as mercurochrome or iodine. This is often best done by a vet, who will treat the fish under anaesthetic. Return the fish to very clean water treated with a proprietary remedy so that the wound can't become infected by bacteria or fungi.

## FISH LOUSE *(Argulus)*

Actually a crustacean like the seaside crab, this parasite swims from host to host, anchoring itself by means of strong suckers, and penetrating the skin with a poison spine.

**Things to look for**

- The fish scrapes itself against objects
- Parasites (5 mm diameter) are visible on the fish

**Treatment**
Remove the fish from the tank (it can survive for 1-2 minutes), hold it in a wet cloth, and pull the adult louse off with tweezers. If stubborn, apply a drop of salt solution (15–30 g to 1 litre of water) to it with a small brush (don't touch the fish's skin). Transfer the affected fish and its companions to a hospital tank (see p. 235), then treat the original aquarium with a proprietary remedy for fish lice to kill any larvae. Return the fishes to the original tank, but watch closely for adult lice or unhatched larvae as the remedy isn't always effective against these forms. If necesssary, repeat the treatments.

## FINROT

Heavily pigmented fishes like Black Mollies and fishes with long, trailing fins like fancy Goldfishes are most prone to this condition. The major cause is a dirty tank; other contributing factors are: □ a poor, vitamin-deficient diet □ outwintering Fancy Goldfishes □ fin-nipping by other fishes □ damage caused by nets □ poor water conditions.

**Things to look for**
- Short or ragged fins

**Treatment**
Give the affected fish a short-term bath (see p. 234) of a proprietary remedy. Make sure that you correct the cause (see pp. 214–9). In severe cases, consult a vet who will trim off the diseased part under anaesthetic, then treat the fish with an antibacterial preparation.

Epidemic finrot in marine aquariums requires specialized care – seek help from a vet, or an experienced marine fishkeeper or dealer.

## FUNGUS

The commonest fungus is *Saprolegnia*. This can affect any type of fish, but is mostly seen on coldwater species. Algae like to grow on the cotton-wool-like fungus, giving it a dirty appearance.

Fungal attack is always a secondary factor to another health problem – for example, a parasitic infestation.

**Things to look for**
- Tufts of a dirty, cotton-wool-like growth on the skin, sometimes covering most of the fish

**Treatment**
Pre-soak cotton wool in water, squeeze it nearly dry, then dip it in povidone iodine or mercurochrome. Take the fish out of the tank (it will survive for 1–2 minutes), holding it in a wet cloth, and use the prepared cotton wool to wipe off the fungus. Then transfer the fish to a hospital tank and treat it with a proprietary broad-spectrum treatment. Find the initial cause of fungal attack and remedy it.

## MOUTH FUNGUS *(Chondrococcus columnaris)*

Despite its common name, *Chondrococcus columnaris* isn't a fungus at all, but a slime bacterium that strings together with its fellows, forming threads. These are finer and shorter than the threads of true fungus. Livebearers are particularly susceptible to this infection.

**Things to look for**
- White, tufty material around the mouth
- White patches on the skin

**Treatment**
Add a proprietary treatment to the aquarium. If particularly stubborn, transfer the affected fish to a separate hospital tank and add antibiotics (prescribed by the vet) to the water. Avoid netting affected fishes; instead, use the net to chase the fish into a plastic bag. As a preventative measure, add a small amount of salt to aquariums containing livebearers.

# Isolation and treatment of sick fishes

When treating a whole tank of fishes, calculating the required dosage of a remedy exactly may be difficult because rocks and decorations will displace an unknown quantity of water. If you quarantine new fishes separately (see p. 214), then you will already possess a second tank which can double as a hospital tank for sick fishes. This approach is always necessary where the treatment would disturb the biological filter system, as in the case of methylene blue and most antibacterial and antibiotic preparations. In certain cases, such as the removal of a large parasite, the fish is best treated by a vet.

**Convalescence**

After a fish has been successfully treated, it shouldn't be returned to the main aquarium straightaway. Over a number of days, gradually replace the water in the hospital tank to re-acclimatize the fish to its original water conditions. Once this has been done, the fish can be returned to the main aquarium. Use a plastic bag *not* a net to transfer it.

**Giving a short-term bath**

To treat an individual fish with certain remedies you will be advised to give it a short-term bath. This involves mixing the treatment with water of the same temperature as the water in the aquarium that the fish comes from, and adding it to a small, watertight container. The water in the container should be aerated. Transfer the affected fish to this bath for the required time (usually 5–60 minutes). If the fish appears to show distress, remove it. Fishes with badly affected gills may not be strong enough to withstand a bath.

*Formalin solution*

A 35 percent formalin solution is used at a dilution of 0.2 ml of formalin per litre of water.

**Warning** : When using formalin and other corrosive poisons, wear rubber gloves or barrier cream (petroleum jelly) to protect your hands.

**Euthanasia and disposal**

Fishes with severe, incurable diseases should be killed painlessly – an approved method is to drop the fish into water cooled with crushed ice. Or the fish can be overdosed with anaesthetic.

*Post-mortems*

If you are in any doubt about the cause of the illness take the sick fish to the vet who will euthanize it and perform a post-mortem on the fresh fish to diagnose the cause of illness. Alternatively, preserve the fish in formalin (1 part formalin plus 9 parts water) and take it to your vet for forwarding to a veterinary laboratory. Or send it to a laboratory direct – your local aquarist's club will provide an address. Such action won't be cheap, but may help to prevent health problems in the rest of your fishes, thus avoiding far greater re-stocking costs. Also, it may provide research material to advance the understanding of diseases.

**Keeping medical records**

It is a good idea to make a written case history for each of your fishes, recording any signs of ill health, behaviour whilst in quarantine, and any treatment given. This will help you to establish which remedies are most successful for your fishes.

# THE HOSPITAL TANK

Because you are unlikely to have the hospital tank in constant use, it is vital to thoroughly disinfect it and any equipment used in it between treatments. Try to create a reassuring environment in the tank by providing "plants" and refuges so that the transfer from the main aquarium will cause the sick fishes as little stress as possible.

Tips for maintaining healthy conditions in a hospital tank:

- Provide a suitable heater and thermostat for tropical species
- Provide extra aeration as many remedies reduce the amount of oxygen in the water
- Don't use filters that include activated carbon as this will remove the treatment from the water
- Light the tank dimly – some remedies are neutralized by bright light and others sensitize the fishes to light, leading to skin disease
- The water should be as similar as possible to the water in the main aquarium, so that fishes aren't subjected to any unnecessary stress as a result of the transfer
- Don't use nets for transferring fishes as they may become contaminated
- A tank for marine fishes should be at least half the size of the main aquarium

***Setting up a hospital tank***
*As the tank shown here contains no heating equipment, it is only suitable for treating coldwater species. A base covering isn't used in a hospital tank, so plastic plants are anchored in weighted trays.*

An overturned flowerpot provides refuge for the fishes
Plastic plants give fishes a sense of security, and unlike real plants, aren't killed by medications
A simple sponge filter keeps the water clean and healthy without removing the medication
An airpump provides vital oxygenation, replacing oxygen that has been removed by medications

# 8

# BREEDING

In order to breed aquarium fishes successfully there are two types of information you must acquire: you should be fully acquainted with the reproductive method of the species you intend to breed, and be aware of the conditions you will need to provide. This section concentrates on the breeding of freshwater fishes. The reason for this is that there is a body of documented experience on freshwater breeding, whilst not so much is known of the reproductive behaviour of marine fishes, either in nature or in captivity.

# The breeding process

Fishes reproduce by fertilization of eggs, from which the young emerge. Depending on the species, the eggs may be fertilized either outside or within the female fish's body. The fry of fishes that fertilize their eggs externally (egglayers) are small and helpless when they hatch, whereas livebearing fishes produce young that are capable of free-swimming (independent swimming at will) at birth and that can fend for themselves.

### Egglaying fishes

The majority of aquarium fishes are egglaying species, with the eggs being expelled by the female and fertilized by the male during spawning (see pp. 246–7). Within this category, the egglaying species commonly found in home aquaria can be divided into five groups according to spawning habit: egg-scatterers, egg-buriers, egg-depositors, mouth-brooders and nest-builders.

## REPRODUCTION AMONGST EGG-LAYING FISHES

Male egg-laying fishes produce sperm or "milt" in their testes. It is passed through tubes to the urino-genital vent, where it is expelled and comes into contact with, and fertilizes, the female's eggs or "ova". These are produced in her ovary and pass along the oviduct to the urino-genital vent where they are ejected, usually after stimulation from the male (see p. 244).

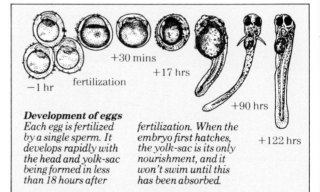

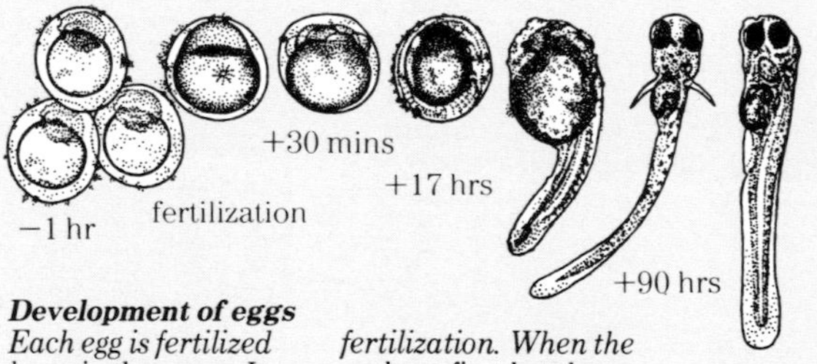


***Development of eggs***
*Each egg is fertilized by a single sperm. It develops rapidly with the head and yolk-sac being formed in less than 18 hours after fertilization. When the embryo first hatches, the yolk-sac is its only nourishment, and it won't swim until this has been absorbed.*

***The sexual organs***
*The difference in sexual anatomy of this pair of Barbs isn't visible externally.*

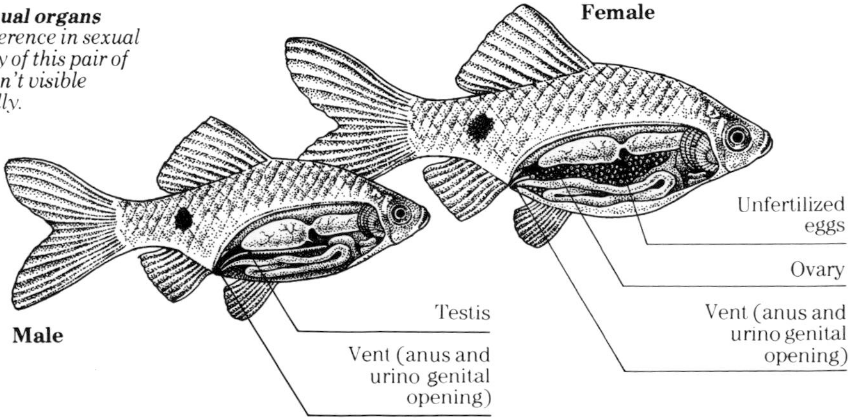

## EGG-SCATTERERS

Little or no parental care is shown by these "promiscuous" fishes. They come together when ready (often once a week), quite spontaneously, and spawn. The eggs may be either non-adhesive or adhesive, falling directly to the riverbed or becoming caught in aquatic plants. In the wild, the eggs from this spawning would probably be hidden in the muddy water or swept away by the current, so there is no need, nor instinct, for parental care.

In the aquarium, however, the eggs have no such protection, and some "egg-saving" measures such as placing rounded stones or marbles on the aquarium floor or filling the breeding tank with plants (see p. 248) must be made to keep them safe from adult fishes.

**Egg-scattering fishes**

☐ Barbs ☐ Danios ☐ some Rasboras
☐ some Characins and related species
☐ Goldfishes

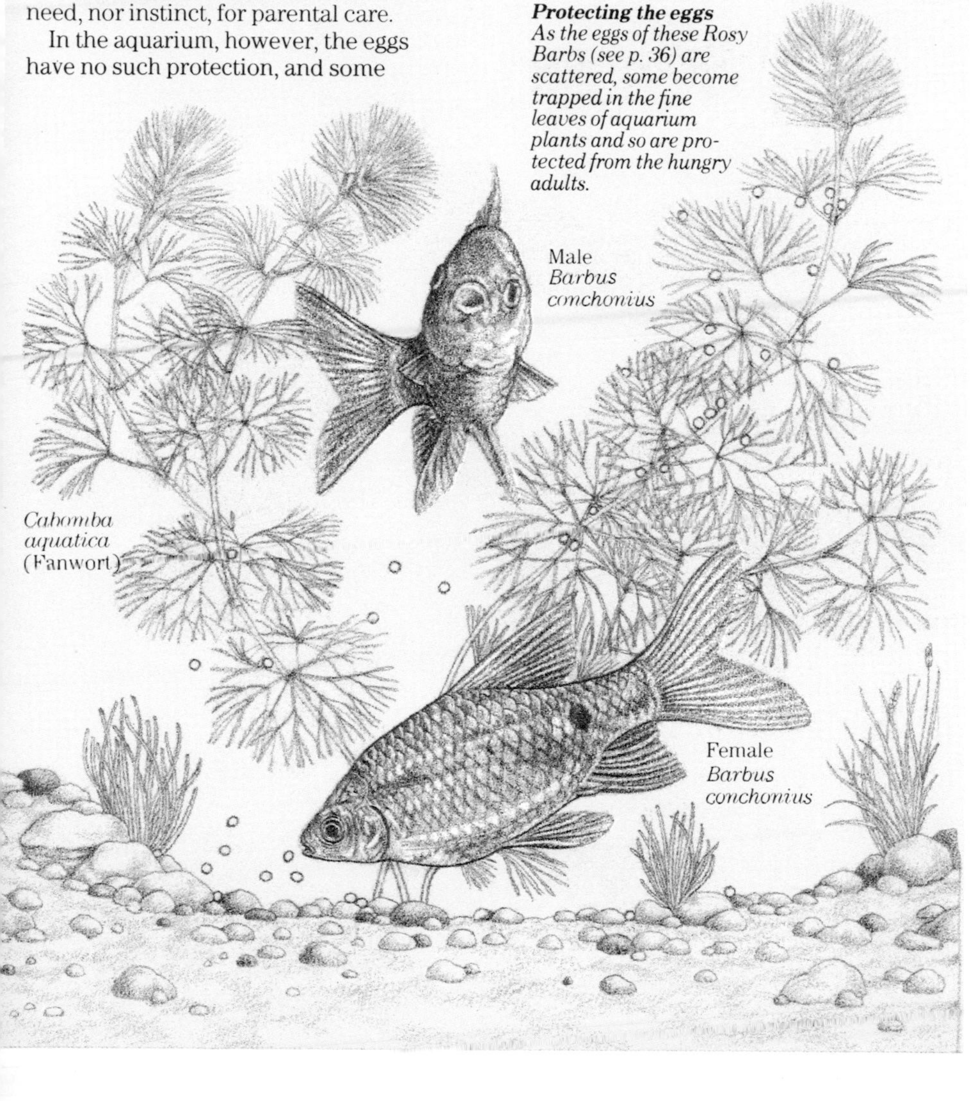


***Protecting the eggs***
*As the eggs of these Rosy Barbs (see p. 36) are scattered, some become trapped in the fine leaves of aquarium plants and so are protected from the hungry adults.*

## EGG-BURIERS

These fishes make an attempt to care for their eggs, even those that never live to see their offspring. This is because the shallow pools and streams in their natural habitat dry up completely once a year, and so if the species is to continue, any spawning must occur before this happens. The adult fishes spawn before the dry season begins, and bury their eggs in the riverbed mud so that they survive the forthcoming drought conditions in which the fishes themselves perish. With the onset of the rainy season, the eggs hatch out. The young fry then have only a few months in which to mature and spawn before they too become victims of the next dry season. The South American egg-buriers generally dive into the mud so that their eggs become buried during spawning whereas African fishes like the *Nothobranchius* genus spawn side by side while the female curls her anal fin to form a channel for the eggs to pass down into the substrate.

Because water doesn't dry out in an aquarium, most annual fishes will live longer than one year. Their eggs can be stored in moist peat – a substitute for riverbed mud (see p. 251).

Some Killifishes will adapt their spawning methods to aquarium life and lay their eggs in dense plant growth or spawning mops (to make these yourself, see p. 250).

**Egg-burying fishes** □ Killifishes (including the Annual fishes)

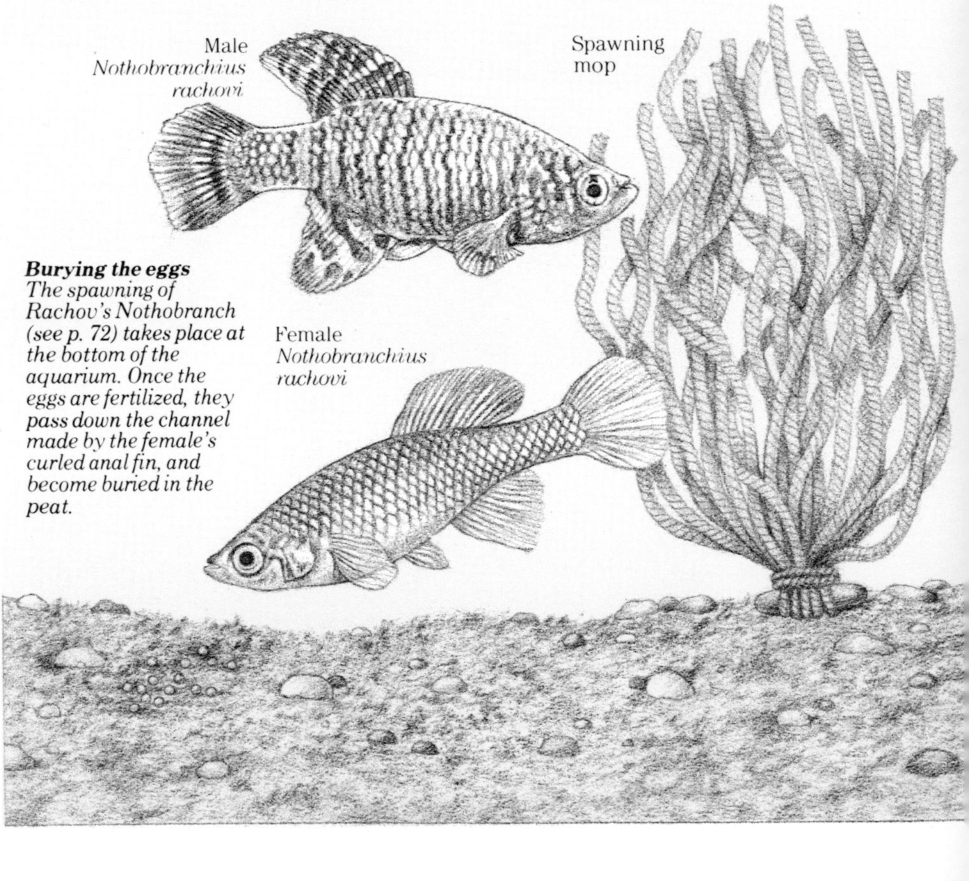


***Burying the eggs***
*The spawning of Rachov's Nothobranch (see p. 72) takes place at the bottom of the aquarium. Once the eggs are fertilized, they pass down the channel made by the female's curled anal fin, and become buried in the peat.*

## EGG-DEPOSITORS

The most significant fishes in this group are the Cichlids, which have complex spawning routines and a high standard of parental care. They select their own breeding partner, then choose and clean a spawning site (for example, pecking off any algae present). The chosen site may be out in the open, on a plant leaf or sloping rockface, or it may be a secret place inside a rocky cave or upturned flowerpot. They will forcibly evict any other fishes from the surrounding area by chasing them away.

Having laid and fertilized the eggs, the pair keep them clean by fanning them to maintain a water current flowing over them, and often by moving them to new, pre-cleaned sites, until hatching occurs. For the first 2–3 weeks after the fry become free-swimming, their parents accompany them around the tank, keeping a watchful eye for predators.

There are other species of fishes that deposit their eggs on plant leaves, on the tank sides, or in depressions in the gravel, but not all of these fishes have developed the same sense of family responsibility as the Cichlids.

One fish in this group takes extreme precautions against predators – *Copella arnoldi* (Splashing Tetra) lays and fertilizes eggs out of the water on the underside of an overhanging leaf. In the aquarium the cover-glass is often a substitute spawning site.

**Egg-depositing fishes** □ Corydoras □ Catfishes □ Cichlids □ some Rasboras □ *Copella arnoldi* □ Sunfishes

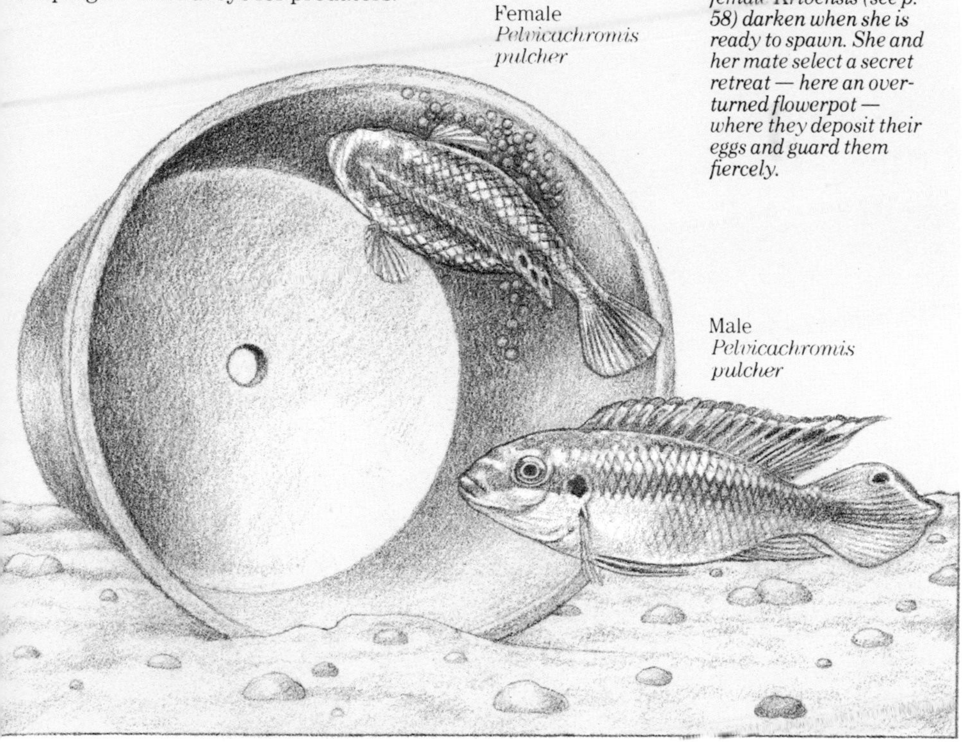


***Depositing the eggs***
*The body colours of the female Kribensis (see p. 58) darken when she is ready to spawn. She and her mate select a secret retreat — here an overturned flowerpot — where they deposit their eggs and guard them fiercely.*

## MOUTH-BROODERS

These fishes carry their eggs in their mouths until they hatch. In many cases, the young are also carried in this way. Mouth-brooders build a nest site, often an elaborate crater in the gravel, and lay their eggs in it, then the female (or, in a few species, the male) picks them up and incubates them in her throat. In some mouth-brooders the eggs are picked up after fertilization, but with others the eggs are actually fertilized during the picking-up process. With this latter type, the female is attracted towards the male's vent by imitation egg markings on his anal fin, and picks up his milt at the same time as the unfertilized eggs. During the incubation period (lasting approximately two weeks) the female takes no food.

**Mouth-brooding fishes** □ Some Cichlids □ some species of *Betta*

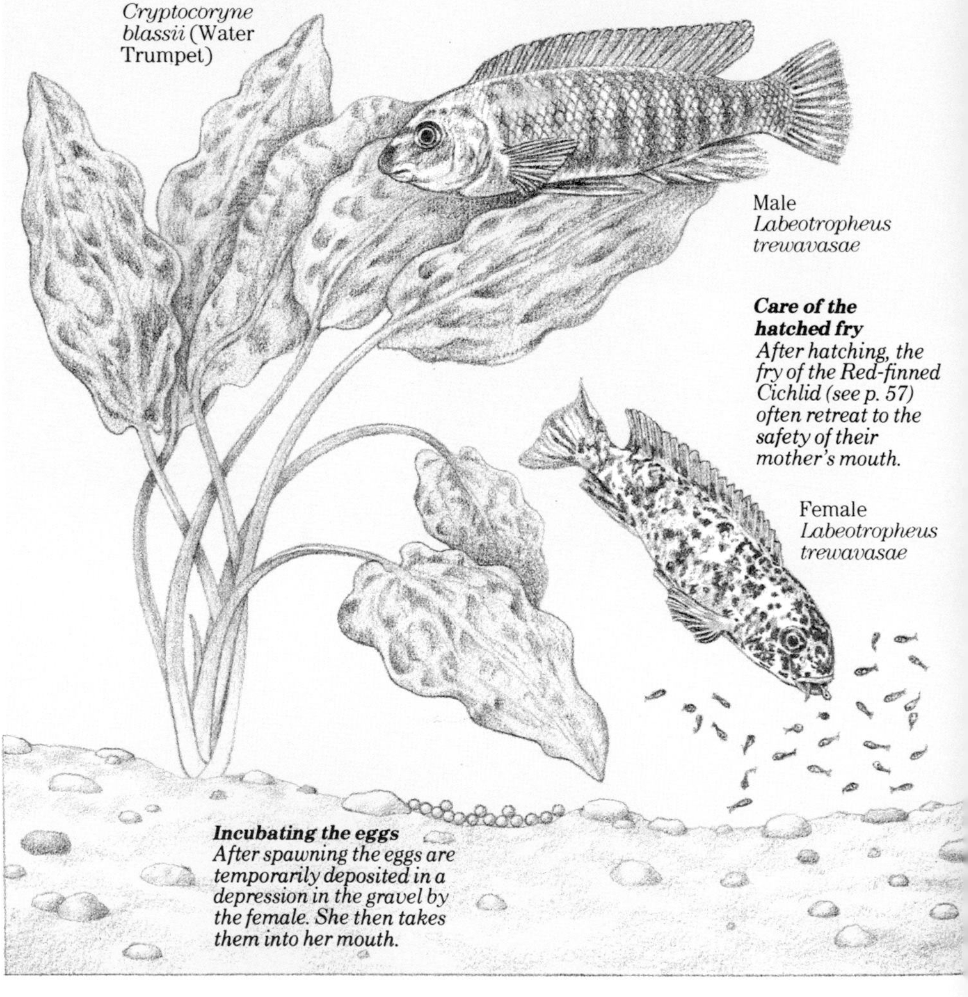


***Care of the hatched fry***
*After hatching, the fry of the Red-finned Cichlid (see p. 57) often retreat to the safety of their mother's mouth.*

***Incubating the eggs***
*After spawning the eggs are temporarily deposited in a depression in the gravel by the female. She then takes them into her mouth.*

## NEST-BUILDERS

These fishes construct nests in which their fertilized eggs are laid and guarded until hatching occurs. Usually built by the male fishes, the nests are either a mass of saliva-blown bubbles floating at the surface or an area for egg deposition prepared at the bottom. As the male fish tends to be aggressive towards the female after spawning, it is best to remove her once she has released her eggs. The male guards the eggs fiercely.

**Nest-building fishes** □ *Betta splendens* □ Gouramis □ some Sunfishes □ some Cichlids □ a few Catfishes

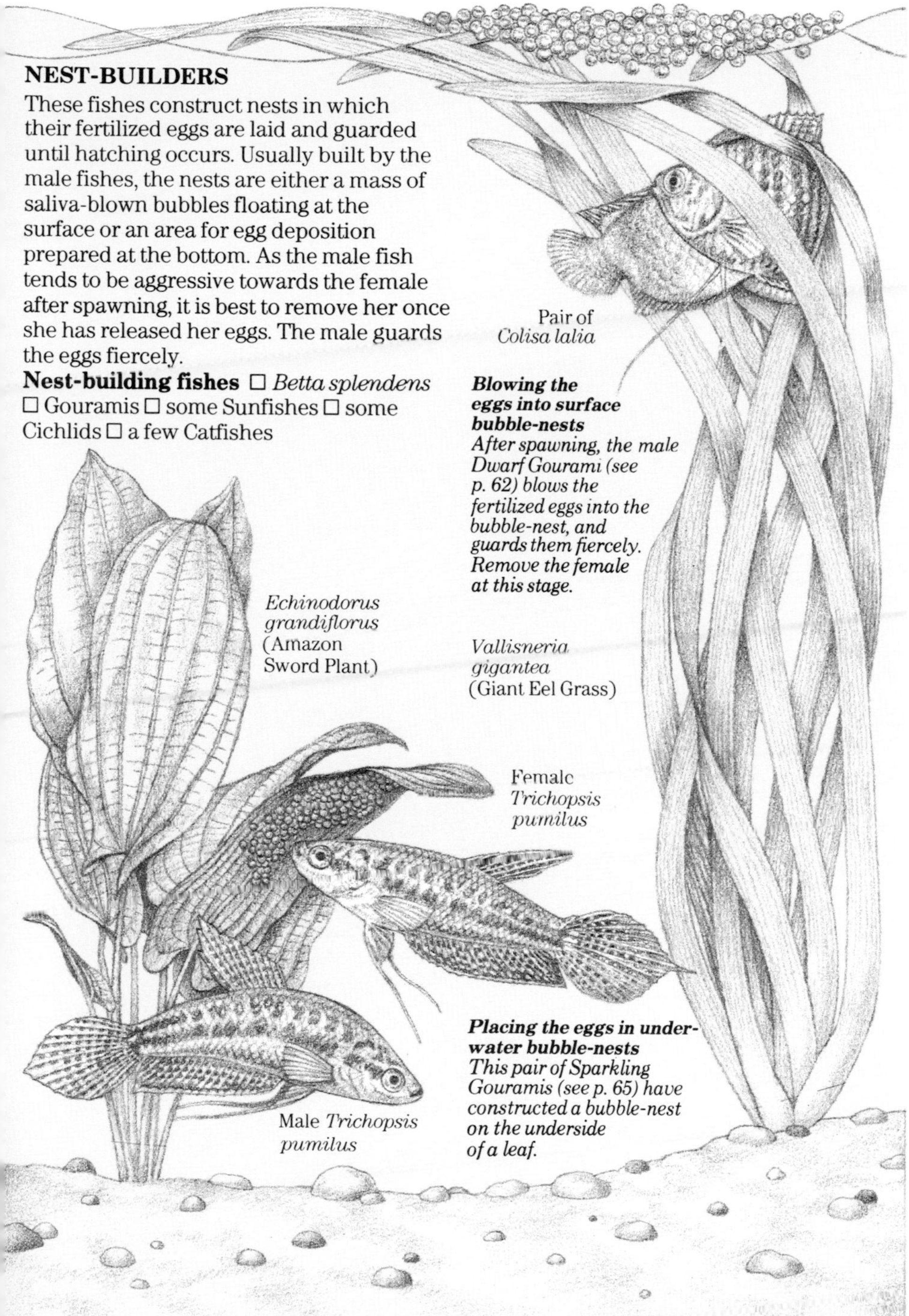


***Blowing the eggs into surface bubble-nests***
*After spawning, the male Dwarf Gourami (see p. 62) blows the fertilized eggs into the bubble-nest, and guards them fiercely. Remove the female at this stage.*

***Placing the eggs in underwater bubble-nests***
*This pair of Sparkling Gouramis (see p. 65) have constructed a bubble-nest on the underside of a leaf.*

## LIVEBEARERS

Livebearing fishes give birth to 30–200 fry at various intervals. Some females are able to store the male's sperm in their body for several months, and so successive broods can be born from one mating. The fry may appear in rapid succession from the female's vent.

**Livebearing fishes** ☐ Guppies ☐ Platies ☐ Swordtails ☐ Mollies ☐ Halfbeaks

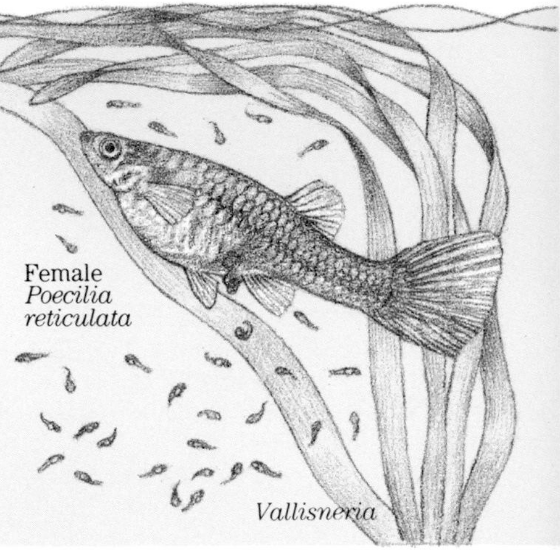


***Free-swimming fry***
*The fry of Guppies (see p. 67) are usually born tail-first. To avoid being eaten, they hide among floating aquarium plants.*

## LIVEBEARERS

Livebearing fishes differ from egg-layers because the eggs develop *inside* the female fish, not outside. The male's anal fin – the gonopodium – is modified so he can introduce sperm into the female. There are two types of livebearers: in "viviparous" species the eggs are nourished through the female's bloodstream, but in "ovoviviparous" types nourishment is provided by the yolk-sacs.

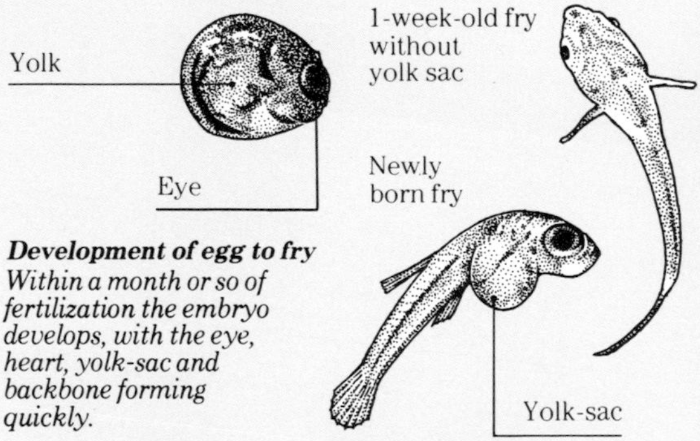


***Development of egg to fry***
*Within a month or so of fertilization the embryo develops, with the eye, heart, yolk-sac and backbone forming quickly.*

***The sexual organs***
*Both the interior and exterior anatomy of the sexes differ. The eggs can be fertilized and will then develop in the ovary of the female.*

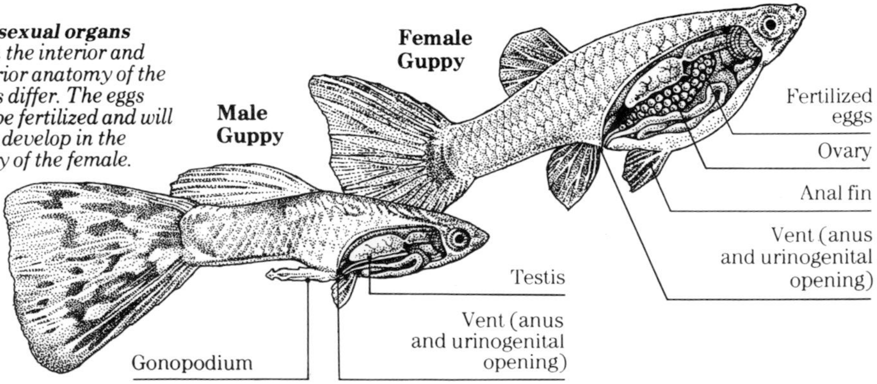

# Preparing for breeding

Begin by selecting a suitable male and female of the species you wish to breed. They must be free from any physical and colour-pattern deficiencies which might be passed on to their fry, and they must also be fully mature and in peak condition. With the majority of egglaying fishes you can select pairs for spawning. However, the Cichlids are an exception as they will choose their own partners.

**Egglaying fishes**

Differentiating between the sexes can be difficult with egglaying species. There is a physical difference in the breeding tube structure, but you need a practised eye to detect this. However, the males are often slimmer, and have brighter colours and longer fins than their intended partners, whilst the females are fatter when viewed from above, due to the build-up of eggs within them. Some male fishes (for example, Goldfishes) develop tiny white pimples (known as tubercles) on the gill covers and head at breeding times, whilst other species change colour. For example, the male *Rhodeus sericeus amarus* (Bitterling) changes from silver-grey to brilliant violet and green during the breeding season.

Observation of the fishes' behaviour may also give some clue which will help you to distinguish between the sexes, although with spontaneously pairing fishes you may have to wait until you see which fish actually lays the eggs to confirm your supposition.

**Livebearing fishes**

Sexing livebearers is easy (see right). Moreover, during pregnancy the female's profile becomes enlarged, the area around her vent may become darker in colour and, towards the end of gestation, she becomes squarer in shape.

## SEXING LIVEBEARERS

In most species the male's anal fin is rod-like and known as the gonopodium, whilst others have a notched fin. The female's anal fin is fan-shaped.

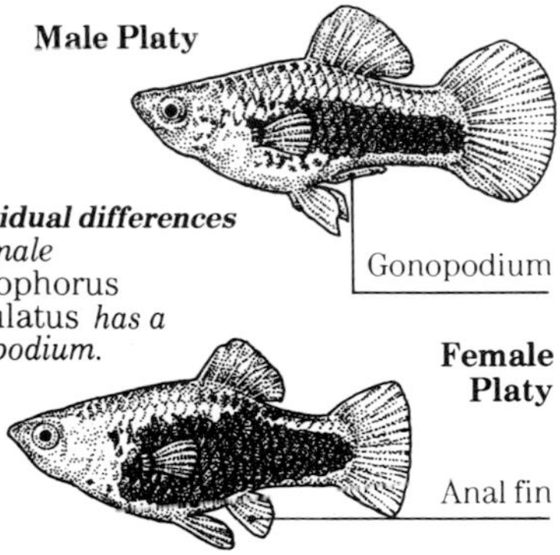


***Individual differences***
*The male* Xiphophorus maculatus *has a gonopodium.*

**Conditioning the adults**

To ensure a successful spawning with a good number of fry, both adult fishes (particularly if they are of an egglaying species) must be ripe for spawning. This can be encouraged in two ways: by feeding them with high-quality, predominantly live foods, and by separating the potential pair for two weeks prior to spawning. The separation can be in a single, divided tank so that the two fishes are still within sight of each other. This ploy also tends to stimulate them further into anticipating spawning. After a fortnight, simply remove the divider. With species that select their own partners, this separation period isn't required. Instead, before you place the fishes in a separate breeding tank, watch the group of fishes from which you intend to breed closely over a period of time, to see whether any particular two fishes show interest in each other. These will make a suitable breeding pair.

# The spawning

In the wild, fishes have a definite breeding season, which may last for about three months and usually follows the wet season when there is plenty of food available. In the aquarium, however, it should be possible to get pairs to breed at any time of the year. They will be stimulated to do this if they are given nutritious food and kept at the appropriate temperature (see pp. 248–9).

The cycle of spawnings varies from species to species. For example, given a good diet and correct conditions, healthy egg-laying fishes (see p. 236) usually become ripe for spawning every 10–14 days (during this time the female will fill up with eggs again). For livebearing fishes (see p. 244), the period of gestation from fertilization to birth is approximately one month, and the female then requires a few days rest before her next spawning.

## Introducing the fishes

If egglaying species have been conditioned in a divided tank (see p. 245), the reunion is a simple matter – just remove the partition. But if totally separate conditioning has occurred, it is wisest to introduce the female into the breeding tank ahead of the male, usually the night before. This not only gives her time to settle down and to find places of refuge, but also means that the male has to court

**Spawning amongst bubble-nest builders**

The spawning sequence of *Trichogaster leeri* (Lace Gourami) is typical of most Anabantoids, and involves elaborate courtship movements. The male instigates and dominates the proceedings.

***Stage one***
*The male begins to take an interest in the female, moving round her body to manoeuvre her beneath the bubble-nest.*

***Stage two***
*Once under the nest, the male wraps his body around the female. As her eggs are ejected, he covers them with milt, and places them in the nest.*

her on her territory, and so is less likely to be aggressive.

With livebearers it is most important to keep only similar colour strains together at all times, as the fishes would otherwise spawn randomly. Livebearers are prolific, and select their own partners.

## The spawning act

It is best to watch the introductory moves during the spawning as some males become very aggressive. If a male attacks the female you should remove her, re-separate the pair or try another conditioned female. Alternatively, it is sometimes a good idea to let the male have two or three females with him, so that one female doesn't take all his attention. Egg-scattering fishes can also be spawned in a shoal, rather than pairs.

## Hand-stripping

It is possible for fishes to spawn without ever meeting as Goldfishes and Koi can be hand-stripped of their ova and milt. In such cases, the water temperature is gradually brought up to the required breeding level over a week to ten days. Once the right temperature has been reached, the fishes are removed from the water and held in a wet towel. Then their flanks are gently pressed so that the ova and milt are expelled from their vents. These are mixed and specially incubated until hatching occurs. This is a complicated procedure, only carried out by experienced fishkeepers and breeders. It has two advantages: by selecting the best parents, the stock can be improved, and by controlling a breeding line new colours can be developed.

### Vertical leaf-spawning

Some species such as *Pterophyllum scalare* (Angelfish) spawn on leaves or slate.

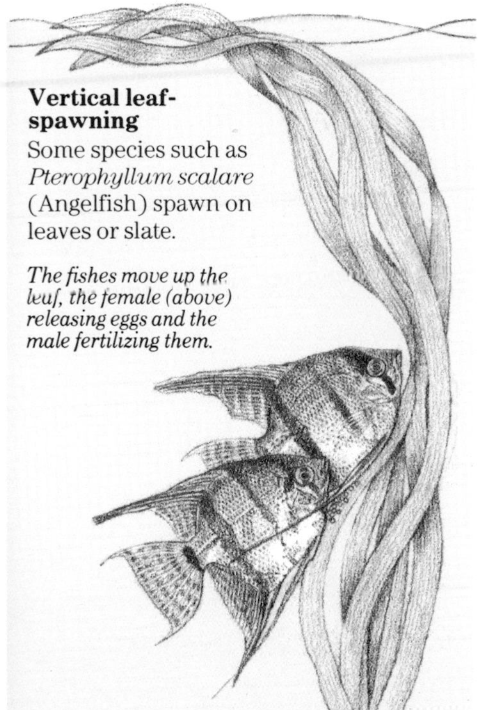

*The fishes move up the leaf, the female (above) releasing eggs and the male fertilizing them.*

### Spawning via a bivalve

A few fishes can only reproduce in the presence of independent organisms. For example, *Rhodeus sericeus amarus* (Bitterling), must have access to certain mussels in order to breed.

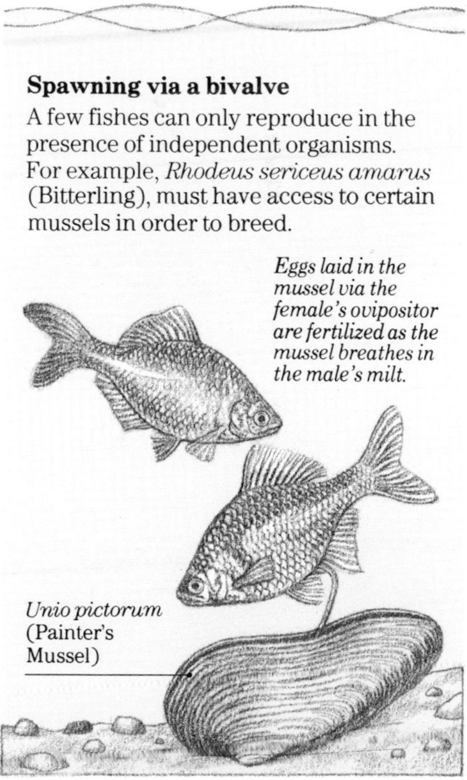


*Eggs laid in the mussel via the female's ovipositor are fertilized as the mussel breathes in the male's milt.*

# Setting up breeding tanks

Although spawning will occur in the main aquarium, it is best to set up a separate breeding tank. The fishes prefer privacy, there is a better chance of the fry surviving away from other fishes, and a separate tank is simpler to control.

**Providing the right conditions**

The breeding tank should be set up in advance of the intended spawning date. You must also make sure that the general conditions in the tank are right.

*Filtration*

It is vital that water conditions remain healthy, but the filtration equipment used must be chosen with care. A strong flow is inadvisable as the tiny fry could easily be sucked into it, and an under-gravel filter can be a problem because

## BASIC BREEDING TANK REQUIREMENTS

When setting up a breeding tank the spawning behaviour of the fishes involved must be considered. For those species that eat their own eggs or fry the tank must be arranged so that this can't happen. Egg-burying fishes need something to bury their eggs in; some nest-builders will appreciate plant material to incorporate in their nest; and secretive egg-depositors need retreats. The size and shape of the breeding tank is important too: a 60—90 cm long tank is necessary for very active or large egg-laying fishes, whilst a 45 cm tank will suffice for a gravid female livebearer or for smaller egg-laying fishes.

***Egg-scatterers' tank design 1***
*Nylon curtain net draped in the tank will allow the newly laid eggs to fall through to the tank floor, where they can't be eaten. Fishes can be spawned in pairs or as a shoal using this method.*
- Barbus conchonius
- *Nylon curtain net*

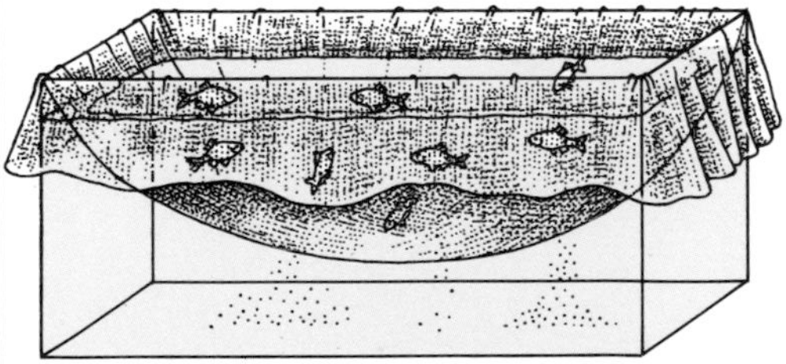

***Egg-scatterers' tank design 2***
*You can protect the eggs by putting a layer or two of glass marbles on the tank floor into which the eggs will fall, beyond the reach of the hungry parents.*
- Barbus tetrazona
- *Glass marbles*

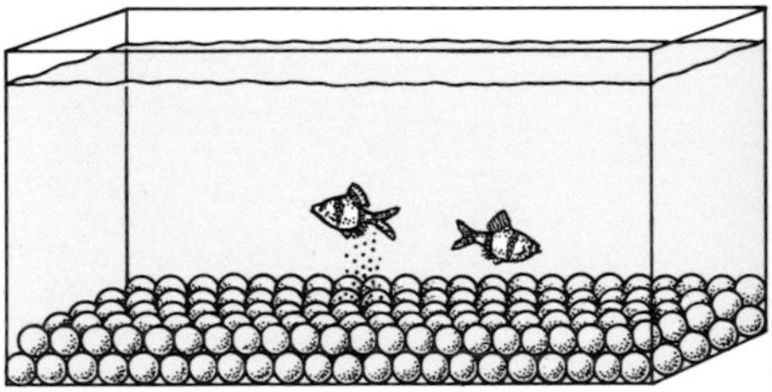

***Egg-scatterers' tank design 3***
*Furnish with artificial spawning mops (see p. 248) or bunches of fine-leaved plants planted in gravel-filled seed-trays to provide traps for the eggs and prevent them from being eaten.*
- Brachydanio rerio
- Ceratophyllum
- *Gravel-filled seed-tray*

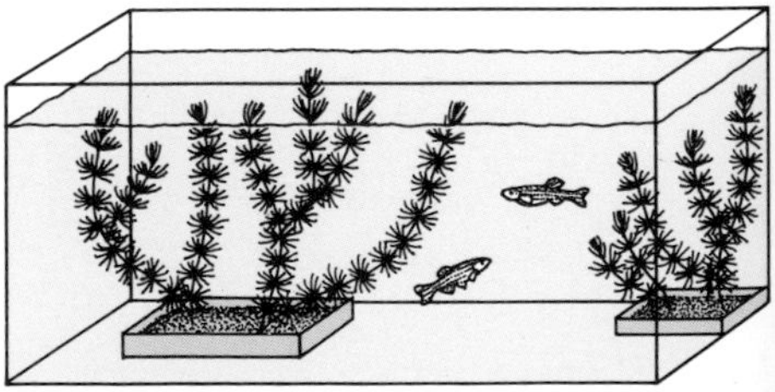

tiny fry could be drawn down into the gravel with the water current. A simple air-operated sponge-type filter (see p. 125) is safer. As the fry grow larger, internal box filters (see p. 126) can be used.

*Aeration*
If the aeration in the breeding tank is too strong, the turbulence will buffet the tiny fry. A gentle flow of air from an airstone is all that is needed. However, you should gradually increase the aeration as the fishes grow larger.

*Water conditions*
You must be able to vary the water conditions (see p. 147), especially the temperature. Most fishes are stimulated into breeding by a slight increase in water temperature, say, 2°C. When setting up a breeding aquarium, initially the water temperature should be the same as that in the main tank, but once the fishes have been introduced you should gradually raise it.

Conversely, a few species, such as the *Corydoras* Catfishes, may be stimulated

***Nest-builders' tank***
*Provide plants to act as refuges or for use as nest-building material.*
- Colisa lalia
- Ceratophyllum
- Hygrophila
- *Any base covering*

***Mouth-brooders' tank***
*These fishes require a 5 cm-deep gravel layer*
- Labeotropheus trewavasae
- Vallisneria
- *5 cm gravel layer*

***Egg-depositors' tank***
*Provide logs, flowerpots and plants as spawning sites.*
- Etroplus maculatus
- Pelvicachromis pulcher
- Pterophyllum scalare
- Echinodorus
- Eleocharis
- Sagittaria
- Vallisneria
- *Logs, flowerpots*
- *Base covering*

***Egg-buriers' tank***
*For egg-burying species, you should line the aquarium floor with a 5 cm-deep layer of peat fibre. Peat should be boiled before use in order to kill any harmful micro-organisms.*
- Nothobranchius rachovi
- *Garden peat*

***Livebearers' tank***
*Although a gravid livebearing female can be placed in a breeding trap, it is better to give her a small well-planted tank. This will provide adequate sanctuary for the fry.*
- Poecilia reticulata
- Cabomba
- *Base covering*

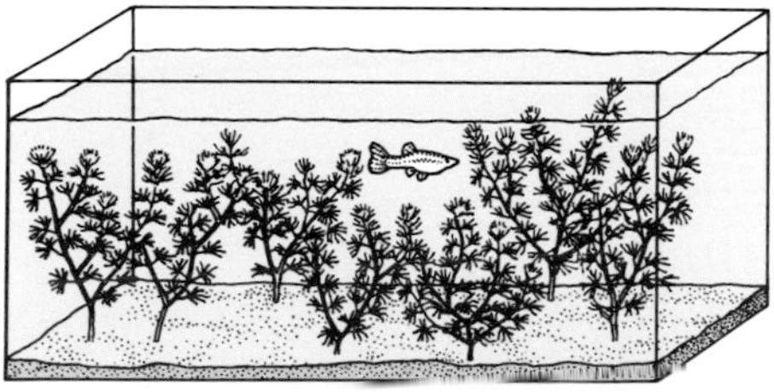

into breeding by siphoning off 2 cms of water and replacing it with cold water.

*Plants*

Where plants are required, place them in a shallow seed tray filled with gravel. No gravel is required in the tank itself.

**Siting the breeding tank**

Although main tanks should be sited to avoid direct sunlight (see p. 120), a breeding tank can be placed so that some morning sunshine reaches it. This is acceptable as the fishes' length of stay in this tank is brief, and because sunlight is said to stimulate breeding.

**Breeding traps for livebearing fishes**

To prevent livebearers eating their fry, breeding traps can be used. These are small tanks placed within the main tank to confine the female, but from which the young can escape. I would advise against using them as many females are ultra-sensitive at "confinement time", and may give birth prematurely if so restricted.

## MAKING ARTIFICIAL SPAWNING MOPS

In their natural habitats, the adhesive eggs of many fishes become caught in the leaves of bushy plants and are thus protected from being eaten. In the aquarium, artificial spawning mops can be used as substitutes. These have the advantage that they make pre-hatching, collection and transferral of the eggs elsewhere easy (see right).

**1** You will need some nylon wool, a pair of scissors and a small book or piece of card (approx. 6–8 cm wide).

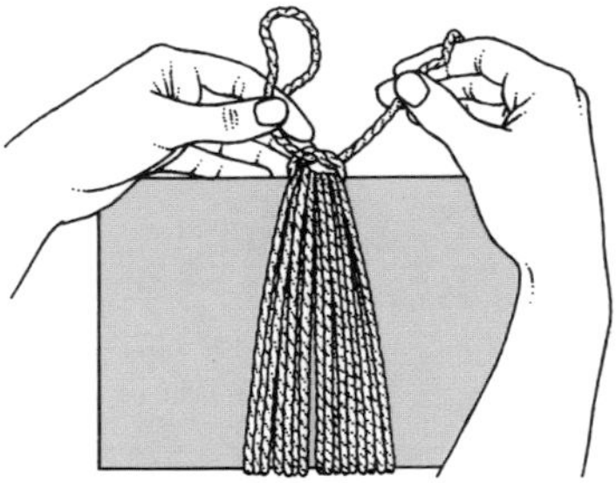

**2** Wind the wool around the book several times, securing with a single strand along the book's spine under the wool turns.

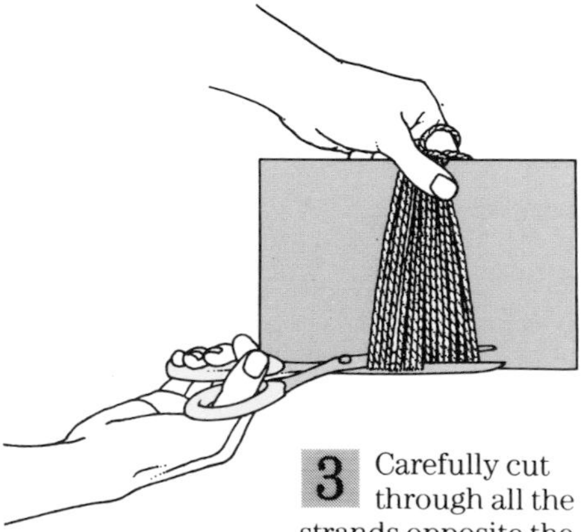

**3** Carefully cut through all the strands opposite the knotted single thread.

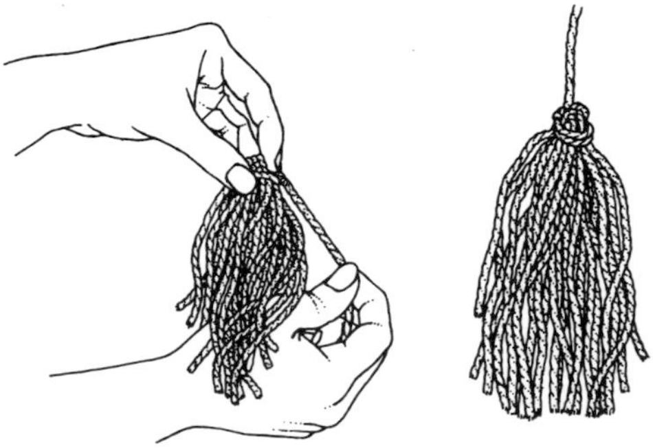

**4** Tease out the strands into a fluffy mass. Suspend the mop by hanging it over the side of the tank or by attaching it to a floating piece of cork.

# After the spawning

Many aquarium species will spawn within an hour of being put into the breeding tank. You must now wait for the young fishes to hatch if they are egg-laying species or to be born if they are live-bearers and prepare to provide for their initial needs.

## Post-spawning care

Depending on the species, you may need to remove one or both of the adult fishes from the breeding tank after spawning in order to protect the eggs or fry, or in cases where the male is inclined to be aggressive, the female. You must also ensure that the general conditions are correct for the eggs to hatch.

*Egg-scatterers*
Once the spawning has taken place, remove egg-scattering adults and shade the breeding tank from direct light. Leave the eggs to hatch.

*Egg-buriers*
These species should be removed from the tank after spawning. The eggs can be gathered and hatched in a number of ways (see below).

*Nest-builders*
The female should be removed immediately after spawning, as the male usually takes over the guarding of the nest and fry, and may be aggressive.

## HATCHING EGG-BURIER'S EGGS

After spawning, collect the eggs. Those from mops are put in a water-filled container, and peat-laid eggs are stored in a bag. Activate hatching by immersing in water of the correct temperature.

**Mop-laid eggs**
*You can see when the eggs are about to hatch as the eyes will show first. (The eggs shown here are five times larger than life-size.)*

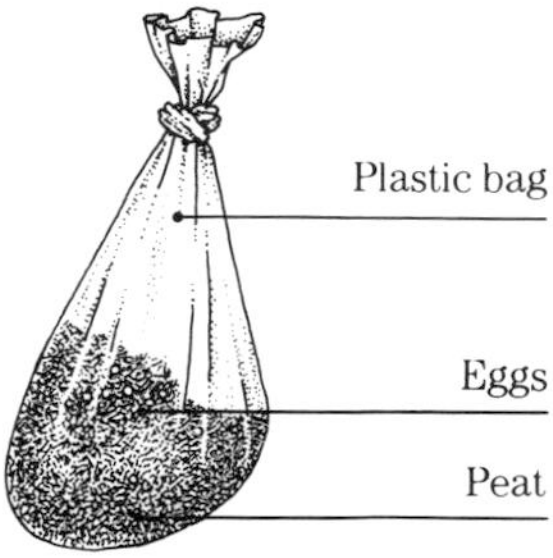


**Hatching mop-laid eggs**
*Pick eggs out of spawning mops with tweezers, then place them in a shallow plastic container with a 2–3 cm depth of water.*

*The egg-filled container can either be floated on the surface of your aquarium, placed on top of it or put in a warm place such as an airing cupboard until hatching occurs (usually within 2–3 days).*

**Hatching peat-laid eggs**
*Gather the eggs and peat and squeeze out the moisture. Seal them in a plastic bag, date it, and store in a warm place. After 2–3 months, immerse the peat in water to hatch the eggs.*

*Cichlids*
It isn't unusual for Cichlids to eat their first spawning. If this occurs regularly, remove the adult fishes after spawning and hatch the young artificially. If not, leave the adults with the eggs so that they can care for them. Don't be alarmed if you see them take the eggs into their mouths – many Cichlids do this to clean them, and will then replace them in the same or a different site.

*Mouth-brooders*
After hatching, the young continue to seek the safety of their mother's mouth at night or when danger threatens. Unfortunately, if the female is alarmed she may eat them. You should therefore keep relatively near the tank immediately after spawning, so that she will get used to you being there and both she and the fry won't be so nervous.

*Livebearers*
Move the female to a well-planted "nursery" tank after spawning, and maintain the temperature at around 24°C. If it drops below this, the embryos will develop slowly and hatching will be delayed. You should be able to tell when the gravid female is about to give birth as a dark mark, known as the gravid spot, will appear close to the urino-genital vent. After giving birth, she should be allowed a few days recuperation before being put back into the main aquarium.

### Egg-hatching times

The length of time it takes for a fertilized egg to hatch will depend on the particular species. However, at ordinary aquarium water temperatures, fertilized eggs from the majority of egg-scattering and nest-building species will hatch in 18–72 hours. The tiny fry will cling to the aquarium sides or hang beneath the bubble-nest like splinters of glass, but they won't swim until they have absorbed the contents of the attached yolk-sac, which may well take another day or two.

Fertilized eggs from egg-depositing species will also hatch in much the same period, but will often remain inactive for a week to ten days before becoming free-swimming. However, the fry of secretive spawners aren't usually seen until they are free-swimming. With mouth-brooders, it may be up to two weeks after fertilization before the young fishes emerge from their mother's mouth.

Killifish eggs may take up to three months to hatch depending on the species. Because the eggs can be kept for long periods, you can regulate the hatching to suit your own requirements. You may, for example, choose to space out the hatchings so that you don't have too many fry to look after at one time, or you may time hatching so that the young fishes will be at a suitable age for entry in a "Breeder's Teams" class at a fish show (see p. 256).

### BREEDING TIPS

- Make sure that the spawning pair are free from physical defects
- Check that the temperature and furnishings in the breeding tank are suited to the species
- Never transfer fishes to a breeding tank without conditioning them slowly beforehand
- Don't disturb the fishes whilst spawning (cover the front glass if necessary)
- Protect the eggs/fry from hungry parents and other predatory fishes
- Don't breed fishes indiscriminately; keep the colour strains pure
- Check your filter box regularly – fertilized eggs from unnoticed spawnings may be drawn into the filter and hatch there

# Raising fry

Vith all young fry, the first few days of ife are the most critical. If you can get hem to start feeding (see p. 209), they tand a good chance of surviving. But lon't expect to raise all of them; it is far etter to raise a few quality fishes than a ot of sub-standard ones. You should be uthless and cull them, throwing out the mall and deformed ones. The fry can be dded to the main tank once they are big nough not to be eaten.

## ighting

f you leave the aquarium lamps on all he time, the fry will be active and feed onstantly, and so will grow quicker. Iowever, it isn't necessary to have full ghting – a 25 watt bulb is sufficient.

## exing livebearer fry

Vith livebearer fry, sex (see p. 243) and eparate them as soon as possible. If in doubt, put the fish in the males' tank not the females', so that it won't interfere with your line-breeding programme.

## Special care for Gouramis

Fry of the Gourami species are prone to cold draughts; a towel draped over the aquarium hood will prevent cold air getting in. If you have a space-heated fish-room, you won't need to do this as there is no risk from cold draughts.

## Keeping breeding records

Always keep written records of your attempts at breeding fishes, particularly if you are spawning a difficult species. If you get it right, you will want to pass on your method to other fishkeepers.

***Looking after fry***
*Like many Cichlid fishes, this pair of Kribensis (see p. 58) display exemplary parental care, guarding their fry fiercely.*

9

# SHOWING FISHES

It may seem strange to you that people take fishes out of their carefully set-up and maintained aquariums, load them into insulated boxes, drive them around the country, and then put them into small, bare tanks, simply to be told how good or bad they are. There aren't any financial gains to be made, since a prize fish can't command high stud fees in the way that a champion dog, cat or horse can, so why should you consider exhibiting at a show? Apart from the pleasure of showing off your pride and joy, the main reason for visiting shows is the opportunity to see species that are quite unfamiliar or unobtainable in your own area. You will also benefit from meeting other fishkeepers and exchanging experiences, breeding hints and advice.

# What is a fish show?

The size of shows and the way that they are organized varies, as shown below. But a basic procedure is common to all shows: exhibitors arrive with their fishes and put them into small, bare tanks, then qualified fish judges assess the entries against the national show standards and select the winners. For show purposes, fishes are split up into groups (known as classes) according to their natural species families: for example, there are separate categories for Barbs, Characins, and Cichlids. The Goldfish group is "segregated" still further into subgroups such as Singletails or Twintails; the number of these classes depends on the size of the show. First, Second and Third prizes are awarded within each class.

**The judging procedure**

At most shows, the exhibitors and the general public are asked to leave the hall while judging takes place. There is a separate judge for each particular class, but the judges may get together to evaluate certain classes such as "Furnished Aquaria" or "Aquascapes". The judge inspects each entry in a class

## HOW IS A SHOW ORGANIZED?

The opportunity for exhibiting fishes begins at local society level, during the regular meetings.

**Bowl shows**

Small shows are held regularly at the meetings of local aquatic societies; these are often referred to as table or bowl shows. The usual practice at these shows is for all the classes of fishes to be covered during the year, so that all members have the opportunity to exhibit, no matter what sort of fishes they keep. In some societies, the different species classes are spread out over the whole year.

**Area shows**

Moving up the scale, the next level of exhibiting is at an area show. This is a one-day inter-society match at which visiting societies take part. In general, you can only enter such shows if you belong to one of the societies concerned.

**Open shows**

Truly "open" shows, which anyone can enter whether they are a member of a society or not, are also one-day affairs. They are much larger than area shows, and are often considered the

**Show classes**

As well as the expected classes for individual aquarium fishes, most shows have several other categories that you can enter. The "Plants" and the "Furnished Aquarium" classes are particularly popular in Europe.

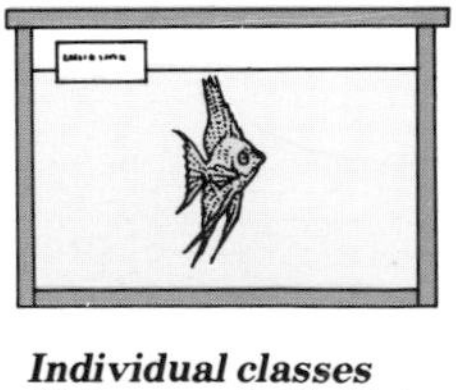

***Individual classes***
*Fishes are divided into classes according to species, and entries are shown singly in small, bare tanks.*

***"Fish Pairs"***
*A matched pair (one male, one female) from a single species are shown together in a small, bare tank.*

***"Breeder's Teams"***
*Entries consist of 4 or 6 fishes less than 14 months old from a single spawning, shown in a small, bare tank.*

individually, and assigns marks to it on an official judging sheet by comparing it to the show standards, *not* to the other fishes present. The fish that gains the highest points in each class becomes the class winner.

At the end of judging, all the class winners are eligible for the "Best in Show" title (see p. 259), and the fish that wins this award is usually selected as a result of a consensus decision between the judges. The "Best in Show" or "Championship Class" award-winning fish will be eligible to compete for "Champion of Champions" or "Supreme Champion" titles at the large national exhibitions (see below right).

**Who organizes the shows?**
Fish shows are organized by aquarist societies, and advertised in the hobby magazines. The show season runs from early spring to late autumn each year. Shows are usually held at weekends, and are one-day affairs: entrants arrive before midday, when judging normally takes place, and the results are known by late afternoon.

Larger shows are often coupled with trade exhibitions, and are promoted by a national aquatic body, often in conjunction with the publishers of one of the hobby magazines. This type of show usually extends over a weekend or national holiday period.

event of the year by the society hosting them. In addition to single fish classes, most open shows include displays of furnished aquariums, and have extra classes for pairs of fishes, home-bred fishes, and aquarium plants.

**Exhibitions**
The largest fish shows are known as exhibitions, and are usually organized by one of the national aquatic bodies, sometimes in conjunction with a hobby magazine. Exhibitions are held over a weekend, and are open to anyone, society member or not. They have the same range of classes as open shows, but with entries in the hundreds, rather than the tens. The makers and suppliers of aquarium equipment and foods are well represented.

***"Aquarium Plants"***
*This class doesn't include fishes. Each entry consists of a single specimen, shown rooted in a pot, as cuttings or floating.*

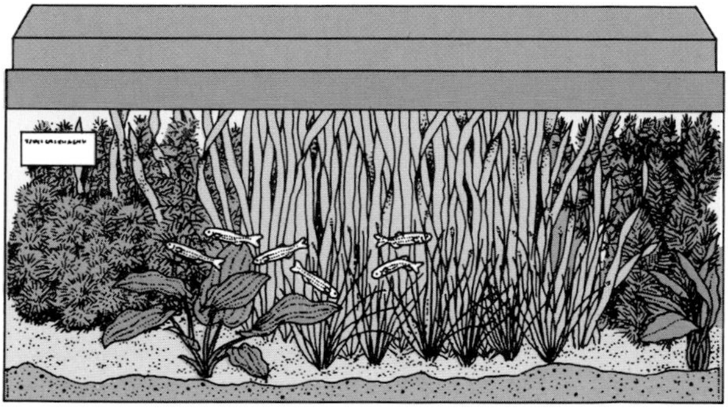

***"Furnished Aquariums"***
*This is the only show class that covers a complete set-up. Healthy, compatible fishes and plants are presented in a medium-sized tank.*

# What makes a good fish?

Assessing a fish's qualities can be very subjective – what seems an attractive fish to one person may look very ugly to another, so how can any fish be adequately evaluated? In exhibition terms, the national aquatic bodies have solved this problem by setting up show standards. Each fish on show is compared to these standards by a qualified judge. The national bodies provide training courses for the judges, who work their way up the proficiency scale in the shows run by small societies, before they get a qualification to judge at national exhibitions.

**A typical fish evaluation system**

A popular judging system is to allocate each fish in a single fish class a theoretical total of 100 points. This total is divided into five groups of 20, with each group covering a specific "quality" which the judge looks for. These may be from a range of characteristics, including size, body, fins, colour, condition and deportment.

**How is size judged?**

+ Points are awarded, up to a maximum of 20, for a fish's size compared to the agreed maximum size laid down in the show rules.
+ Points are awarded on a rising scale which slopes steeply upwards as the final growth size is approached. The reason for this is to reward the skilful individual who can care for a fish so that it grows to its full size, as almost anyone can grow a fish to half size.

**How is the body judged?**

+ Points are awarded for a body shape that compares well to that expected of the natural species.
— Points are deducted if the fish's body is deformed or out of proportion.
— Points are deducted if there is any deviation from the "standard" required for that particular fish. For example, the absence of barbels around the mouth of a Tinfoil Barb.

**How are the fins judged?**

— Points are deducted if any fins are missing. Occasionally, fish are born without pelvic fins, and it is surprising how many people don't notice this. (N.B. Some species of fishes lack these fins naturally.)
— Points are deducted if the fins aren't in good condition (for example, if they are split or frayed).
— Points are deducted if there is any deviation from the "standard" required. For example, in some cultivated varieties such as Fancy Goldfish or Siamese Fighting Fish, fins must be carried in a certain manner, or be of certain sizes in proportion to each other or to the fish's body size.
— Points are deducted if the single, filament-like fins are bent or branched.

**How is colour judged?**

— Points are deducted if the fish's colour deviates either from that found in nature, or from the rigorous standards set down for aquarium-developed varieties.
— Points are deducted if the colour isn't dense and evenly distributed. And if the colour is expected to spread from the fish's body onto its fins, then the fins must also be densely coloured with no signs of fading.
— Points are deducted if the colour patterns in cultivated fishes aren't clearly marked, or if adjacent colours that should be distinct blur together.
— Points are deducted from fishes whose colours are artificially-heightened –

brought on by the feeding of colour-intensifying hormonal foods, rather than by good fishkeeping and the feeding of nutritious live foods.

**How is condition judged?**
— Points are heavily deducted or the fish is disqualified if it is diseased. A note from the judge giving the reason will be provided.
— Points are deducted from a female livebearer if she is heavily pregnant or gives birth during the show.

**How is deportment judged?**
+ Points are awarded for fishes that show off their good points naturally and fearlessly.
— Points are deducted if a fish sulks or cowers in the corner of the tank, rather than swimming freely around in the centre. After all, if the fish doesn't show itself to the judge, how can it be successfully evaluated?
— Points are deducted if a fish isn't behaving according to its natural disposition. For example, the judge wouldn't expect a bottom-dwelling fish to be cavorting around the upper levels of the tank, or a top-swimming fish to be lying on the bottom.

**Other considerations**
In certain classes, the individual "20s" may be awarded for different characteristics:

- "Size" may be omitted in Goldfish classes, to be replaced by "Characteristics", which evaluates the differences between the varieties more accurately.
- In a "Fish Pairs" class, a degree of matching is required between the male and the female.
- In the "Breeder's Teams" class, points are awarded for "Achievement and Difficulty", "Size for Age", and "Matching", as well as "Colour" and "Condition and Deportment".
- Obviously, a different points system is used in the "Aquarium Plant" and "Furnished Aquarium" classes, where factors other than fish qualities are involved in the judging.

**How are plants judged?**
The typical evaluation system used for judging plants divides a potential maximum of 100 points into five 20-point categories, these are: "Size", "Colour", "Leaves", "Difficulty" and "Condition".

**How is a furnished aquarium judged?**
Plants and fishes used in this class must come from the same geographical area. The fishes only account for 20 out of the possible 100 marks. Another 20 marks are allotted to the plants, and 20 to the rockwork and gravel. The last two sets of 20 points are given for "Design" and "Technique".

**How is the "Best Fish in Show" class judged?**
Once all the fishes in the show have been judged, a further selection is made from the winners of the Single Fish classes to determine the overall Best Fish in Show. This fish won't necessarily be the fish that gained the highest number of points as some judges mark higher or lower than others. So each judge puts forward his or her own preferences for the Best Fish prize, and by a process of voting and discussion the judges gradually eliminate nominated fishes until one mutually agreed fish emerges as the winner of the Best in Show category.

When you view the fishes at a show, you might not always agree that the chosen fish is the best in the show, but remember that the fishes on display may have improved or deteriorated in their appearance and condition since the judges evaluated them.

## A POOR SHOW FISH

Most self-respecting exhibitors wouldn't consider entering the sorry specimen below in a show; it is included here to illustrate a wide range of common faults that affect a judge's marking. This fish falls short in all of the five categories – Size, Body, Fins, Colour, Condition and Deportment – that the judges look at. Most fishes have a few faults in one or two categories, and the fish that scores full points in every category, like the specimen shown opposite, is very rare indeed. In general, prizewinning fishes score between 70 and 85 points.

You should *never* exhibit a diseased fish. It isn't a good idea to advertise the fact that you aren't a very good fishkeeper, but more importantly, you risk spreading an infection to other entries. And avoid exhibiting a female livebearer when she is heavily pregnant as this can bring on a premature birth.

**POINTS AWARDED: 56**
(maximum 100)

- **Size:13 points**
(maximum 20) At 75 mm long, this fish isn't full size.
- **Body:10 points**
(maximum 20) The fish's shape doesn't conform to the required standard: its head is snouty, and has an uneven outline.
- **Fins:10 points**
(maximum 20) The fish's fins don't conform to the required standard: the pelvic fins are bent and the caudal fin is split and lacks extension filaments.
- **Colour:13 points**
(maximum 20) The fish looks too pale, and its dark bars have smudged edges.
- **Condition and deportment:10 points**
(maximum 20) The fish's condition is poor: its scales are damaged, its body looks thin, and its fins are folded. And its behaviour is wrong: it was hanging at the surface of the water when judged.

**Faulty *Pterophyllum scalare* (Angelfish) on an uneven keel**

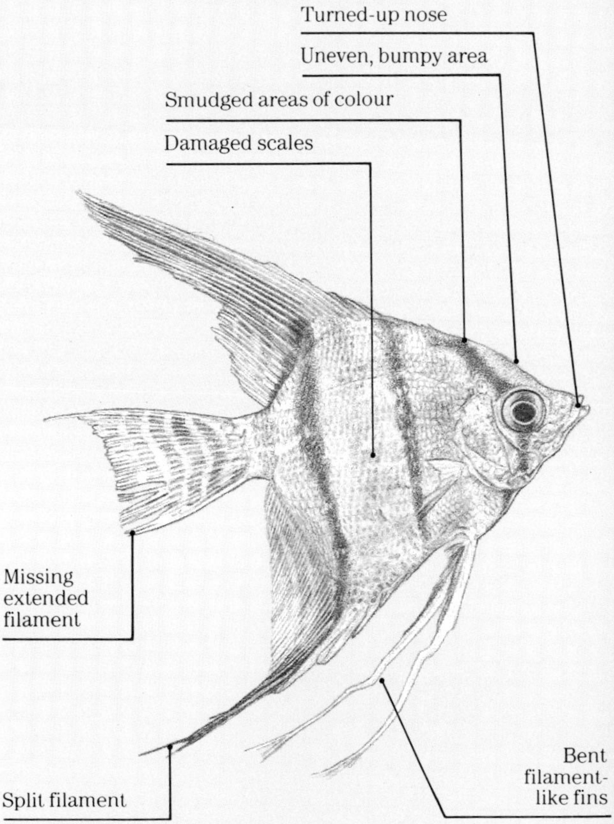

## AN IDEAL SHOW FISH

The "ideal" show fish is rarely found, although most exhibitors claim that they have plenty at home. However, it is easy to imagine just what it would look like. An ideal fish would be full size, with attractive, undamaged fins and good colour. It would have obvious health and vigour, it would be fully confident on show days, and would stand out amongst its fellow competitors.

If you are aiming to own such a fish, you must select a good specimen from the start, and then feed a full and varied diet, including plenty of live foods. And, of course, its aquarium conditions must always be the best.

**Ideal *Pterophyllum scalare* (Angelfish) on an even keel**

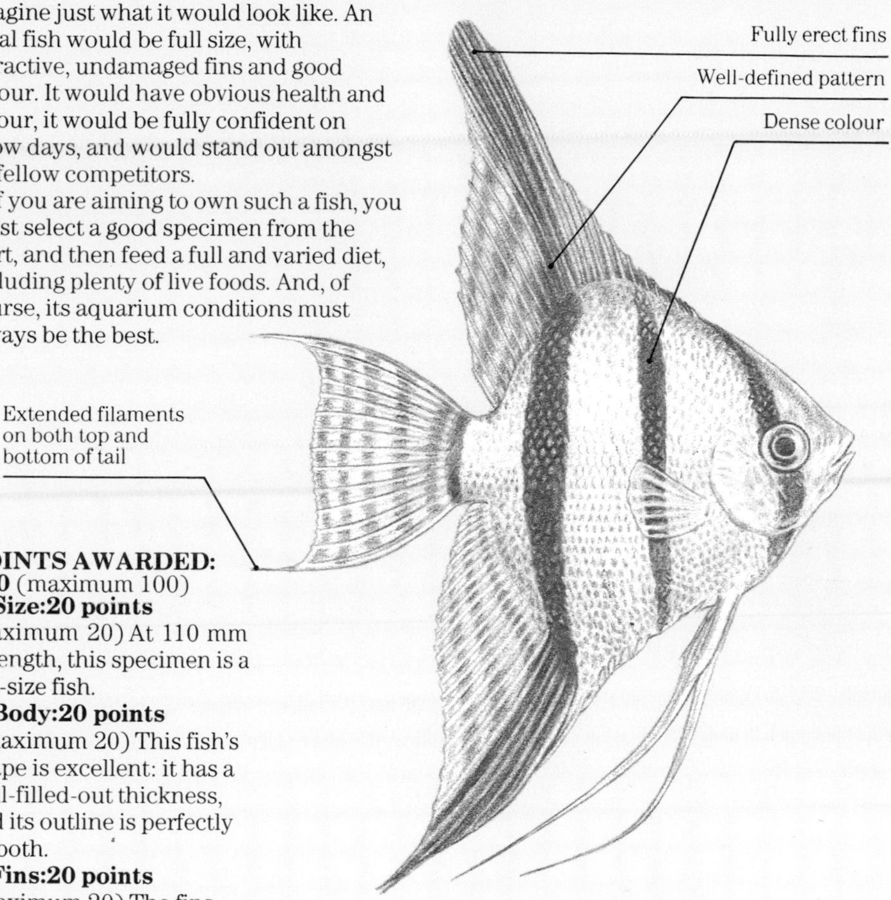


**POINTS AWARDED: 100** (maximum 100)

- **Size:20 points** (maximum 20) At 110 mm in length, this specimen is a full-size fish.
- **Body:20 points** (maximum 20) This fish's shape is excellent: it has a well-filled-out thickness, and its outline is perfectly smooth.
- **Fins:20 points** (maximum 20) The fins match the required standard: they spread well, and the pelvic fins look excellent. In addition, this fish has beautifully developed extensions to its caudal fin. (The total lack of fin damage suggests a plastic bag was used to catch it rather than a net, see p. 215.)
- **Colour:20 points** (maximum 20) The colour matches the required standard: it has good, dense black bars with edges that stand out well against the silver body colour. And the patterning shows well into the fins.
- **Condition and deportment: 20 points** (maximum 20) The fish's behaviour and condition are both excellent: when judged, it displayed itself fearlessly, and its colour patterns didn't fade or intensify

# Guidance on exhibiting fishes

If you want to take part in a show, your first step is to select your best fishes and send off for the necessary entry forms. You should study the rules, and make sure that you understand the procedures required. You must submit your entry in good time (most shows have a closing date for entries, and this is usually several weeks before the show date). The next step is to get your fishes to the show safely, and exhibit them so that they stand a good chance of gaining points.

**Selecting a show fish**

Naturally, any fish worthy of entering a show must be of good quality and size. It should therefore be pampered and well-fed. Many fishkeepers isolate their "specimen" fishes from other fishes in order to preserve their good qualities and to prevent them from being damaged in the hurly-burly of community-tank life.

Under the mistaken impression that big fishes get more points, some exhibitors deliberately enter heavily gravid (pregnant) livebearer females. This isn't a good idea, as the stress of transportation to the show and confinement in a small exhibition tank will often encourage such fishes to give birth prematurely.

## SHOW TIPS

There are many legitimate "tricks" that seasoned exhibitors use to make sure that their fishes are given every chance to gain good points:

- To avoid damaging a fish when catching it, use a large, clear plastic bag rather than a net. Because this type of bag is invisible underwater, you should be able to capture the fish without a chase around the tank. And, unlike a coarse net, a plastic bag won't damage the fish's scales, pelvic fins or barbels.
- To get a show fish used to people passing the tank and peering in, give it regular hour-long sessions in a small, bare tank, sited in a place where people are constantly moving about.
- A fish's colours depend on how it feels. Therefore, you should try to get your fish to the show ahead of schedule so that it has plenty of time to settle down and "colour up" before the judging starts.
- A fish should be exhibited in the water it is used to, so take along a supply of its aquarium water and use this to fill the exhibition tank.
- The water quality will affect the fish's colour: a pinch of salt added to the water will often make a fish exhibit its colours more strongly.
- Water temperature will affect a fish's behaviour: using water that is a degree or two warmer than usual can make it "show off" well.
- One way to make a male fish "show off" is to enter two females in the same class as him so you can bench him between them. He will spend the day trying to impress them, and he may catch the judge's eye as a result!

Catch fish in
a plastic bag
Use a plastic bonded
tank to move a
small fish
Use a snap-top bucket
to transport a
large fish

***Moving procedure***
*You must catch a show fish carefully, using a plastic bag rather than a net. Then transfer it to a suitable tank or bucket (above) or box (see p. 29).*

**Transporting fishes to a show**

Often, you will have a long journey ahead of you when you attend a show, and if you are taking fishes to exhibit you will have to devise a safe way of transporting them that minimizes any harmful effects. You can use a bucket or a small tank, as shown on the right, or a purpose-built box (see p. 29).

There is always a risk of chilling tropical species en route, so make sure that any transportation tank used for a tropical fish is well insulated against heat loss. Many shows have a supply of hot water available so that when you arrive you can add some to the fish's transportation water to bring it up to the right temperature again.

The best way to transport a large fish is in a plastic bucket with a snap-on lid. If your journey lasts more than half-an-hour, you should lift the bucket's lid occasionally to let fresh air in and to check on the passenger.

**Setting up a tank for a show fish**

Once you have got your fish to the show safely, you should put it in its show tank. Transfer the fish with its transportation water, then top up the tank from the show supply if necessary. A show tank should be completely bare; plants or decorations are only allowed if you are entering the "Furnished Aquarium" or "Aquarium Plants" class.

*Gravel and gravel substitutes*

Some show rules allow the use of aquarium gravel in the bottom of the exhibition tanks, but don't rely on this. You must check that gravel is allowed before you install it. One good reason why many organizers ban gravel is that exhibitors clog up the drains with it when disposing of the water after the show.

If you aren't allowed to use gravel you will have to install a dark-coloured substitute base as fishes feel uncomfortable swimming over a pale-coloured background, and this may affect performance. The simplest way to "modify" the bare tank bottom is to place the tank on a dark mat that will show through the glass and reassure the fish. A sheet of black plastic, cut to the same size as the tank base, will do the job. Another advantage of using a dark mat is that the fish's colours won't be "washed out" by the reflection of light from the bare tank base.

**Feeding**

Fish shows are usually one-day affairs, and so your fishes won't need feeding. A short fast won't harm them, so even if a show runs over several days it is best to avoid feeding, and thus prevent any pollution of the tank water.

FIRST

# 10

# PHOTOGRAPHING FISHES

If you have a suitable camera and you are a competent photographer you can combine this interest with fishkeeping. You will find your camera very useful for taking "portraits" of particular fishes, for shooting a sequence of pictures to "log" a home-bred fish's growth, or for recording a well-furnished tank that you have set up. Also, some aspects of fish behaviour such as spawning are well worth preserving on film for reference.

# The camera and lenses

The best type of camera to use for photographing fishes is a 35 mm single lens reflex (SLR) model. There are several reasons for choosing a 35 mm SLR:

- It allows you to see through the viewfinder exactly what will appear on the film.
- It takes auxiliary lenses, allowing you to fit a telephoto lens in order to take close-up shots.
- It is capable of electronic flash synchronization (see p. 270).

Of these requirements, the first is probably the most important. If you want to take a shot of an aquarium without including the rest of the room then the single lens reflex design, which allows you to look through the lens that views the subject, is essential. A camera with a separate viewfinder will give a very unsatisfactory result. The reason for this is that what you see through the viewfinder won't be exactly what the lens sees, and this discrepancy becomes more of a problem the closer you get to a subject. This effect is known as "parallax error" (see below), and results in a photograph of only part of what you see in the viewfinder frame. So, for example, a shot in which you thought you had captured all of your prize Guppy may only show its rear end.

## CAMERA TYPES AND PARALLAX ERROR

A camera with a separate viewfinder will only give satisfactory results with subjects more than 1.5 m away. At closer distances, the difference between the image you see in the viewfinder and what the lens records may result in only part of the picture you composed appearing on the film.

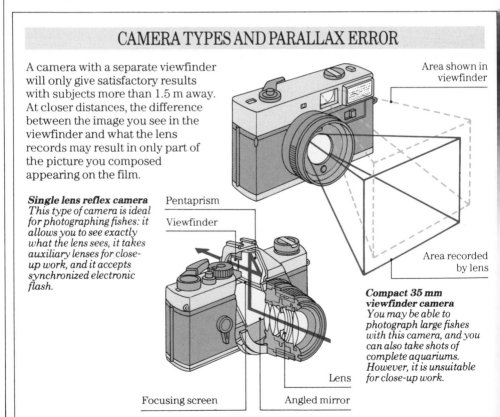

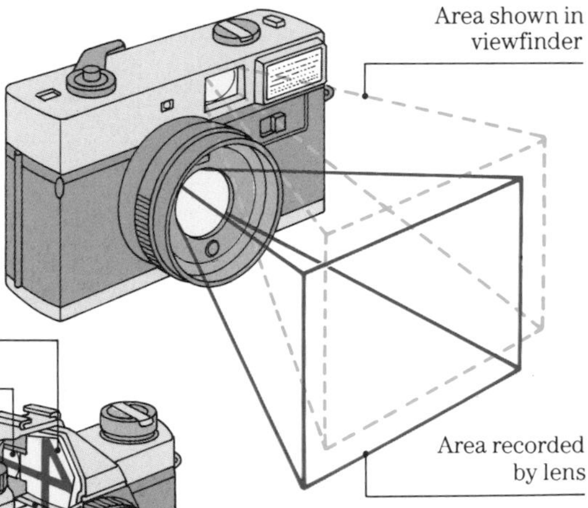


***Compact 35 mm viewfinder camera***
*You may be able to photograph large fishes with this camera, and you can also take shots of complete aquariums. However, it is unsuitable for close-up work.*

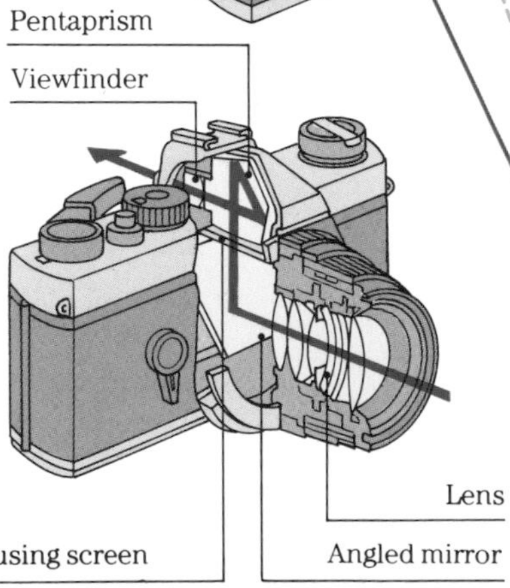


***Single lens reflex camera***
*This type of camera is ideal for photographing fishes: it allows you to see exactly what the lens sees, it takes auxiliary lenses for close-up work, and it accepts synchronized electronic flash.*

## Supporting the camera

If you can't hold the camera still while the exposure is made, your photograph will be blurred. To avoid any problems with blur, the model of camera you use shouldn't be too heavy or awkward for you to hold. Ideally, you should set your camera up on a tripod, as a rigid support will prevent camera shake. However, this may not always be possible, and as an alternative you can support the camera on a nearby object such as a table or chair back, using either a soft gadget bag or a photographer's "bean bag" to cushion the lens.

## Lenses

If you want to photograph fishes such as large Barbs or Cichlids (15 cms or longer) the standard 50 mm lens that comes with your camera will be adequate. But to obtain successful shots of smaller specimens like Tetras or Rasboras, and for taking close-up details, you will need to change to a telephoto or macro lens or fit extension tubes (see below). Alternatively, close-up attachments (see p. 268) which fit over the lens can be used to increase the image size.

*Zoom lenses*

These lenses have a variable focal length, allowing you to compose your picture carefully and quickly – ideal for moving subjects like fishes. Image quality is now very good, and you will find a telephoto zoom such as a 100–200 mm or 135–250 mm very useful.

## LENS TYPES

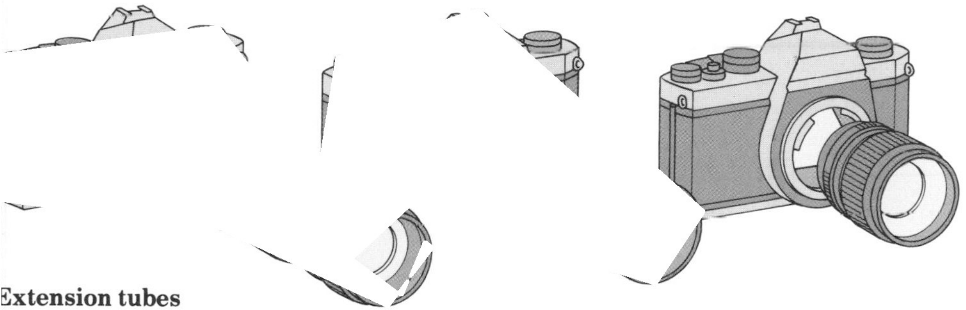

### Extension tubes

These tubes usually come in a set of three, and are fitted, singly or in combination, between the camera body and a standard lens. They are an inexpensive way of taking close-ups, but unless you can "stop down" (set a high f number), the quality ("resolution") at the picture edges may suffer. You should buy the automatic type, otherwise the mechanism that operates your lens diaphragm won't work.

### Telephoto lenses

A telephoto lens works rather like a telescope, enlarging a distant image. The advantage of this type of lens is that it enables you to work some way away from the aquarium and thus avoid disturbing the fishes. There are two main disadvantages: greater risk of camera shake (because these lenses are larger and heavier than the standard type), and shallower depth of field (see p. 269). A 135 mm or 200 mm are good choices.

### Macro lens

A macro-focusing lens is specially designed for taking close-ups, and gives a good-quality image. A focal length of between 50 and 90 mm will be very useful. Often, this type of lens will give 1:1 ratios (i.e., the subject will be life-size in the picture).

## THE EFFECT OF CHANGING LENSES

By using different lenses, you can vary the size of the fish's image in your picture.

**Close-up attachments**
These inexpensive fitments give a slight image enlargement, but the picture quality isn't as good as with a prime lens. They are available in several sizes, giving magnifications from 0.05 to 1 (life- size).

**Extension tubes**
These tubes are very versatile and inexpensive. The magnification given will vary depending on the combination of lens and tube(s).

**Macro lenses**
Specially designed to give images up to half life-size without additional attachments, macros produce good-quality images, but are more costly and less versatile than close-up attachments or extension tubes.

***Standard 55 mm lens***
*This type of lens gives a top-quality image, and has fast aperture settings, but won't focus any closer than 0.6 m.*

***55 mm lens with a 0.7 diopter close-up attachment***
*This gives a slight image magnification, but reduces picture quality.*

***55 mm lens with a 3 diopter close-up attachment***
*More powerful than the 0.7, this gives a larger image, but a greater quality reduction.*

***50 mm macro lens on standard setting***
*At this setting, the macro lens gives an image that is 1/10th life-size.*

***50 mm macro lens racked out***
*The lens was set to its closest focusing distance, giving a half life-size result.*

***135 mm telephoto lens***
*A long lens magnifies the image but has the disadvantage of producing shallow depth of field.*

***8 mm extension tube fitted to a standard 55 mm lens***
*This combination produces an image that is 1/7th life-size.*

***14 mm extension tube fitted to a standard 55 mm lens***
*A quarter life-size image is given by this combination.*

***27 mm extension tube fitted to a 55 mm standard lens***
*The largest extension tube gives a half life-size image.*

# ᶠocusing

Vhen focusing on a fish, especially in lose-up, there is very little margin for rror. In order to get as much of the fish ı focus as possible, and, conversely, to nsure that any distracting background lements are out-of-focus, you must nderstand the effect of different aperıre settings on focus. This is particularly nportant with close-up work because ıe zone of sharp focus (the depth-ofeld), which is determined by the aperıre, decreases as image size increases.

## Vhat is aperture?

'he aperture is simply a hole that estricts the amount of light entering the ɔns and reaching the film. It is usually djustable by means of an iris diaphragm ·ithin the lens, and is calibrated in "f umbers" or "f stops" on the aperture etting ring around the lens.

## ·perture and focus

. small aperture (high f number) will ive the largest zone of sharp focus, known as depth-of-field. A wide aperture (low f number) will confine the zone of sharp focus to a very small area. If you want to set a small aperture to get as much as possible in focus, you will need plenty of light. With some subjects, you can overcome lack of light by setting a slow shutter speed. But using a slow speed is out of the question when photographing fishes because it gives a long exposure, and as your subject is most likely to be moving, the result will be blurred. In general, when working with aquarium fishes, particularly close-up, to get enough of the subject in focus you will need to use a flashgun (see p. 270).

## Bracketing

To make sure you get a sharp picture allow a margin for error by bracketing aperture settings. That is, if you calculate that you need to set the aperture to f8, take three shots: one at f5.6, one at f8, and the third at f11, keeping all other factors constant.

## DEPTH OF FIELD TABLE

When taking a close-up shot, the extent of the depth of field is slightly different to that found in general photography, extending *equally* either side of the point of focus. To find the magnification of your image, refer to the table supplied by your lens manufacturer. In this chart the zone of sharp focus is given in millimetres.

| **Aperture** | **Magnification** | | | | | | | | |
|---|---|---|---|---|---|---|---|---|---|
| | **× 0.1** | **× 0.15** | **× 0.2** | **× 0.25** | **× 0.33** | **× 0.5** | **× 0.75** | **× 1** | **× 1.5** |
| **f5.6** | 0.3 | 0.32 | 0.34 | 0.34 | 0.38 | 0.42 | 0.5 | 0.56 | 0.7 |
| **f8** | 0.44 | 0.46 | 0.48 | 0.5 | 0.54 | 0.6 | 0.72 | 0.8 | 1 |
| **f11** | 0.6 | 0.63 | 0.66 | 0.68 | 0.74 | 0.82 | 0.99 | 1.1 | 1.38 |
| **f16** | 0.88 | 0.92 | 0.96 | 1 | 1.08 | 1.2 | 1.44 | 1.6 | 2 |
| **f22** | 1.22 | 1.28 | 1.32 | 1.38 | 1.48 | 1.64 | 1.93 | 2.2 | 2.74 |

# Lighting

The aquarium usually has its main light source directly overhead in the hood. This arrangement probably won't provide sufficient light for photography and may give unacceptable light casts (green if your lamps are fluorescent, orange if they are tungsten). It is better to turn off the lamps, remove the hood and cover-glass and use electronic flash. This has the additional advantage of freezing movement, thus avoiding blur.

**Avoiding light reflection**

When photographing an aquarium, light will often reflect off the glass back of the tank into the camera, causing small "flare spots" on your photograph. The best way to prevent this is to make a mask from black cardboard and fit it round the lens.

**Flash devices**

There are many accessories available which help to create a good result.

## FLASHGUN ARRANGEMENTS

The methods shown here are fairly simple, as long as you use one or more "dedicated" flashguns, each linked to the camera with its own cord. With this type of flash, the amount of light from the gun is metered by a sensor mounted on the camera and exposure is set automatically. However, very light or dark backgrounds may "fool" the sensor, and so you will have to compensate for this.

If you don't have a dedicated flashgun which works with your TTL meter to measure the light during exposure, you can achieve similar results with a sensor flashgun. However, with this type of gun the sensor must face the subject in order to get a correct exposure, so with off-camera flash you will need to mount a remote sensor accessory in the hot shoe, connect it to the gun(s) with extension cord(s), and use a transmitter and slave trigger to fire all the heads simultaneously.

***Using three flashguns*** *For the photograph below, one flashgun was positioned facing down into the tank, and the other two directed into the tank from the side, slightly above it, and pointing at an angle of about 45 degrees downwards (see right).*

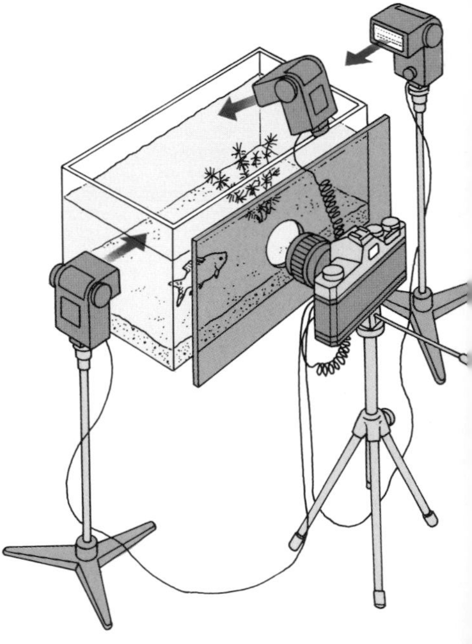

*Multiple flash connector*
This device plugs into the flash synchronization socket on the camera body or hot shoe adaptor, and takes extension leads to link several guns to the camera. Using more than one flashgun enables you to provide deeper, more even illumination.

*Zoom attachment*
Used with a telephoto lens, this device fits over the flash head to ensure that your flashgun's light beam covers the range of your lens.

*Bounce reflector*
This angled holder slots on top of the camera and takes a piece of white card. It is used to bounce flash – a technique which reduces harsh shadows and gives a more natural, even effect.

A bounced flash will lose about half to one-third of its intensity, so you should take this into account when calculating exposure.

In order to bounce flash you will need to have a flashgun with a tilting head.

***Bounced flash***
*For the shot below, a flashgun with a tilting head was mounted on the camera, and pointed vertically upwards onto an angled white reflector or a white ceiling (see right), thus "bouncing" light forward into the tank.*

***Using a single gun***
*A tilt-head flashgun held off-camera above and in front of the tank and directed downwards (see right) can give an adequate result (see below).*

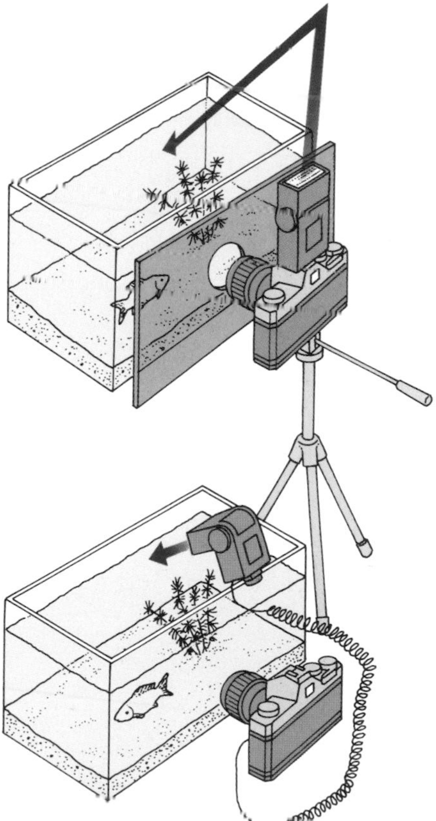

# Setting up a good picture

When photographing fishes, a little preparation will go a long way towards ensuring a good result. All the usual rules of composition apply, and you should pay particular attention to the elements in the background.

**Backgrounds**

Often, an attractive shot of a fish is marred by distracting elements in the picture, for example, a slotted-steel bench support showing through from the other side of the tank. Even if you have avoided such pitfalls, the photograph may still look too "busy" because the background is too dominant, distracting the viewer's eye from the main subject – the fish. It is therefore important to arrange a suitable, uncluttered background before you take your picture.

*Backgrounds for identification shots*

If you are taking shots purely for identification purposes, a blank, one-colour background installed behind the rear glass of the tank is probably best. It is a good idea to have a selection of coloured cards, including both light and dark shades, so that you can select a colour that suits the fish you want to photograph.

**Composition tips**

- A zoom lens allows you to close in on your subject, filling the frame with it, without changing your camera position.
- To soften a chosen part of the background, smear part of a UV filter with petroleum jelly and fit it over your lens.
- Use a fish-confining device (see opposite) to restrict the fish you are photographing to a small area of the tank, lessening your depth-of-field problem.
- You may find a polarizing filter helpful for reducing reflections from the glass.

## FISH PHOTOGRAPHY DONT'S

- Don't leave the hood and aquarium lamps in place for your photograph
- Don't point the flashgun directly at the aquarium – it will reflect light directly back into the lens
- Don't take photographs through dirty water or a dirty aquarium glass
- Don't forget to set your shutter speed to the correct setting for flash photography
- Don't use too large an aperture setting as you will restrict the depth of field

***Abnormal "growths"***
*The metal side of the tank seems to be growing out of this fish's tail! To avoid such ugly extras, take notice of what else is in the frame when you compose a shot.*

***Distracting background***
*Before you press the shutter, make sure that none of the room furnishings are visible through the rear glass. If necessary, use a plain card behind the tank.*

# ARRANGING A TANK FOR PHOTOGRAPHY

You can either set up a small tank specially for photography, or use your regular aquarium. In the latter case, you will have to use a fish-confining device – a sheet of glass or a special three-sided inner tank – otherwise your subject will probably lurk behind plants or rocks, refusing to pose for the camera.

***Comparison tank***
*It is a good idea to set aside a small tank with a simple gravel base and a single plant for fish photography. You can use this tank for a scale-size comparison in two main ways: you can compare different fishes or take a comparative sequence of photographs of the same fish over a period of time. A ruler along the bottom edge of the tank will give you an accurate comparison of size.*

***Spawning tank***
*To photograph breeding behaviour when you set up a breeding tank (see pp. 248–51) furnish it so that the most likely spawning locations are visible. For your first attempt, try Cichlids or Gouramis as they are relatively slow-moving, even in the excitement of spawning. Once you know where the fishes' pre-selected site is (see p. 241), set up and focus your camera in advance. You may have to wait an hour or so, but the result will make your effort worthwhile.*

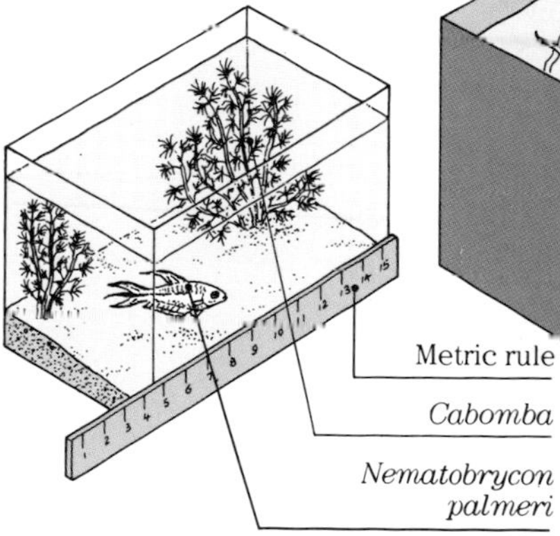


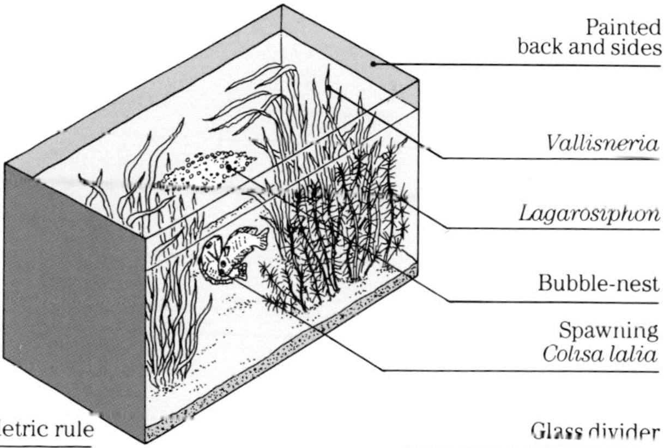


***Fish confining devices***
*These devices range from a sheet of glass placed in the tank to confine the fish's movements to a pre-selected focusing zone, to a complex, three-sided "inner" tank with a layer of gravel glued into place with resin.*

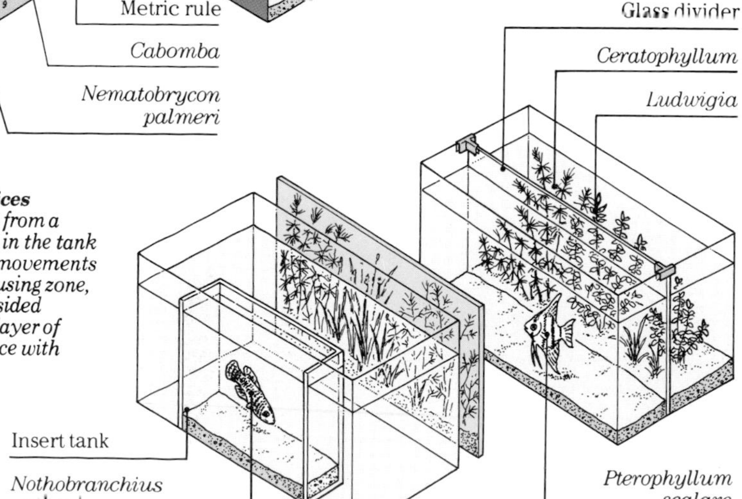

# Glossary

## A

**Acid**
A term used by aquarists in connection with water conditions. Water can be termed acid if it contains dissolved hydrogen ions. Acidic water can be high in carbon dioxide. Opposite of ALKALINE.

**Activated carbon**
A FILTER MEDIUM used to ADSORB dissolved waste substances from the water.

**Adipose fin**
A small extra FIN found on some fishes between the DORSAL and CAUDAL FINS.

**Adsorption**
The process of accumulating other substances on the surface of a solid. In the aquarium, FILTER MEDIUMS with this ability may be used to clean the aquarium water.

**Aeration**
The introduction of air into the aquarium in order to agitate the water surface to facilitate absorption of oxygen.

**Aerobic**
Oxygen-rich.

**Airlift**
Device that facilitates the movement of clean water out of the FILTER MEDIUM by introducing airbubbles into the water. Being lighter than water, the mixture of airbubbles and water rises, and so is forced out of the FILTER.

**Airline**
Synthetic tubing through which air is fed into the aquarium.

**Airpump**
An electric vibrator-diaphragm device, or piston pump, which provides a continuous supply of air to the tank.

**Airstone**
A porous block fitted at the end of the AIRLINE to produce a fine stream of air bubbles.

**Albino**
Lacking in pigmentation.

**Algae**
Primitive aquatic plants ranging from microscopic UNICELLULAR types to sea-weeds.

**Alkaline**
A term used in connection with water conditions. Water can be termed alkaline if it contains hydroxyl ions (carbonates). These carbonates will neutralize the acidity caused by the respiration of water plants and animals. Opposite of ACID.

**All-glass tank**
A frameless tank made of five pieces of glass stuck together with silicone SEALANT.

**Ammonia**
A compound of nitrogen and hydrogen ($NH_3$).

**Anabantoids**
A group of fishes which are able to breathe atmospheric air. See LABYRINTH FISHES.

**Anal fin**
Single, keel-like FIN beneath the fish's body.

**Annuals**
Fishes that in nature only live one year, dying as their river habitats dry up.

## B

**Barbel**
Whisker-like growth at the corner of the mouth on some fishes. Used for locating food. (From the Latin *barbus* meaning beard.)

**Biological filter**
A filter plate installed under the GRAVEL as part of a BIOLOGICAL FILTRATION system. Also known as an undergravel filter.

**Biological filtration**
A system that uses bacteria to break down toxic nitrogenous compounds into harmless ones.

**Brackish water**
Estuarine water containing a small proportion of SALT WATER and a larger proportion of FRESH WATER.

**Breeding-trap**
A small tank placed within the aquarium, in which a GRAVID female LIVEBEARER is confined to give birth. The fry are able to swim out of the trap to safety.

**Brine-shrimp hatcher**
A small tank used for breeding *Artemia salina* (brine shrimp).

## C

**Cable-tidy**
Junction box with control switches for aquarium electrical equipment connections.

**Calcareous**
Chalky, containing calcium.

**Caudal fin**
FIN at the rear end of a fish. Also known as the tail.

**Caudal peduncle**
Narrow part of the fish's body joining directly onto the CAUDAL FIN.
**Chemical filtration**
Process of removing dissolved waste products from the aquarium by a chemical reaction.
**Class**
A major biological division of life forms, such as mammals, fishes.
**Coldwater**
Term applied to all fishes kept in unheated tanks.
**Community tank**
Aquarium in which many different SPECIES of fishes are kept together.
**Compost**
One name for the medium covering the aquarium floor. Also known as substrate.
**Conditioning**
Preparing an adult pair of fishes for breeding.
**Coral sand**
Pulverized coral with particles similar in size to sand.
**Cover-glass**
Sheet of glass placed on the tank to prevent evaporation and to stop fishes from escaping. It also protects the lamps from water splashes, and the plants from scorching by the lamps.
**Crown**
Junction of plant stem and ROOT.
**Crustacean**
CLASS of segmented animals with jointed limbs, and, often, hard outer coverings.
**Culling**
Quality control of young fishes by discarding the inferior specimens.
**Cuttings**
Method of PROPAGATION of plants by re-rooting a severed leaf or top part of the stem.

## D

**Denitrification**
Process of removing NITROGEN and its compounds.
**Density**
Ratio of mass to volume.
**Detritus**
SEDIMENT on the bottom of the aquarium.
**°DH**
Measurement of the HARDNESS of the water.
**Diatomaceous earth**
A FILTER MEDIUM made from the powdered, fossilized remains of diatoms (UNICELLULAR ALGAE with two hard interlocking shells).
**Division**
Method of reproduction whereby the original organism splits into two or more independent parts.
**Dorsal**
Top surface of a fish.
**Dorsal fin**
FIN on top of the fish. Usually single; some SPECIES have two.

## E

**Egglaying fishes**
Fishes whose eggs are fertilized and hatched outside the female's body.
**Electric impeller**
An electrically driven rotor in a power filter which pumps water.

## F

**Fertilization**
The combining of the male's MILT with the female's ova. It can occur either internally in livebearers or externally in egg-laying species.
**Filaments**
Long, thread-like extensions to the FINS, usually single RAYS.
**Filter**
Device for cleaning the aquarium water.
**Filter medium**
Material used to trap suspended matter in the aquarium water as it passes through the FILTER.
**Filtration**
Process of removing waste materials from the aquarium water to prevent pollution.
**Finrot**
Bacterial disease of the FINS.
**Fins**
External "limbs" of the fish used to provide propulsion or stability.
**Fresh water**
Water that contains no SALT.
**Fry**
The young of a fish.
**Fungus**
A cotton wool-like organic growth sometimes found on a fish's body.

## G

**Genus**
Scientific name for a group or family of closely related life forms. It can be thought of as a surname (with SPECIES being the Christian name). It is the first part of any scientific name, and begins with a capital letter.
**Gill cover**
A flap of skin which protects the GILLS. Also known as the operculum.
**Gills**
Organs which form part of the fish's respiratory system through which it extracts oxygen from the water.

**Gonopodium**
A rod-like modification of the ANAL FIN of male LIVEBEARER fishes, used to assist FERTILIZATION of the female.
**Gravel**
Type of aquarium base-covering material consisting of very small, round stones. Also known as compost or substrate.
**Gravid**
Condition of a female LIVEBEARER when carrying young internally.
**Guanin**
A waste product that fishes store internally beneath the SCALES, and which gives rise to the FRESHWATER fishes' IRIDESCENCES when light is reflected from it.

# H

**Hand-stripping**
See STRIPPING.
**Hardness**
Amount of dissolved minerals in water.
**Hardness test kit**
Device used to determine either the specific HARDNESS (i.e., the amount of calcium carbonate) of water or the overall hardness (i.e., the total amount of minerals) by the introduction of a chemical agent into a sample of the water.
**Heater**
Small electrical heating element encased in a watertight tube. May be combined with a THERMOSTAT as a single unit.
**Hood**
Tank cover which houses the lighting equipment. Also known as the reflector.
**Hydrometer**
Floating device for measuring the density (SPECIFIC GRAVITY = S.G.) of SALT WATER in marine tanks.

# I

**Ichthyology**
Study of fish. (From the Greek *ichthys* = fish.)
**Infertile**
Incapable of breeding; eggs which have not been fertilized.
**Invertebrate**
Animal without a vertebral column (spine).
**Ion-exchange resin**
Chemical used to soften water.
**Iridescence**
Glittering, changing colour.

# L

**Labyrinth fishes**
Alternative name for ANABANTOID fishes.
**Labyrinth organ**
Additional anatomical development in ANABANTOID fishes which enables them to breath atmospheric air.
**Lateral line**
Visible row of pierced SCALES along the fish's side through which vibrations in the water are transmitted to the nervous system.
**Length**
Standard measurement of a fish from the SNOUT to the end of the CAUDAL PEDUNCLE, excluding the CAUDAL FIN.
**Line-breeding**
Maintaining a pure colour strain of a SPECIES through a controlled, selective breeding programme.
**Livebearers**
Fishes whose eggs are fertilized and hatched internally in the female's body.

# M

**Marine**
Pertaining to SALT WATER.
**Matt**
Non-reflecting SCALES.
**Mechanical filtration**
Process of removing suspended waste products from the aquarium water by means of a mechanical "sieve".
**Metallic**
Highly reflective SCALES.
**Milt**
The male fish's fertilizing fluid.
**Mops**
Bundles of nylon wool which are substituted for plants as SPAWNING receptacles, particularly in breeding tanks for egg-scattering SPECIES.
**Mouth-brooder**
Fishes where the female incubates fertilized eggs in her throat.

# N

**Nacreous**
Semi-reflective, "mother-of-pearl" SCALES.
**Natural pairing**
The spontaneous selection of mating partners by the fishes themselves.
**Nauplius**
Newly hatched stage of brine shrimp.
**Nitrate**
Final compound produced by DENITRIFYING bacteria in BIOLOGICAL FILTRATION systems.
**Nitrite**
Intermediate compound (between AMMONIA and NITRATE) produced in the DENITRIFYING process.

**Nitrite test kit**
A device used to measure the NITRITE levels in the water. A chemical agent is added to a water sample and any colour change is observed.
**Nitrobacter bacteria**
Cultured in the aquarium to convert NITRITE into NITRATE.
**Nitrosomonas bacteria**
Cultured in the aquarium to convert AMMONIA into NITRITE.

## O

**Operculum**
See GILL COVER.
**Ova**
Fish eggs.
**Oviparous**
Egglaying.
**Ovipositor**
A tube extended, at breeding time, by egg-depositing SPECIES through which eggs are laid and fertilized.
**Ovoviviparous**
Producing eggs which hatch inside the female's body.

## P

**Pectoral fins**
Paired FINS situated immediately behind the GILL COVER.
**Pelvic fins**
Paired FINS located on the bottom of the fish immediately in front of the ANAL FIN. Also known as the ventral fins.
**pH**
Measurement of the acidity or alkalinity of water.
**pH test kit**
Means of determining the aquarium water's acidity and alkalinity by use of a chemical agent and colour comparison.
**Pharyngeal teeth**
Teeth situated in the throat of Cyprinid fishes.
**Photosynthesis**
Process by which plants assimilate food under the action of light, consuming carbon dioxide and giving off oxygen.
**Plug**
Nutrient-rich, preformed fibrous material planted in the GRAVEL to nourish plants.
**Polyps**
Live animals whose dead skeletons form coral reefs.
**Power filter**
A filter of high WATERFLOW rate, usually operated by an ELECTRIC IMPELLER.
**Propagation**
The process of growing new plants from established ones.
**Protein skimmer**
Air-operated device for filtering waste products out of the water.

## Q

**Quarantine**
The holding in a separate tank of any new stock, before being introduced into the main collection, in order to eliminate the spread of disease.

## R

**Rays**
Supports for the FIN tissues. Usually preceded by stiffer spines.
**Reflector**
See HOOD.
**Reticulated**
A net-like pattern.
**Reverse-flow filtration**
Process of pumping water under the aquarium base-covering (usually GRAVEL) and then up through the BIOLOGICAL FILTER bed (WATERFLOW is usually the opposite way).
**Root**
The part of a plant that anchors it in place, and through which terrestrial and semi-aquatic plants are nourished.
**Runners**
Exposed, root-like plant growths that carry new plantlets.

## S

**Salinity**
Relative amount of salts (minerals) in the water.
**Salt mix**
Proprietary dry mixes of minerals used to make up synthetic sea-water for the marine aquarium. Also known as "sea-mix".
**Salt water**
Water containing a high proportion of salts, usually sodium chloride (Na Cl).
**Scales**
Thin, bony plates that cover the fish's skin.
**Scalpels**
Erectile spines on the CAUDAL PEDUNCLE of Surgeonfishes.
**Scutes**
Large, bony, plate-like covering found on some fishes such as various Catfishes instead of SCALES.
**Sealant**
Usually silicone rubber. Used as an adhesive to stick glass panels together.
**Sediment**
Settled rubbish such as dead leaves or waste products on the aquarium floor. Also known as detritus.

**Shimmies**
Symptom shown by fish suffering from chilling. Consists of undulating from side to side without forward movement.
**Shoal**
Group of fishes of the same SPECIES.
**Shoaling**
Action of swimming together in groups.
**Singletail**
A fish with an undivided CAUDAL FIN.
**Sintered glass**
Material that is formed at a very high temperature without becoming liquified. Used as AIRSTONES.
**Siphon tube**
Device used to transfer liquids from one level to another level.
**Snout**
Extreme point (tip) of the head.
**Spawning**
Breeding.
**Spawning tank**
Separate aquarium in which a pair, or SHOAL, of fishes can breed.
**Species**
A group of similar animals capable of interbreeding and producing fertile young. Closely related species are grouped in a GENUS; each species within a genus is distinguished by the second part of its scientific name.
**Species tank**
An aquarium containing a single SPECIES of fishes.
**Specific Gravity**
(S.G.) Ratio of a liquid's DENSITY compared to that of pure water.
**Specimen plant**
A plant with outstanding foliage or shape that is planted singly in the aquarium.
**Strain**
A stock of fishes which have one particular inherited trait, such as colour.
**Stripping**
The removal, by hand, of OVA and MILT from fishes. Often used in LINE-BREEDING large Fancy Goldfishes whose natural mating may be hampered by their exaggerated FIN forms. Also known as hand-stripping.
**Substrate**
See COMPOST.
**Surface area**
Expanse of water surface in contact with the atmosphere.
**Swim-bladder**
Internal gas-filled organ, which automatically expands or contracts so that a fish can maintain itself at any depth in the water.

# T

**Tail**
See CAUDAL FIN.
**Terminal**
Position at extreme front end of a fish. Used to describe mouth location in mid-water swimmers whose mouths are neither upturned nor underslung.
**Territory**
An area a fish decides to occupy to the exclusion of others, especially at breeding times.
**Thermostat**
Device for controlling the amount of heat to the tank. May be combined with the HEATER as a single unit.
**Tropical**
Term applied to all fishes that require heated aquariums.
**Tubercles**
Small pimple-like protruberances which some male coldwater fishes, especially Goldfishes, develop on the SNOUT, GILL COVERS, along the body, and on the PECTORAL FINS at breeding times.
**Twintail**
A fish, especially a Goldfish, whose CAUDAL and ANAL FINS are divided.

# U

**Undergravel filter**
See BIOLOGICAL FILTER.
**Unicellular**
Consisting of a single cell.

# V

**Variety**
Belonging to one SPECIES, but distinguishable by differing colouring, patterning or FIN development.
**Ventral fins**
See PELVIC FINS.
**Ventral**
Fish's lower surface.
**Viviparous**
Livebearing.

# W

**Waterflow**
Rate (litres per hour) or direction of water movement.
**Water-turnover**
Rate at which water flows through a FILTER. It is normally taken as an indication of how frequently the water is filtered (litres per hour).

# APPENDIX 1: Classifying fishes

For most purposes, it is only necessary to know which family a particular fish belongs to and its individual scientific name.

A *family* is a biological classification which groups together fishes that share common features. Characins, for example, are all tropical freshwater fishes and most have adipose fins. Within a family, fishes are divided into *genera* and then into *species*. Each fish has a scientific name comprising both a generic and specific name by which it is known worldwide. The genus (which is always spelt with an initial capital letter, and may be regarded as the surname) is applied to a number of closely related fishes, e.g., the *Corydoras* Catfishes. The species (which is like a forename) is used to define a particular fish, e.g., *Corydoras reticulatus* (Reticulated Corydoras). Occasionally, these names are updated as new information is discovered about fishes.

A fish's names may refer to its place of origin, e.g., *Jordanella floridae* (American Flagfish) which is found in Florida; or be a dedication to its discoverer, e.g., *Cheirodon axelrodi* (Cardinal Tetra) after Dr H. Axelrod; or describe its appearance, e.g., *Chaetodon octofasciatus* (Eight-banded Butterflyfish).

# APPENDIX 2: Garden ponds

Outdoor ponds used for keeping coldwater freshwater fishes in should receive some sunshine and mustn't be sited beneath overhanging trees. The size will depend upon the species of fishes you keep. For example, Koi require a depth of at least 1.5–2 metres for overwintering, whilst Goldfishes can survive in as little as 0.6 metres.

**Special requirements**
A pond intended for Koi should have a filtration system. Your local dealer will be able to suggest suitable internal or external types. Plants will facilitate the oxygenation of the water and will enhance the pond's appearance. You should also provide some rocks as shelter for the fishes.

**Introducing livestock**
Always wait until the plants have established themselves and the water has "settled" before you put fishes into the pond. Float them in a container until the water temperatures have equalized (see p. 29).

**Maintenance**
Leaves falling into the water will gradually decay, creating toxic substances that could harm the fishes. And, if the leaves settle on the surface they cut down the amount of oxygen entering the water. To avoid this, lay nets across the pond during the autumn so that the leaves can be gathered and disposed of.

Another serious problem may occur if a layer of ice forms on the pond. If the pond is sufficiently deep, the fishes can move around in the unfrozen water beneath the ice. But if toxic gases build up under the ice, the fishes may suffocate. You must therefore maintain a hole in the ice so that the gases can escape. But *never* smash the ice as the shock waves will alarm, and may even kill, the fishes. A pond heater will serve as a convenient means of keeping one area ice-free.

**SUITABLE POND-KEPT FISHES**
**All year** □ Singletail Goldfishes □ Koi □ Bitterling
**Summer months only**
□ Fancy Goldfishes □ Pigmy Sunfish □ Black-banded Sunfish □ Pumpkinseed □ Red Shiner □ Eastern Mudminnow □ Pale Chub

# APPENDIX 3: Tank facts and figures

Throughout this book metric measurements are quoted. However, you may occasionally come across products in imperial units. These guides will show you how to convert from one system to another, and how to calculate measurements.

**Tank measurements**
The important statistics for tanks relate to length, weight and volume.

*Length* Linear dimensions are measured in metres.

METRIC/IMPERIAL CONVERSIONS:

1 metre (100 cms) = 39 ins (approx.)
1 foot = 30 cms (approx.)
To convert centimetres to inches multiply by 0.40.
To convert inches to centimetres multiply by 2.54.

*Weight* Measured in kilograms, the weight of water in a tank is equal to the volume measured in litres. But bear in mind that water becomes lighter in weight above and below 4°C, which is why a pond freezes from the top down, not from the bottom up.

METRIC/IMPERIAL CONVERSIONS:

To convert kilogrammes to pounds multiply by 2.2.
To convert pounds to kilograms multiply by 0.453.

*Volume* This is measured in litres, and is obtained by multiplying depth by surface area. For rectangular tanks, the surface area is the length multiplied by the width, and the volume is therefore length × width × height.

For irregularly shaped tanks, divide the surface area into regular shapes (rectangles, circles or triangles) and treat each section as a separate tank, then add the total volumes together. To calculate the surface area of a triangle, multiply $\frac{1}{2}$ base by the perpendicular height. To calculate the surface area of a circle, multiply the radius by $\frac{22}{7}$.

METRIC/IMPERIAL CONVERSIONS:

1 litre (1000 cc) = 1.75 pints (approx.)
5 litres = 1.1 gallons (approx.)
To convert litres to gallons multiply by 0.22.
To convert gallons to litres multiply by 4.56.

## TANK BACKDROPS

One way of making your aquarium look more attractive and of giving the impression of a greater depth in the tank is by using decorative backcloths or plastic "3D" dioramas. Built in a three-sided open box and fitted to the *outside* of the aquarium, they can be seen through the tank. Alternatively, an internal backdrop made of cork bark can be fitted in the tank. Make sure that it is fitted flush with the rear glass so that fishes can't be trapped behind it.

Sidelighting will show dioramas and some fishes to the best advantage. The source can either be natural or, more usually, an exterior spotlight angled down onto the side wall of the tank.

This chart shows the volume and weight of water that certain tank sizes can hold, and the total length of fishes (excluding caudal fins) that each aquarium can house. No figures are given for keeping tropical marine fishes in the two smallest tanks because a 90-cm-long tank is the

| **Size (1 × d × w)** | **Volume** |
|---|---|
| 45 × 25 × 25 cms | 28.1 litres |
| 60 × 38 × 30 cms | 68.4 litres |
| 90 × 38 × 30 cms | 102.6 litres |
| 120 × 38 × 30 cms | 136 litres |

**Water measurements**
Water statistics relate to temperature, hardness, and acidity/alkalinity.

*Temperature* This is measured in degrees Celsius.

METRIC/IMPERIAL CONVERSIONS:

To convert °Fahrenheit to °Celsius:
°C = (°F – 32) × 5/9.
To convert °Celsius to °Fahrenheit:
°F = (°C × 9/5) + 32.

*Hardness* This is measured in degrees DH.

*Alkalinity/acidity* The pH scale is used to measure this. If neutral, the reading will be 7, if acidic below 7, and if alkaline above 7.

## WATER HARDNESS

| Water type | Suitable fishes |
|---|---|
| **FRESHWATER** | |
| Very soft 0–6° DH | Characins, Barbs, Killifishes, Discus |
| Soft/medium-hard 7–15° DH | Catfishes, Loaches, Goldfishes, most Cichlids, Gouramis |
| Hard 15° DH+ | African Lake Cichlids, Livebearers, Australian Rainbowfishes |
| **BRACKISH** | |
| 1 tsp salt/5 litres and above | Monodactylus, Mollies, Australian Rainbowfishes |
| **MARINE** | |
| S.G. 1.023 | All tropical and coldwater saltwater fishes, invertebrates |

## TANK CAPACITY

smallest recommended size for such species. However, smaller tanks can house coldwater marine invertebrate collections. For tanks with different dimensions, multiplying the length by the depth by the width will give the volume, the weight in kilograms will be equal to the volume in litres, and each centimetre of a fish's body will require 30 $cm^2$ of water surface in a tropical freshwater system, 75 $cm^2$ of water surface in a coldwater freshwater system and 120 $cm^2$ of water surface in a tropical marine system.

| Weight of water | Total body lengths of tropical freshwater fishes | Total body lengths of coldwater freshwater fishes | Total body lengths of tropical marine fishes |
|---|---|---|---|
| 28.1 kg | 38 cms | 15 cms | – – – |
| 68.4 kg | 60 cms | 24 cms | – – – |
| 102.6 kg | 90 cms | 36 cms | 23 cms |
| 136 kg | 120 cms | 48 cms | 30 cms |

# General Index


## A

**acclimatization** of fishes, 28-9, 178, 182-3, 234
**acid water**, 145, 274
***Acorus gramineus*** (Japanese Dwarf Rush), 159, 163
***Acropora pulchra*** (coral), 191
**activated carbon**, 122, 274
**adipose fin**, 17, 274
**adsorption**, 122, 274
**aeration**, 121-3, 274
  coldwater freshwater species, 83, 91, 216
  marine species, 95, 188, 194
**aerobic**, (def.) 274
**airlift**, 124, 274
  making, 193
**airline**, 274
**airpump**, 121-3, 274
  installation, 175
  maintenance, 217
**airstone**, 123, 175, 179, 194, 217, 274
**albino**, (def.) 274
**algae**, 150-1, 206, 217, 219, 274
**alkaline water**, 145, 274
**Amazon Sword**, 159, 165
***Ambulia***, 168
**ammonia**, 128-9, 158, 274
**anal fin**, 17, 274
**anchor worm**, 221-3, 226, 232
***Anubias lanceolata***, 178
***Aponogeton***
  *crispus*, 178
  *madagascariensis* (Madagascar Lace Plant), 159, 163, 173, 179
**aquarium**
  aquascaping, 154-97, 256
  costs of, 9-11
  cover-glass, 150-1, 176, 189, 217
  "furnished", 256-7, 259
  heating, 130-5
  layouts, 173-97
  lighting, 120, 148-51, 160, 215-17, 253
  loss of heat, 216
  maintenance, 216-19
  plants, 24, 154-5, 158-71, 250, 257, 259
  siting, 11, 118-20
  space utilization, 31
  *and see* fish, plants, systems, tank
***Argulus*** (fish louse), 222-3, 232
**Arrowhead**, 159, 171, 219
**Axelrod**, Dr H., 47, 279

## B

**Baby's Tears**, 163, 173
***Bacopa monnieria*** (Baby's Tears), 159, 163, 173
**bacteria and fungae**, 227, 229, 275
**barbel**, 36, 73, 78, 89, 274
**bath**, formalin, 234
**Best Fish in Show**, 259
**biological filter**, *see* undergravel
**body fluid levels**, 19-20
**body shape**, 14-15, 27, 258
***Bolbitis heudelotii***, 178
**box filters**, 126
**breeding**, 30, 237-53
  *and see* individual families (Fishes index)
**breeding tank**, 250, 278
**breeding trap**, 250, 274
**Brine Shrimp** hatching, 123, 125, 209, 274
**bubble-nests**, 60, 243
  *and see* nest-builders (Fishes index)

## C

**cable tidy**, 177, 274
***Cabomba*** (Fanwort)
  *aquatica*, 159, 164
  *caroliniana*, 178
**calcareous**, (def.) 274
**Canadian Pondweed**, 159, 166
**cannister filters**, 127
**carbon**, activated, 122, 274
**caudal fin**, 17-18, 274
**caudal peduncle**, 275
***Ceratophyllum demersum*** (Hornwort), 159, 164, 173
**chemical filtration**, 122, 275
***Chilodonella*** (slime disease), 232
**class**, (def.) 274, 281
**coldwater fishes**, (def.) 275
**colour** in fishes, 16, 19, 27, 258-9, 262
**Common Eel Grass**, 179, 184, 219
**community tank**, 34, 275
**compatibility** grouping, 30, 172
**compost** (substrate), 275
**conditioning**, 245, 252, 275
**coral**, 24, 154, 188, 190-1
**coral fish disease**, 231
**coral sand**, 275
**cork bark**, 157
***Costia*** (slime disease), 232
**cover-glass**, 150-1, 176, 195, 217, 275
**crustacean**, (def.) 275
***Cryptocaryon*** (marine white spot disease), 230
***Cryptocoryne blassii*** (Water Trumpet), 159, 164, 173, 179
**culling**, 253, 275
**culturing live foods**, 206
***Cyclochaeta*** (slime disease), 232

## D

***Dactylogyrus*** (gill flukes), 231
**denitrification**, 146, 275
**density** of water, 147, 189, 218, 275
**detritus**, 146, 275
**DH** of water, 145-6, 275, 281
**diatomaceous earth**, 126, 275
**diseases/**disorders of fishes, 220-35
  prevention, 213-19
**division**, (def.) 275
***Doitsu***, 89
**dorsal fin**, 17, 275
**dragonfly** larvae, 205
**dropsy**, 220-1, 228, 229
**Duckweed**, 159, 168

## E

***Echinodorus grandiflorus*** (Amazon Sword), 159, 165, 173
**ecological rules** for collecting, 192
***Egeria densa*** (Waterweed), 159, 165, 184
**egg-hatching**, 251-2
**electric impeller**, 126, 275
***Eleocharis acicularis*** (Hairgrass), 159, 165, 173
***Elodea canadensis*** (Water Thyme/Canadian Pondweed), 159, 166, 184-5
***Elodea crispa***, 167
**equalization** of temperatures, 28-9, 178, 182-3, 279
**European Curly Pondweed**, 159, 170
**euthanasia**, 229, 234
**evaporation**, 218
**excretion**, 19
***Exopthalmia*** ("pop-eye"), 221, 228
**eye**
  disorders, 221, 228
  structure, 20-1

## F

**family**, (def.) 279
**Fanwort**, 159, 164, 178
**feeding**, 199-209, 263
  habits, 14-15, 31, 36, 45, 73, 84, 91, 98, 200-1, 208
  overfeeding, 146-7, 207, 214, 219
**fertilization**, 238, 244, 275
**filaments**, 18, 275
**filters**, 124-9, 275
  air-operated, 124-5
  cleaning, 216-17
  power, 126-7, 181-2, 218, 277
  undergravel (biological), 128-9, 156, 160, 174, 186, 193, 216-17, 274
**filter wool**, 187
**filtration**, 121-2, 124-9, 146-7, 155
**fin-nipping**, 212, 233
**fin rot**, 229, 233, 275
**fins**, 17-18, 258
**fish**
  anatomy, 13-21
  breeding, 237-53
  classification, 279
  compatibility grouping, 30, 172
  diseases/disorders, 220-35
  feeding, 199-209
  feeding groups, 14-15
  health care, 211-19
  photography, 265-73
  selection criteria, 26-8
  shows, 255-63
  sources, 28
  transporting, 29, 194, 263
  -watching, 212, 220-1
**fish louse**, 222-3, 232
**fluid levels** in fishes, 19-20
**flukes**, 221, 224-5, 231
**foam**, use of, 122
**foil**, metal, and ventilation, 149
***Fontinalis antipyretica*** (Willow Moss), 159, 166
**formalin bath**, 234
**fossils**, powdered, 126
**freshwater fishes**
  body fluid levels, 20
  breeding, 237-53
  **freshwater fishes** (cont.)
  species, 38-94 (*see* Fishes index)
**freshwater white spot disease**, 231
**fry**, 209, 253, 275
  dangers to, 205
**fungus**, 227, 275
**fungus disease**, 221-3, 233

## G
**genus**, (def.) 275, 279
**Giant Eel Grass**, 159, 171, 184, 219
**Giant Indian Water Star**, 159, 170
**gill flukes**, 221, 231
**gills**, 14, 227, 232, 275
***Gin-rin***, 89
**glass** thickness, 118-19
**gonopodium**, 244, 276
**gravel** as base, 155-6, 174, 180, 193, 263, 276
**gravid**, 67, 276
**Great Diving Beetle larvae**, 205
**green water**, 209
**guanin**, 19, 155, 276
***Gyrodactylus*** (skin flukes), 231

## H
**Hairgrass**, 159, 165
**hand-stripping**, 247, 276
**header-tank**, 209
**heating**, 130-5
fitting heaters, 175, 187
loss of, 216
**hood**, reflector, 149, 177, 276
**Hornwort**, 159, 164
**hospital tank**, 234-5
**"humped-back"** fishes, 28
**Hydra**, dangerous to fry, 205
***Hydrilla verticillata***, 159, 166
**hydrometer**, 147, 188-9, 276
***Hygrophila***, 159
*difformis* (Water Wisteria), 167, 179
*polysperma* (Water Star), 167

## I
**ichthyology**, (def.) 276
***Ichthyophthirius*** (freshwater white spot disease), 230
***Ichtybodo*** (slime disease), 232
***Infusoria*** (green water), 209
**invertebrates**, 24, 276, 281
**ion exchange resin**, 146, 276
**iridescence**, (def.) 276

## J
**Japanese Dwarf Rush**, 159, 163
**Java Fern**, 159, 169
**Java Moss**, 159, 171

## K
***Kin-rin***, 89

## L
**labyrinth organ**, 19, 60, 276
***Lagarosiphon madagasericuse***, 159, 167
**lamp types** and positioning, 149, 151
**lateral line**, 21, 46, 276
**leeches**, dangerous to fry, 205
***Lemna minor*** (Duckweed), 159, 168
***Lernaea*** (anchor worm), 222-3, 232
**lifespan** of fishes. 30
**lighting**, 120, 148-51, 160, 253
changes in, 215-16
maintenance, 216-17
***Limnophila aquatica*** *(Ambulia)*, 159, 168, 173
**line-breeding**, (def.) 276
**live foods**, collecting and culturing 204-6
***Ludwigia repens***, 159, 168, 173

## M
**Madagascar Lace Plant**, 163, 179
**Malawi bloat**, 228
**marine**, (def.) 276.
**marine fishes**
body fluid levels, 19
species, 95-111 (*see* Fishes index)
**marine white spot disease**, 230
***Matsuba***, 89
**matt scales**, 84, 276
**mechanical filtration**, 122, 276
**metallic deposits**, removal of, 144
**metallic scales**, 84, 276
**Microeels** as food, 209
***Microsorium pteropus*** (Java Fern), 159, 169
**Microworms** as food, 209
***Millepora complanata*** (hydrozoan coral), 190
**milt**, 238, 276
**mops**, spawning, 250-1, 276
**mouth fungus**, 222-3, 233
**mouth structure** of fishes, 14-15
***Myriophyllum aquaticum*** (Water Milfoil) 159, 169, 184

## N
**nacreous scales**, 84, 276
***Najas guadelupensis***, 159, 169, 173
**"naked" fishes**, 16
**natural pairing**, 245, 276
**nauplius**, 209, 276
**Neon Disease** (Plistophora), 52
**net usage**, 214-15
**nitrate/nitrite levels**, 129, 146, 158, 186, 219, 277
***Nomaphila stricta*** (Giant Indian Water Star), 159, 170
**nutritional requirements** of fishes, 200-2

## O
***Oodinium***
*limneticum* (velvet disease), 231
*ocellatum* (coral fish disease), 231
**operculum** (gill cover), 275
**ova**, 238, 277
**overfeeding**, 146-7, 207, 214, 219
**ovipositor**, 93, 277
**ovliviparous species**, 244, 277
**oxygenation** *see* aeration

## P
**parasites**, 213, 227, 230-2, 236
**pectoral fins**, 17, 277
**pelvic fins**, 17-18, 277
**pharyngeal teeth**, 36, 277
**photosynthesis**, 158, 277
**pH testing** of water, 145, 217-18, 277
***Pistia stratiotes*** (Water Lettuce), 159, 170, 173
**plants, aquatic**, 24, 154-5, 158-71, 250, 257, 259
crown, 275
cuttings, 162, 275
filtration, 129
floating, 162
lighting, 148, 151, 158, 160
maintenance, 217, 219
nourishment, 161
planting techniques, 160-1, 176
plug, 161, 277
rooted, 162, 277
sources, 163
specimen, 161, 278
and undergravel filtration, 160
**plastic bags**, use of 215, 262-3
***Plistophora*** (Neon Disease), 52
**plug**, plant, 161, 277
**poisoning**, prevention of, 214, 224-5
**pond**
fishes, 64, 84, 89, 279
predators, 205
**"pop-eye"**, 221, 228-9
**post-mortems**, 234
***Potamogeton crispus*** (European Curly Pondweed), 159, 170
**power filter**, 126-7, 277
**predators**, 205
**processed foods**, 202-3
**protein skimming**, 125

## Q
**quarantine**, 35, 214, 234, 277

## R
**rays**, 18, 277
**record-keeping**, 234, 253
**reflector-hood**, 149, 177, 217, 276
making, 195
**respiration**, 19
**reverse-flow filtration**, 128, 277
**rockpools**, 192
**rocks as decoration**, 157, 175, 181, 188

## S
**safety advice**, 132, 146, 150, 174, 192, 216, 217
***Sagittaria graminea*** (Arrowhead), 159, 171, 173, 210
**salinity**, (def.) 277
**salt mix**, *see* "sea-mix"
**sand** as base, 156
***Saprolegnia*** (fungus disease), 233
**scales**, 16, 84, 89, 277
**scalpels**, 103, 277
**scutes**, 16, 73, 277
**"sea-mix"**, 24, 144, 147, 188, 194, 218, 277
**seawater**, use of, 144
**seaweed**, 173
**selection criteria** for fishes, 26-8
**senses** of fishes, 20-1
**septicaemia**, 221-3, 229
**sexing livebearers**, 30, 245, 253
**shimmies**, 224-5, 278
**show groups**, 34
**shows**, 255-63
**show tank**, 263
**shrimp culture**, 206
**sick fishes**, treatment of, 226, 234-5
**sintered glass**, 123, 278
**sizes** of fishes, 28, 34, 258
**skin** of fishes, 16
problems, 222-3, 227, 231
**"sleep"**, 19
**slime disease**, 232

**sources** for fish, 28
**spawning**, 246-50, 278
mops, artificial, 250-1
**species**, (def.) 278, 279
**species tank**, 34, 278
**specific gravity** of water, 147, 189, 194, 217, 278
**specimen plants**, 161, 278
**standard mix** of synthetic sea-water, 34, 95
**strain**, (def.) 278
**stress**, 211, 214-16, 222-3, 235
**stripping**, 247, 276, 278
**substrate** (compost), 276
**swim-bladder**, 21, 278
problems, 221, 228
**swimming levels**, 14-15, 31, 36, 45, 66, 70, 73, 96, 98, 259
**swimming problems**, 224-5
***Synnema triflorum*** (Water Wisteria), 167
**systems**
comparison of, 24-5, 115-16, 137, 155
**systems, freshwater**, 147, 217
coldwater, 9, 83, 173, 180-5
tropical, 8, 35, 130-6, 173-9
**systems, saltwater**, 138, 217
coldwater, 9, 109, 147, 173, 192-7
tropical, 9, 95, 130-6, 147, 173, 186-91, 217

# T

**tail shapes**, 17-18, 256
**tank**
aeration and filtration, 121-9
backdrops, 280
breeding, 250, 278
buying and siting, 118-120
capacity, 280-1
community, 34, 278
filling, 176, 181
/fish ratio, 116
glass thickness, 118-19
header, 209
heating, 130-5
hospital, 234-5
layouts, 173-97
maintenance, 216-19
measurements, imperial/metric, 280
moving, 120
secondhand, 119, 193
show, 263
space utilization, 31
special requirements of coldwater freshwater fishes, 83, 91, 114, 180
species, 34, 278
surface area of water, 115-16, 122, 278
temperature measurement, 134
types, 115-17
water quality and sources, 144-7, 216
*see also* aquarium, systems, water
***Taonia atomaria*** (Seaweed), 173
**temperature**
equalization, 28-9, 178, 182-3
maintenance, 216, 217
measurement, 134
**terminal mouth**, 15, 278
**thermometers**, 134, 176, 182, 190, 216
**thermostats**, 133-4, 278
**transporting fishes**, 29, 194, 263
***Trichodina*** (slime disease), 232
**tropical fishes**, (def.) 278
**tubercles**, 245, 278
**tuberculosis**, 221, 229
***Tubipora musica*** (Organ-pipe coral), 191
**twintail fish**, 18, 86, 278

# U

**ulcer disease**, 222-3
**undergravel filters**, 128-9, 156, 217, 274
fitting, 129, 174, 186
maintenance, 216
making, 193
and plants, 160

# V

***Vallisneria***
*gigantea* (Giant Eel Grass), 159, 171, 173, 184, 218
*spiralis* (Common Eel Grass), 179, 184, 218
**variety**, (def.) 278
**velvet disease**, 231
**ventilation** and metal foil, 149
***Vesicularia dubyana*** (Java Moss), 159, 171
**vibration** and stress, 216
**vitamin requirements** of fish, 200, 214-15
**viviparous species**, 244, 278

# W

**wasting disease**, 228-9
**water**
brackish, 138, 274
changes, 215, 217-18
chemical composition, 122
chlorine and metallic deposits, 144
conditions and breeding, 147
density, 147, 189, 218, 275
flow/turnover, 126-8
and habitat, 138-42
hardness (DH), 145-6, 275, 281
maintenance, 216
nitrate/nitrite levels, 129, 146, 158, 186, 219, 277
pH testing, 145, 217-18, 277, 281
rainwater, 144, 146
requirements of plants, 160
seawater, 144
surface area, 115-16, 122, 278
temperature, 134, 281
**Water Boatmen,** dangerous to larvae, 205
**Water Lettuce**, 159, 170
**Water Milfoil**, 159, 169, 184
**Water Star**, 167
**Water Thyme**, 159, 166, 184-5
**Water Trumpet**, 159, 164, 179
**Water Weed**, 159, 165, 184
**Water Wisteria**, 167, 179
**Whirligig Beetle** larvae, 205
**white spot disease**, 221, 230
**Willow Moss**, 159, 166
**worm culture**, 206
**worm feeder**, 208

# Index of Fishes

*alic page numbers refer to Species Guide entries*


## A

***budefduf oxyodon*** (Blue-velvet Damselfish), 30, *96*
***canthopthalmus kuhli*** (Coolie Loach), *78*
***canthurus leucosternon*** (Powder Blue Surgeonfish), *103*
***ctinia equina*** (Beadlet Anemone), 196-7
***equidens***
*curviceps* (Sheepshead Acara/Flag Cichlid), *53*
*maronii* (Keyhole Cichlid), *54*
**frican Catfish**, 16
**merican Flagfish**, *71*, 279
***mphiprion***
*ocellaris* (Common Clownfish), *97*, 142
*sebae* (Black Clownfish), 191
**nabantoids**, 19, 60-5, 246, 274, 276
**nemonefishes**, 96-7
**ngelfish**, 16, 18, 31, 59, 98-100, 140, 142, 178, 260-1
breeding of, 247
**nnual Fishes**, 70, 274
***nostomus anostomus*** (Striped Headstander), *45*
***phyosemion gardneri*** (Steel-blue Aphyosemion), 15, *70*, 141
***pistogramma ramirezi*** (Ram), *54*
***plocheilus dayi*** (Ceylon Killifish), *71*
***pogon nematopterus*** (Pyjama Cardinalfish), 28, *104*
**rmoured Catfish**, 16
***sagi*** (Koi), *90*
***sterias rubens*** (Common Starfish), *111*, 192
***styanax mexicanus*** (Blind Cave Fish), 21, *46*

## B

***alistapus undulatus*** (Undulate Triggerfish), 18, *104*
***arbus*** (Barbs), 36-9, 138-9, 239, 256
*conchonius* (Rosy Barb), *36*, 178, 239
*cumingi* (Cuming's Barb), *37*
*nigrofasciatus* (Black Ruby Barb/Purple headed Barb), *37*, 178
***Barbus*** (cont.)
*oligolepis* (Checker Barb/ Island Barb), *38*
*schwanenfeldi* (Tinfoil Barb/Goldfoil Barb), *38*
*tetrazona* (Tiger Barb), 31, *39*, 139
*titteya* (Cherry Barb), *39*
**Beadlet Anemone**, 196-7
***Bedotia geayi*** (Madagascar Rainbowfish), *80*
***Belontia signata*** (Combtail), *60*
***Betta splendens*** (Siamese Fighter), 30, *60-1*, 141, 243
**Bitterling**, 91, *93*, 245, 247, 279
**Black-banded Sunfish**, *92*
**Black Clownfish**, 191
**Black Molly**, *67*, 232, 244
**Black Ruby Barb**, *37*, 178
**Black-spotted Corydoras**, *75*
**Black-tailed Humbug**, 190
**Bleeding-heart Tetra**, 31, *49*
**Blennies**, *110*, 192, 197
***Blennius gattorugine*** (Tompot Blenny), *110*, 192, 197
***Blennius pholis*** (Blenny/ Shanny), 192
**Blind Cave Fish**, 21, *46*
**Blue-banded Sergeant Major**, *96*
**Blue Devil**, 190
**Blue-girdled Angelfish**, *99*
**Blue Streak**, *106*, 142
**Blue-velvet Damselfish**, 30, *96*
***Bodianus rufus*** (Spanish Hogfish), *105*
***Botia***
*macracantha* (Clown Loach/Tiger Botia), 31, *79*, 139
*sidthimunki* (Chain Loach/Dwarf Loach), *79*
***Brachydanio*** (Danios), 15, 31, 36, 40-1, 138-9, 239
*albolineatus* (Pearl Danio), *40*, 139
*frankei* (Leopard Danio), *40*
*rerio* (Zebra Danio), 31, *41*, 179
**Bridled Beauty**, *106*, 142
**Bristol Shubunkin**, *85*
**Brocaded Carp**, *89*
***Brochis splendens*** (Short-bodied Catfish), *73*
**Bronze Catfish**, 14, *74*, 179
***Buccinum undatum*** (Common Whelk), 197
**Butterflyfishes**, 98, 100-2, 142

## C

**Cardinal Tetra**, *47*, 279
***Carnegiella strigata*** (Marbled Hatchetfish), 14, *46*, 140
**Catfishes**, 16, 18-20, 31, 73-7, 140-1, 154
breeding of, 241, 243
**Celestial**, *88*
***Centropyge loriculus*** (Flame Angelfish), *98*
**Ceylon Killifish**, *71*
***Chaetodon***
*lunula* (Red-striped Butterflyfish), *100*, 142
*octofasciatus* (Eight-banded Butterflyfish), *101*, 142, 279
**Chain Loach**, *79*
**Characins**, 17, 45-52, 138, 239, 256, 279
**Checkerboard Cichlid**, *56*
**Checker Barb**, *38*
***Cheirodon axelrodi*** (Cardinal Tetra), *47*, 279
***Chelmon rostratus*** (Copper-banded/Long-nosed Butterflyfish), *101*, 142
**Cherry Barb**, *39*
***Chilodus punctatus*** (Spotted Headstander), *47*, 140
***Cichlasoma***
*festivum* (Flag/Festive Cichlid), *55*
*meeki* (Firemouth Cichlid), *55*
**Cichlids**, 53-9, 138, 154, 228, 256
breeding, 241-3, 245, 252-3
**Cleanerfish**, 103, 105, *106*, 142
**Clownfishes**, 96-7, 142, 191
**Clown Loach**, 31, *79*, 139
***Colisa*** (Gouramis), 18, 61-5, 141
*chuna* (Honey Gourami), *61*
*labiosa* (Thick-lip Gourami), *62*
*lalia* (Dwarf Gourami), 31, *62*, 178, 243
**Combtail/Combtail Paradise Fish**, *60*
**Comet**, *85*
**Common Clownfish**, *97*, 142
**Common Goldfish**, *84*, 184-5, 232
**Common Prawn**, 192, 197
**Common Starfish**, *111*, 192
**Congo Tetra**, *50*
**Coolie Loach**, *78*
***Copella arnoldi*** (Splashing Tetra), *48*, 241
**Copper-banded Butterflyfish**, *101*, 142
***Corydoras*** (Catfishes), 18, 19, 140-1, 147, 154, 241, 279
*aeneus* (Bronze Catfish), 14, *74*, 179
*julii* (Leopard Catfish), *74*, 140, 179
*melanistius* (Black-spotted Corydoras), *75*
*reticulatus* (Reticulated Corydoras), 31, *75*, 279
***Crenicara filamentosa*** (Checkerboard Cichlid), *56*
**Croaking Gourami**, *65*, 243
***Ctenopoma***
*acutirostre* (Spotted Climbing Perch), *63*, 141
*ansorgei* (Ornate Ctenopoma), *63*, 141
**Cuming's Barb**, *37*
**Cyprinids**, 36-44
***Cyprinus carpio*** (Koi), *89*

## D

**Dahlia Sea-anemone**, *110*, 192
**Damselfishes**, 97-7, 190-1
**Danios**, 15, 31, 36, 40-1, 138-9, 239
***Dascyllus***
*albisella* (Hawaiian Humbug), *97*
*melanurus* (Black-tailed Humbug), *190*
*trimaculatus* (Three-spotted Damselfish), *191*
***Dermogenys pusillus*** (Wrestling Halfbeak), *66*
**Diamond Tetra**, *50*
***Dianema urostriata*** (Stripe-tailed Catfish, *76*
**Discus**, 14, *59*, 147
**Dwarf Gourami**, 31, *62*, 178, 243
**Dwarf Loach**, *79*

## E

**Eastern Mudminnow**, *94*, 279
**Egg-buriers**, 240, 251, 252
**Egg-depositors**, 241, 252
**Egg-layers**, 238-53, 275
**Egg-scatterers**, 239, 247, 251-2
**Eight-banded Butterflyfish**, *101*, 142, 279

***Elassoma evergladei*** (Pigmy Sunfish), *91*
**Emperor Angelfish**, 16, *100*, 142
**Emperor Tetra**, 18, *51*
***Enneacanthus chaetodon*** (Black-banded Sunfish), *92*
***Epalzeorhyncus kallopterus*** (Flying Fox), *44*, 139
***Etroplus maculatus*** (Orange Chromide), *56*, 220
***Euxiphipops navarchus*** (Blue-girdled Angelfish), *99*

# F

**Fairy Basslet**, *105*
**Fantail**, *86*
**Festive Cichlid**, *55*
**Fingerfish**, 80, *82*
**Firemouth Cichlid**, *55*
**Flag Cichlid**, 53, 55
**Flame Angelfish**, *98*
**Flying Fox**, *44*, 139
**Forcepsfish**, *102*
***Forcipiger longirostris*** (Long-snounted Coralfish/Forcepsfish), *102*

# G

**Giant Danio**, *41*
**Glass Catfish**, *77*
**Gobies**, 21, 192
**Goldfishes**, 6, 18, 24, *83-8*, 184-5, 232, 256, 279
breeding 239, 245, 247
**Goldfoil Barb**, *38*
**Gouramis**, 18, 61-5, 141
breeding, 243, 253
***Gramma loreto*** (Royal Gramma, Fairy Basslet), *105*
**Guppies**, 31, 66, *67*, 179, 244

# H

**Halfbeaks**, *66*, 244
***Hariwaki Koi***, *89*
**Harlequin Fish**, *42*, 178-9
**Hatchetfishes**, 45, 140
**Hawaiian Humbug**, *97*
**Headstanders**, 45, 47, 140
***Helostoma temmincki*** (Kissing Gourami), *64*
***Hemigrammus rhodostomus*** (Rummy-nosed Tetra/ Red-nosed Tetra), *48*
***Heniochus acuminatus*** (Wimplefish, Pennantfish, Poor Man's Moorish Idol), *102*
***Hippocampus kuda*** (Yellow Seahorse/ Oceanic Seahorse), *106*
***Holacanthus tricolor*** (Rock Beauty), *99*, 220
**Honey Gourami**, *61*
***Hyphessobrycon***
*erythrostigma* (Bleeding-heart Tetra), 31, *49*
*pulchripinnis* (Lemon Tetra), *49*
**Hypostomus species** (Suckermouth Catfish), 19, 20, *76*

# I

**Invertebrates**, 24, 194-7
**Island Barb**, *38*

# J

**Jack-in-the-Box**, *107*, 142
***Jordanella floridae*** (American Flagfish), *71*, 279
***Julidochromis marlieri*** (Marlier's Julie), *57*

# K

**Keyhole Cichlid**, *54*
**Killifishes**, 70-2, 138, 240, 252
**Kissing Gourami**, *64*
***Kohaku Koi***, *89*
**Koi**, 6, 83, *89-90*, 247, 249
**Kribensis**, *58*, 141, 241, 253
***Kryptopterus bicirrhus*** (Glass Catfish), *77*

# L

***Labeo bicolor*** (Red-tailed Black Shark), *44*
***Labeotropheus trewavasae*** (Red-finned Cichlid), *57*, 242
***Labridae*** (Wrasses), 19, 103, 105-6
***Labroides dimidiatus*** (Cleanerfish/Blue Streak/ Bridled Beauty), *106*, 142
**Labyrinth fishes**, 19, 60-5, 246, 274, 276
**Lace Gourami**, *65*, 246
**Lacetails**, 18
***Lamprologus brichardi*** (Lyretail Lamprologus), 18, *58*, 154
***Leander serratus*** (Common Prawn), 192, 197
**Leeri Gourami**, *65*, 246
**Lemon Tetra**, *49*
**Leopard Catfish**, *74*, 140, 179
**Leopard Danio**, *40*
***Lepomis gibbosus*** (Pumpkinseed), *92*, 279
**Limpet, Common**, 196
**Lionfishes**, 18, 103, *107*
**Lionhead**, *88*
***Littorina littorea*** (Edible Periwinkle), 197
**Livebearers**, 8, 17, 25, 30, 66-9, 138, 233, 259-62, 276
breeding, 238, 244-7, 250-3
sexing, 245, 253
**Loaches**, 31, 78-9, 139
**London Shubunkin**, 84
**Long-nosed Butterflyfish**, *101*, 142
**Long-snouted Coralfish**, *102*
**Lyretail Lamprologus**, 18, *58*, 154

# M

***Macrognathus aculeatus*** (Spiny Eel), *81*
***Macropodus opercularis*** (Paradise Fish), *64*
**Madagascar Rainbowfish**, *80*
**Malayan Angelfish**, *82*
**Mandarinfish**, *108*
**Marbled Hatchetfish**, 14, *46*, 140
**Marlier's Julie**, *57*
***Mastacembelus argus*** (Spiny Eel), *81*
***Melanotaenia nigrans*** (Australian Rainbowfish), *82*
***Micralestes interruptus*** (Congo Tetra), *50*
**Millions Fish**, *67*
***Moenkhausia pittieri*** (Diamond Tetra), *50*
**Mollies**, 66-8, 138, 232, 244
**Mongrel Koi**, *90*
***Monodactylus argenteus*** (Monofish/ Malayan Angelfish/ Fingerfish), *82*
**Monofish**, 80, *82*, 138
**Moorish Idol**, *108*
**Mosaic Gourami**, *65*, 246
**Mouth-brooders**, 243, 246, 251, 276

# N

**Naked Catfish**, 16, 73
***Nannostomus unifasciatus*** (One-lined Pencilfish), *51*, 140
***Nematobrycon palmeri*** (Emperor Tetra), 18, *51*
**Neon Tetra**, *52*, 140
**Nest-builders**, 243, 246, 251
**"Nishiki Koi"**, *89*
***Nothobranchius rachovi*** (Rachov's Nothobranch), *72*, 155, 240
***Notropis lutrensis*** (Red Shiner), *93*, 279

# O

**Oceanic Seahorse**, *106*
**One-lined Pencilfish**, *51*, 140
***Opistognathus aurifrons*** (Yellow-headed Jawfish/ Jack-in-the-Box), *107*, 142, 155
**Oranda**, *87*
**Orange Chromide**, *56*
**Orange-green Triggerfis** *104*
**Ornate Ctenopoma**, *63*, 141

# P

***Pachypanchax playfai*** (Playfair's Panchax), *72*, 220
**Pale Chub**, *94*, 279
***Papiliochromis ramire*** (Ram), *54*
***Paracheirodon***
*axelrodi* (Cardinal Tetra), *47*, 279
*innesi* (Neon Tetra), *52*, 140
**Paradise Fish**, *64*
***Patella vulgata*** (Comm Limpet), 196
**Pearl Danio**, *40*, 139
***Pelvicachromis pulcher*** (Kribensis), *58*, 141, 241, 253
**Pencilfishes**, 16, 45, 51, 14 147
**Pennantfish**, *102*
**Periwinkle, Edible**, 197
**Pigmy Rasbora**, *43*
**Pigmy Sunfish**, *91*
**Platys**, 66, 69, 138, 178-9
breeding, 244-5
**Playfair's Panchax** *72*, 2
***Poecilia***
hybrid (Black Molly), *67*, 232, 244
*reticulata* (Guppy), 31, *67*, 179, 244
*velifera* (Sailfin Molly), 244
**Polka Dot Cardinalfish**, *104*
***Pomacanthus imperat*** (Emperor Angelfish), 16 *100*, 142
***Pomacentrus caeruleu*** (Blue Devil), 190
**Pompadour Fish**, *59*
**Poor Man's Moorish Ido** *102*
**Powder Blue Surgeonfi** *103*
**Prawn, Common**, 192, 19
***Pterois volitans*** (Lionfis Scorpionfish/Turkeyfish 18, *107*

***Pterophyllum scalare*** (Angelfish), 31, *59*, 140, 178, 247, 260-1
**Pumpkinseed**, *92*, 279
**Purple-headed Barb**, *37*
**Pyjama Cardinalfish**, 28, *104*

# R

**Rachov's Nothobranch**, *72*, 155, 240
**Rainbowfishes**, 80, 82
**Ram**, *54*
***Rasbora***, 36, 138-9, 238
*heteromorpha* (Harlequin Fish), *42*
*maculata*, Pigmy Rasbora/ Spotted Rasbora), *43*
*trilineata* (Scissortail), *43*, 139
**Red-finned Cichlid**, *57*, 242
**Red-nosed Tetra**, *48*
**Red Piranha**, *52*
**Red Shiner**, *93*, 279
**Red-striped Butterflyfish**, *100*, 142
**Red-tailed Black Shark**, *44*
**Reticulated Corydoras**, 31, *75*, 279
***Rhodeus sericeus amarus*** (Bitterling), 91, *93*, 245, 247, 279
**Rock Beauty**, *99*, 220
**Rooseveltiella nattereri** (Red Piranha), *52*
**Rosy Barb**, *36*, 178, 239
**Royal Gramma**, *105*
**Rummy-nosed Tetra**, *48*

# S

**Sailfin Molly**, *68*, 244
**Scissortail**, *43*, 139
**Scorpionfish**, *107*
**Sea-anemones**, 96-7, *110*, 142, 192, 194-7
**Seahorses**, 103, *106*
***Serpula vermicularis*** (Tubeworm), *111*, 142, 192
***Serrasalmus nattereri*** (Red Piranha), *52*
**Shanny**, 192
**Sheepshead Acara**, *53*
**Short-bodied Catfish**, *73*
***Showa Sanke***, *90*
**Siamese Fighting Fish**, 30, *60-1*, 141, 243
**Spanish Hogfish**, 105
**Sparkling Gourami**, *65*, 243
***Sphaeramia nematopterus*** (Pyjama Cardinalfish), 28, *104*
**Spiny eels**, *80*, *81*
**Splashing Tetra**, *48*, 241
**Spotted Climbing Perch**, *63*, 141
**Spotted Headstander**, *47*, 140
**Spotted Rasbora**, *43*
**Starfish, Common**, *111*, 192
**Steel-blue Aphyosemion**, 15, *70*, 141
**Striped Headstander**, *45*,
**Stripe-tailed Catfish**, *76*
**Suckermouth Catfish**, 19, 20, *76*
**Sunfishes**, 91-2, 241, 243, 279
**Surgeonfishes**, *103*
**Swordtail**, 18, 66, *68*, 138, 179, 244
***Symphysodon discus*** (Discus, Pompadour Fish), 14, *59*, 147
***Synchiropus splendidus*** (Mandarin-fish), *108*
***Synodontis nigriventris*** (Upside-down Catfish), *77*, 141

# T

***Taelia felina*** (Dahlia Sea-anemone), *110*,, 192
***Taisho Sanke***, *90*
**Tancho Orandas**, *87*
***Tanichthys albonubes*** (White Cloud Mountain Minnow), *42*
**Telescope Moor**, *87*
**Tetras**, 18, 31, 45, 47-52, 154
**Thick-lip Gourami**, *62*
**Three-colour Koi**, *90*
**Three-Spotted Damselfish**, 191
**Tiger Barb**, 31, *39*, 139
**Tiger Botia**, 31, *79*, 139
**Tinfoil Barb**, *38*
**Toby**, *108*
**Tompot Blenny**, *110*, 192, 197
***Trichogaster leeri*** (Lace Gourami/Leeri Gourami/ Mosaic Gourami), *65*, 246
***Trichopsis pumilus*** (Croaking Gourami/ Sparkling Gourami), *65*, 243
**Triggerfish**, 18, *103-4*
**Tubeworm**, 111, 142, 192
**Turkeyfish**, *107*
**Two-colour Koi**, *89*

# U

***Umbra pygmaea*** (Eastern Mudminnow), *94*, 279
**Undulate Triggerfish**, 18, *104*
**Upside-down Catfish**, *77*, 141

# V

**Variatus Platy**, *69*, 179
**Veiltail**, 18, *86*

# W

**Whelk, Common**, 197
**White Cloud Mountain Minnow**, *42*
**Wimplefish**, *102*
**Wrasses**, 19, 103, 105-6
**Wrestling Halfbeak**, *66*

# X

***Xiphophorus***
hybrid (Swordtail), 18, 66, *68*, 138, 179, 244
*maculatus* hybrid (Platy), *69*, 138, 178-9, 245
*variatus*, hybrid (Variatus Platy), *69*, 179

# Y

**Yellow-headed Jawfish**, *107*, 142, 155
**Yellow Seahorse**, *106*

# Z

***Zacco platypus*** (Pale Chub), *94*, 279
***Zanclus cornutus*** (Moorish Idol/Toby), *108*
**Zebra Danio**, 31, *41*

# Acknowledgements


## Author's acknowledgements
I wish to thank Alan Buckingham for believing this book was possible in the first place, and then the production team that made it so: Judith More and Janice Lacock, two very "word-wise" ladies for making the transition of text from floppy disk to paper so successful; Julia Goodman, Jo Martin and Tina Vaughan for original design ideas, artwork and photographic coordination and for fitting everything together.

## Dorling Kindersley would like to thank:
Bernice Brewster for checking the anatomical illustrations: Lesley Davy for picture research; Peter Scott for veterinary advice; Eric Crichton for his hard work in producing much of the commissioned photography; David Ashby for his major contribution to the illustration work; Gillian della Casa for her design help on the Species guide; Biddy Martin for the index; Chelsea Aquarist, Chelsea Farmers Market, Sydney Street, London SW3; David Quelch of Water World, Turnford, Roxbourne, Herts; Stapeley Water Gardens Ltd, 92 London Road, Stapeley, Nantwich, Cheshire; and Robbie Somers for supplying photographs, equipment and livestock; Ron Bagley, Mike Hearne, and Ken Hone for photographic services.

## Illustrators
David Ashby, Andrew Macdonald, Kuo Kang Chen.

## Photographic credits
Heather Angel: pp. 8 b l, 9 t r, 40 t, 48 t, 81 t, 111 b
The Bridgeman Art Library/British Museum: p. 6 b l
Ron Boardman/Bruce Coleman Ltd.: p. 111 t
Jane Burton/Bruce Coleman Ltd.: pp. 9 t l, 12, 21, 32, 42 t, 52 t, 55 b, 59 b, 64 t, and b l, 66, 68 b, 72 b, 74 t, 78, 85 b, 101 b, 110 b, 192
Alain Compost/Bruce Coleman Ltd.: p. 35
Eric Crichton: pp. 2-3, 22, 152, 162, 174-5, 176-7, 178-9, 180-1, 182-3, 184-5, 186-7, 188-9, 190-1, 193, 194-5, 196-7, 210, 235, 254, 264, 268, 270-1
Stephen Dalton/NHPA: p. 61 l
Thomas Dobbie: pp. 202-3
Martin Dohrn/Bruce Coleman Ltd.: p. 109
Mary Evans Picture Library: pp. 6 r, 7
Laurence Gould/Planet Earth Pictures: p. 105 t
J. M. Labat/Ardea London: p. 92 b
Jan-Eric Larsson: pp. 1, 40 b, 45, 56 b, 69 b, 70 b, 71 b, 72 t, 77 b, 80, 86 t, 91, 98, 99 b, 100 t, 102 t, 103, 104 t, 105 b, 108 b, 213, 221 t l, 221 b l, 226
Lacz Lemoine/NHPA: pp. 41 b, 42 b, 51 b, 67 b, 77 t, 79 t l, 84 b, 106 t
Ken Lucas/Planet Earth Pictures: p. 97 b
Harold Metcalf/NHPA: p. 102 b
P. Morris/Ardea London: pp. 16, 93 b, 97 t, 99 t, 108 t
Arend van den Nieuwenhuizen: pp. 67 t, 71 t, 86 b, 92 t, 100 b, 101 t, 236, 253 b
M. Timothy O'Keefe/Bruce Coleman Ltd.: p. 95
Dick Mills: pp. 93 t, 94 b, 107 t and b, 272 t and b
Paulo Oliveira/Planet Earth Pictures: pp. 38 t, 62 r, 64 r, 221 b c r
Vincent Oliver: p. 90 b
Laurence E. Perkins: p. 88 t and b
Christian Petron/Planet Earth Pictures: p. 110
Hans Reinhard: pp. 37 b, 38 b, 39 b, 41 t, 43 t, 44 t and b, 46 b, 47 t, 48 b, 49 b, 51 t, 52 b, 54 t and b, 61 r, 68 t, 69 t, 76 b, 81 b, 85 t, 87 b, 89, 90 t, 94 t
Hans Reinhard/Bruce Coleman Ltd.: pp. 36, 39 t, 49 t, 55 t, 57 b, 60, 62 l, 63 r, 65 l, 82 t
David D. Sands: pp. 73, 74 b, 75 t
Mike Sandford: pp. 43 b, 46 t, 50 t and b, 53, 58 b, 59 t, 63 t and c l, 75 b, 76 t, 87 t
Peter Scott: p. 221 t r, c l, b r
Peter Scoones/Planet Earth Pictures: p. 104 b
John Shaw/Bruce Coleman Ltd.: p. 83
Spectrum Colour Library: p. 198
Stapeley Water Gardens Ltd.: p. 10
William Tomey: pp. 9 b r, 11, 37 t, 47 b, 56 t, 57 t, 58 t, 65 r, 79 r, 82 b, 96 b, 112, 220, 221 t c r
Bill Wood/NHPA: p. 106 b

## Cover
Illustrations: David Ashby
Photographs: Jan-Eric Larssen, Lacz Lemoine/NHPA, Eric Crichton

Typesetting by Gedset, Cheltenham
Reproduction by Mondadori, Verona